LAW AND SOCIETY IN MODERN INDIA

LAW AND SOCIETY
IN MODERN INDIA

MARC GALANTER

Edited with an introduction by
RAJEEV DHAVAN

DELHI
OXFORD UNIVERSITY PRESS
BOMBAY CALCUTTA MADRAS

Oxford University Press, Walton Street, Oxford OX2 6DP

New York Toronto
Delhi Bombay Calcutta Madras Karachi
Kuala Lumpur Singapore Hong Kong Tokyo
Nairobi Dar es Salaam
Melbourne Auckland

and associates in
Berlin Ibadan

This book is published in association with
THE BOOK REVIEW LITERARY TRUST

Printed at Rekha Printers Pvt. Ltd., New Delhi 110020
and published by Neil O'Brien, Oxford University Press
YMCA Library Building, Jai Singh Road, New Delhi 110001

To the memory of my mother
Mary Linett Galanter 1898–1988

פִּיהָ פָּתְחָה בְחָכְמָה
וְתוֹרַת־חֶסֶד עַל־לְשׁוֹנָה

She openeth her mouth with wisdom;
And the law of kindness is on her tongue

PROVERBS 31:26

Preface

Academics are often prey to the illusion that the appearance of their work in print is tantamount to its effective communication with their intended readers. Most of the papers in the present volume were published earlier, but I am now painfully aware that their appearance in scattered publications, most not readily available in India, means that they reached few among those for whom they were written. I am grateful for the opportunity to set them forth once more in the hope that in each other's company they will find an easier journey to their intended destination.

A second look has enabled me to catch a few egregious lapses, but it has not been possible to rewrite and update these essays. The various styles of the original publications regarding transliteration and other matters have not been disturbed, except that citations occurring in the text have been placed in footnotes and the original lists of references combined into a unified list.

Acknowledgements

I am indebted to the various publishers, journals, individuals and learned societies who graciously permitted these essays to be reprinted here, namely:

The University of Chicago Press for permission to reprint Chapter 1, which originally appeared in *Languages and Areas: Studies presented to George V. Bobrinskoy* (1967)

The Society for the Psychological Study of Social Issues, for permission to reprint Chapter 2, which originally appeared in Volume 24 of the *Journal of Social Issues* (1968)

The Cambridge University Press for permission to reprint Chapter 3, which originally appeared in Volume 14 of *Comparative Studies in Society and History* (1972)

Professor Mauro Cappelletti and Martinus Nijhoff Publishers to reprint Chapter 4, which originally appeared in Volume III of Cappelletti and Garth, editors, *Access to Justice* (1979) and Professor Upendra Baxi with whom the article was co-authored

The Social Science Research Council for permission to reprint Chapter 5, which originally appeared in Volume 32 of *Items* (1978)

The *Journal of Asian and African Studies* for permission to reprint Chapter 6, which originally appeared in Volume 2 of that journal in 1967

The Wenner-Gren Foundation for Anthropological Research for permission to reprint Chapter 7, which originally appeared in *Structure and Change in Indian Society*, Milton Singer and Bernard S. Cohn, editors, Viking Fund Publications in Anthropology, No. 47 (1968)

Sage Publications for permission to reprint Chapter 8 which originally appeared in *Social and Economic Development in India: A Reassessment*, Dilip K. Basu and Richard Sisson, editors (1986)

Philosophy East and West for permission to reprint Chapter 10 which originally appeared in Volume 21 of that journal in 1971

Rajeev Dhavan, S. Khurshid and R. Sudarshan for permission

to reprint Chapter 11, which was published in *Judges and the Judicial Power*, R. Dhavan, S. Khurshid and R. Sudarshan, editors, (1985).

Finally, I owe an inestimable debt to Rajeev Dhavan whose unfailing energy and intellectual fellowship brought this book into being. Everyone who writes, dreams of a reader who reads one's work with an intensity that matches its writing and who delights in friendly argument that reveals to the writer more than he knew was there. I am extraordinarily fortunate to have found such a reader and friend in Rajeev Dhavan.

Contents

Introduction

by Rajeev Dhavan

I

Although the history and development of political philosophy and legal thought is often represented and characterized as a series of quantum leaps emanating from the contribution of distinguished individuals, it is not easy to assess the contribution of any one scholar. Scholarship does not exist in isolation. It draws from the work of those that precede it and is contextualized by its participation in contemporary controversies of the day. Often, the contribution of any one thinker or jurist seems unique. Yet the strands of this uniqueness can be traced to traditions of scholarship within which it was trapped and from which the scholar in question — along with others — liberated it. Significantly, the scholarship of the ancient Hindus sought to de-personalize the contribution of individual scholars and de-contextualize the scholarship from the milieu that produced it.[1] In this way, the basic paradigmatic notions that underlay the scholarship were anonymously and silently reviewed and revised. While it is only right that the growth of political ideas should be traced to the contribution of individual scholars, such contributions can be best understood as part of the wisdom of the life and times that created them.

In this essay, we will *first* examine the various strands of scholarship on modern Indian Law in recent years. This must, perforce, include an examination of the contribution of Marc Galanter. Having identified Galanter's position on the landscape, we will proceed to evaluate and critique his juristic sociology.

THE JURISPRUDENCE AND SOCIOLOGY OF MODERN INDIAN LAW

The Transformation of Indian Law

Over the last two hundred years, Indian Law has attracted a considerable amount of foreign scholarship. This interest has taken various shapes and forms. Administrators and judges of the Raj had a direct interest in the legal system even if they represented varied interests and maintained varying standpoints. Those interested in the administration concerned themselves with institutional reform—careful attention being given to the development of judicial institutions to service the needs and responsibilities of the ever-expanding Raj.[2]

Historically, the first of these modern colonial jurists consisted of people who were interested in native law.[3] Spurred on by a curiosity for the indigenous and aware that it would be politically and socially cumbersome to administer English or Western Law to supplant an already complex set of native rules, English Law decreed that the vast area of residuary law to be administered by the British courts would be according to 'justice, equity, and good conscience'.[4] It has now been convincingly demonstrated that this much misunderstood clausula became a vehicle for recognizing some version of native law rather than imposing Western Law in India.[5] A group of British scholars set about the task of 'discovering' native law. There were a large number of '*sanskrit* treatises compiled at the instance of the British'.[6] Pandits and *moulvis* assisted British courts in the process of discovery.[7] However, this process of discovery was as inexact as it was purposive. Its inexactness led to what Gandhi later described as the 'egregious blunders' committed by the British in their interpretation of native law.[8] Some of these errors were deliberate even though they appeared to be causal. British administrators wrote their accounts of Indian Law to assist the courts. These accounts were later treated as authoritative by British Indian courts. British Indologists sought to re-write native law while claiming a traditional authenticity for their efforts. The most celebrated of these efforts is Colebroke's *Digest* — a translation of a contemporary *pandit's* account of Hindu Law. But the *Digest* is more than a translation and represents an effort to fuse *shastric* principles with Chancery jurisprudence.[9] Yet another group of scholars — led by the idiosyncratic J.H. Nelson who, in turn, had his own reasons for writing what he did — seriously questioned the official policy of taking ancient texts, rather than customary practice, as its basis for understanding Indian Law.[10] Nelson pointed to the disparity between local customs and *shastric* law and felt that the former should be enforced rather than the declarations of the *rishis* and *moulvis* which had been accorded an *ex cathedra* status through, and in, the scriptures. While these controversies abounded — and amidst fresh social and political controversies — native law was being transformed at the hands of British judges. Hindus and Muslims were made subject to the new judicially transformed personal laws which can be termed Ango-Hindu and Anglo-Muslim law respectively.[11] These new creations are, too often, interpreted and critically evaluated by reference to the ancient texts from which they were supposed to derive authority. No one should doubt the dexterity with which the Anglo-Indian judges of the Raj approached intricate questions of *shastric* antiquity. Most of the changes that emerged did not result from errors of understanding but constituted deliberate and adaptive re-creations. To that extent, they are a part of a general effort to codify Indian Law which was otherwise statutorily achieved through the efforts of Macaulay, Maine, Fitzjames Stephen, and others.[12] By the turn into this century, legal scholarship turned to servicing the lawyering and litigational needs of Anglo-Indian law and paid a lesser attention to the deeper pre-British tradition that governed decision making in civil society.

This earlier tradition has never really been subjected to systematic sociological exposition. The *rishis* had created very detailed and sophisticated doctrines and carefully re-constructed those doctrines into a coherent ideology which formed an intellectual basis from which to reason and develop a comprehensive and unified view of Indian society.[13] What were the *dharmasastris* up to? What was their influence? What were the asymmetrical social relations created, consolidated, and where necessary, disguised by their prescriptions? The *rishis* had a very clear understanding of what they were doing as they carefully re-interpreted seemingly monolithic doctrine to meet the demands of competing inequalities. Despite their seemingly orthodox learning, they were men of the world reacting to the social changes and challenges of their society. Their rigour, ingenuity, and scholarship were matched by the profound influence of their writing, an influence which survived in unbroken waves till modern times. This is not to repudiate Nelson's powerful thesis that the *dharmasastra* was not the actual law which governed local communities in their day-to-day folkways. The *dharmasastra* sought to create an enduring set of attitudes and beliefs rather than to actually supplant customary practices with the imperatives ordained by the *rishis*. It tried to house in its mansions a variety of practices which were confronted with similar problems and resolved them on the basis of a unified, even if seemingly amorphous, set of beliefs. The efforts of the *rishis* have been commented on, but not been sufficiently critically examined and evaluated. Kane's *History of the Dharmasastra* stands over twentieth century Hindu Law scholarship like a colossus.[14] But, given the enormity of Kane's task, it is not surprising that this gigantic work did not develop sociological insights into the development of Hindu law.[15] While a large number of scholars continue to write about pre-British Indian law, some of the work is mechanical, uninspired and demonstrates a rote learning rather than imaginative interpretation. Some of the work is marred with a 'wonder-that-was-India'[16] or a modernity of tradition approach.[17] A very small band of scholars — Indian, English, German, and American — continue to sustain an interest in the development of earlier Indian law and the socio-economic, political and ideological dimension in which it is set.[18]

Maine's dismissive remark that much of the *dharmasastric* wisdom consisted of 'dotages of brahmanical superstition' has been rightly criticized.[19] It represents attempts made by the Raj to discredit the ideological foundations of the Hindu hegemony of ideas. But such an off-hand dismissiveness of the earlier law should not detract from the importance of Maine's examination of the Hindu and Indian law in the wider context of sociological explanation. *Ancient law*, *Village Communities* and several of his *Minutes*[20] display an incisive assessment of Indian law, and erect new typologies to further both legal reform and socio-legal understanding. Although Maine did not provide a sophisticated understanding of whom Indian law serviced and why, there is no doubt that Indian scholarship remained trapped in Maine's metaphors for a long time. His characterizations of Indian law were not only the basis of nineteenth and twentieth century law reform but also remain a vehicle for socio-legal analysis to this day.[21] Maine's intuition rather than empirical evidence structured future law reform and scholarship. It was assumed that pre-British *shastric* jurisprudence was static, linked to highly structured patterns of social stratification and· possessed only marginal capabilities for re-adaptive use. Modern law, on the other hand, was seen as constituting a new liberating instrument of change. An insufficient range of questions were — and are — asked about the ideological and material nature of native law.

Too much has been made of Maine's 'status to contract' metaphor in analysing the juxtaposition between imposed English law and Indian social life.[22] More often than not, notions of 'status' and 'contract' have been used to identify the different forms in which Anglo-Indian law was expressed. This emphasis on the form of law has often obscured our understanding of how law and legal procedures are appropriated to social, economic, and political use. The emergent classes who exploited Anglo-Indian law appropriated the so-called 'status' provisions as well as the 'contract' provisions to their use with equal facility.[23] Aspects of the Hindu law relating to the joint Hindu family may well have been conceived in terms of status. This did not stop the Hindu bourgeoisie from using these laws to assert the emerging demands of commercial and landed interests.[24] But quite apart from the transforming effect of litigation upon Anglo-Indian 'status' and 'contract' based law, Indian law did explicate a formal tension between the orientalists who favoured using native Indian legal idioms to serve emerging Indian and colonial needs and the modernists whose work can be seen in the reports of the Indian Law Commissions of the nineteenth century which led to the enactment of various statutes on property, contract, commercial, procedural, and penal law. These enactments, known as the Anglo-Indian Codes, were enacted in the heyday of codification and survive as the basis of Indian statute law today.[25] The modernism of the Law Commission was matched by the vigour with which some judges — like Justice Holloway — vigorously championed the introduction of Western law in India.[26] This, in turn, led to a curious scholarship which compared the law of the *dharmasastra* with Roman law without references to differences in context.[27] In actual fact, many of these controversies about the 'status/contract' or 'tradition/modernity' tensions in Indian law and Indian society are about form rather than substance. Even the discussions about form do not pay enough attention to the relationship between the ideological importance of form and its relationship to the political economy. An insufficient amount of work has been done on the relationship between law, state, and the political economy. The signal achievement of the Raj was to bring the entire social and political economy within the shadow of its law and legal institutions and processes. This was achieved and legitimated by a dual reference to both the *shastra* and the new dispensation. Disabused by Maine's analysis, scholarship on Indian law has to this day not found sufficient insights into the origins and social foundations of either traditional or modern Indian law.

The 'Black Letter Law' Tradition

As the legal and administrative system imposed by the British began to exercise an influence on social affairs and took over the working of crucial aspects of the political economy, legal scholarship was self-protectively encapsulated in a 'black letter law' tradition.

The 'black letter law' tradition seeks to interpret law as a distinct, relatively autonomous reality. Within this tradition, 'law' is separated from morality. It is understood and interpreted by esoteric rules known only to the initiated and critiqued on the basis of self-constituted legal principles and concepts. Such a tradition exists at two distinct levels. At a nobler and more insidious level it concerns itself with legal doctrine — seeking to redefine and reconstitute peoples' understanding of their social, political, and economic relations. This reconstruction of society is guaranteed exclusiveness and derives strength from the notion of the 'rule of law'. Presenting itself as a 'fair' arrangement, the 'rule of law' draws support for the legal reconstruction of social reality while at the same time providing full power to the State to contain transgressions of the letter and spirit of the 'rule of law'. At a less sophisticated level, the 'black letter law' tradition is not concerned with doctrine but with servicing the narrow, technical day-to-day needs of the administrator and legal practitioner. As litigation in Indian courts developed, legal literature concentrated on producing practitioners' books which found cases for lawyers rather than developed jurisprudence. With all this, we see a virtual collapse of Indian legal scholarship both *within* and *about* Indian law. Scholarship on Indian law reached its nadir when India became independent. Indian publishing houses produced stacks of case digests. Standard editions of famous Indian practitioners' books sometimes survived as credible scholarship; but, in the main, deteriorated to become incomplete and improperly classified collections of an ever increasing mountain of case law.[28]

Although many lawyers and judges were prominent in the national movement, the 'black letter law' tradition in India survived virtually unscathed during the troublesome years that led to India's independence. The early lawyer and judge reformers emphasized the importance of the 'rule of law' and propagated the need for an ordered and disciplined social reform to precede political and economic change in India.[29] From 1909, following the granting of representation in the advisory political institutions of the Raj, a new breed of liberal constitutional lawyers concerned themselves with the mechanics of legal change and put forward plans for alternative constitutional structures for India.[30] The advent of Gandhi did not alter the centrality of the 'black letter law' tradition. While the law and administration were criticized, the central thrust of the movement was to achieve legal and constitutional change. In its own curious way, civil disobedience urged respect for legal institutions and processes while making political capital of their misuse.[31] Gandhi's own solutions in dealing with peasant and other uprisings smack of legal compromise and seek — as in the case of Champaran — to provide an administrative and legal *status quo* response to the problem.[22] Ultimately, the aims and objectives of the national movement were narrow legal ones. Once independence was attained, lawyers were left virtually free to play around with the spoils of politics. Independent India continued to espouse the 'black letter law' view that once the Constitution was established and the legal and administrative system reconstituted under its auspices, 'law' would develop within its own relatively autonomous field.

India's Constitution was the largest ever devised in the world. When the draft Constitution was discussed on the floor of the Constituent Assembly, most members were overawed by the document's complexity. Some members regarded the new Constitution as imitative and eclectic whilst others were appreciative of the detail of attention given to it. Even though the discussions on the Constitution seem neat and impressive to the undiscerning reader, the formal record does not do justice to the varying standpoints which were accommodated during the making of the Constitution. While the forces of emergent capitalism have diluted and dissolved the basic political framework of the Constitution, lawyers have used the new legal and constitutional processes for both money and social and political mischief.

The new Constitution generated a new literature.[33] Many of the ideas and idioms used in the Constitution were new. They were imported from various jurisdictions. The new literature concentrated on servicing the needs of administrators, lawyers, and judges. The earlier constitutional law generated by this literature and the courts contains a great deal of scissor and paste jurisprudence. To be harsh on these imitative developments is to undervalue the enormity of the interpretative tasks assigned to lawyers and judges. The scissor and paste jurisprudence which was characteristic of the early fifties represents both genuine attempts to come to terms with the newly posed constitutional questions as well as an intellectual camouflage to reinforce personal and class interpretations of the Constitution by reference to cosmopolitan scholarship. Some of the genuine attempts to reassess doctrinal aspects of Indian law are impressive, but these are rare and were developed more by practitioners rather than the emerging scholars on Indian law. The fact remains that the passing of the Constitution did not stimulate a revival in Indian studies of law. To the practitioner the new idiomatic phrases used in the Constitution were a means, and an opportunity, to reinforce their declining status. To the judge, the new constitutional doctrines provided an opportunity to display his cosmopolitan understanding and learning. A new comparative literature was developed which mechanically cross-referenced English, American, and Australian citations for analogous provisions in the Indian Constitution. While Indian law was caught in its 'craze for foreign things',[34] it was also being manipulated by the more advantaged sections. The result was as piquant as it was purposive. Dressed in Western garments, Indian public and private law served emerging interests with an untidy integrity.

Indian legal scholarship seemed to enter into its mechanical phase.[35] After the *rishis*, the nineteenth century had witnessed a creative, even if patently false, re-interpretation of indigenous law, a carefully worked-through induction of Western law into India and a spate of litigation which gave this law social meaning. The *rishis* became reformers; and the reformers compilers of indexes. Connected more with the market place of litigation rather than the market place of ideas, Indian publishing houses produced a plethora of law reports, legal digests, and practitioners' books. To this was added a whole series of student crib books. But that came later. As it happened, there were not that many law schools in India in the beginning of the nineteen fifties. The more famous law colleges in the presidency towns served the ever-expanding corps of lawyers with a kind of gross learning which impatient young law students tolerated only as a means of access to the profession. Apart from some notable exceptions, Indian law schools were, and are, not the source of a new-found theoretical nourishment.

Mature versions of the 'black letter law' tradition were kept alive by contacts with English jurisprudence and English legal education.[36] Even after independence, there was always a large stream of students who went to England to read law in English universities and be called to the English bar. Most of these went into law practice and many became judges. The quality of their imported learning is visible in their advocacy and their judgment writing. It must not, however, be readily assumed that these 'England returned' lawyers — as they are sometimes derisively or appreciatively called — represent either the cream of the Bar or its leading intellectual talent. The India educated — or 'home grown' — Bar lawyers are more often than not infinitely more imaginative and adroit in their use of legal doctrine and legal processes than their 'England returned' counterparts. Perhaps, the latter were less taken in by the beatitudes of English learning and quite prepared to develop a more responsive — even if less symmetrical — body of jurisprudence.[37] Much of the 'England returned' student's exposure to English scholarship lay either in undergraduate courses at English universities or through the more practical Bar courses. This limited exposure provided incomplete insights into the true nature of English law but entrenched the 'black letter tradition' in Indian law more securely.[38] The few students who did dissertations at English universities — principally at the School of Oriental Studies, London — stayed, by and large, within the 'black letter law' tradition.

The impetus for a renewed Indian scholarship in law came from exchange programmes with the United States.[39] Unlike their English counterparts, these programmes were pitched at post-graduate level and took a much more mature research-oriented look at Indian law.[40] In time, some American scholars also began to take an interest in Indian law.[41] The first and most direct impact of this traffic of scholars was at the very elementary level of legal research methods. Simple matters like citing all the case law, looking up a more comprehensive range of cases, reading background material, and writing more adequately referenced papers came to be given a more considered attention in India. This affected undergraduate teaching. New law schools — one or two with American help — were established. The creation of the Indian Law Institute represents a move to establish a more adequately referenced, even though not necessarily more imaginative, legal scholarship.[42] At post-graduate and law teacher level, more work on legal doctrine and public policy became visible. Even where this work was competent and thorough, it explored American paradigms on the basis of an ill-thought theory of comparative equivalence that culture-specific legal doctrine from foreign jurisdictions could, without further ado, be cross-fertilized with nineteenth-century Anglo-Indian law and adapted for use in contemporary India. The second — less direct but more significant — impact of the American connection came from the writing and work of American scholars who were funded by American sources and who sought to impose American ideas on Indian jurisprudence. One wave of this scholarship concerned itself with public law and questions of public policy. Broader speculations about public policy were replaced by more incisive case studies. Another wave came to be concerned with legal education and legal research. In time, Indian law came to be studied by Americans not just as something unique and mysterious, but for its own sake. The emerging pattern of American legal studies in India had something of a 'Marshall' plan flavour to it. But it also reflected America's experiment with the New Deal. While some of the research was thoughtful, most of the American initiatives reflected distinct patterns of legal studies which had been invented by American law schools and which were being experimented upon in Africa, Latin America, and elsewhere. This was the 'law and development' approach. It produced its *pandits*, its *gurus*, its followers, and its students.[43]

Mainstream scholarship in Indian law continues to espouse the narrow 'black letter law' tradition. It is concerned with a knowledge of statute and case law. A leading constitutional law theorist has insisted that constitutional interpretation should limit itself to the consideration of doctrine.[44] This approach has received support from judges, lawyers, academics, and, above all, the Law Commission.[45] But given the social rhetoric of the Constitution, government planning and the increasing differential between the rich and the poor, there is a considerable amount of pressure to break out of the 'black letter law' tradition to ask a new range of questions about the task and purposes of Indian law.

The Imperatives of Planning

The declaration of India's five-year plans to implement Nehru's socialism had ideological, instrumental, suspicious, and pathological dimensions.[46]

At the *ideological level*, it was a powerful statement of a new kind of ideological framework which would lead India towards modernization, industrialization, secularism, and socialism. Some of this ideology was articulated in the Constitution itself. The 'State' has been given powers to restrict Fundamental Rights — ostensibly to achieve these stated aims. Federal governments at the Centre were given increased powers to implement planning and other goals.

New instrumentalities were created to plan, review, and monitor programmes for the disadvantaged — and especially certain tribes and castes (who came to be called Scheduled Tribes and Castes because a Schedule of the Constitution earmarked them for special attention) and Backward Classes. One part of the Constitution called the Directive Principle of State Policy, which has been variously presented both as the soul of the Constitution[48] as well as a 'veritable dustbin of sentiment',[49] listed certain priorities: the national ownership of all resources, a uniform civil code to replace existing personal laws, the creation of new but indigenous based forms of dispute settlement and a new economic policy for a living wage for low (and, presumably, no) income earners. All this was really at the level of ideological declaration. The parallel political declaration of creating a 'mixed' economy with State Socialist and capitalist components set the stage for adjustments, compromises, entropy, and corruption.

At the *level of instrumentalities*, new laws had to be designed which had to be fitted into the legal framework of the Constitution and guided by a socio-political understanding of the uses and abuses which any such programme would both engender and encounter. New regulatory organizations and bureaucracies had to be entrusted with powers to match the gigantic responsibilities ascribed to these agencies. These instrumentalities had to be created by a 'democratic' legislature and implemented by a politically 'accountable' executive. This left open a great number of possibilities. Were these new instrumentalities constitutional? What kind of powers should the new bureaucracies have? What would constitute the framework of limitations within which the power of these new instrumentalities would be contained? Would the simple English 'rule of law' model, distrustful even of the delegation of 'wide discretionary authority' suffice? Would there be a need for an improved 'anti-arbitrariness' jurisprudence? What were the 'process' requirements that would, or should, be imposed on the new regulatory State?[50] And, if all this failed—as, indeed, some, if not all, were expected to—*quis custodiet ipsos custodes*: what kind of instrumentalities would have to be created to watch over the custodians? How would all these new instrumentalities work as they struggled to accommodate various plural pressures in society (including efforts designed at their own corruption)?[51] In the Indian context, some of these questions came to the forefront in a controversy between Nehru and the law courts.[52] Nehru — amazed that law courts had interpreted some aspects of the 'eminent domain' and 'just compensation' provisions of the Constitution in favour of the traditional landlords who had fallen out of favour with the ruling Congress—chastised the courts for standing in the way of socialism.[53]

At the *level of suspicion*, there was some concern about what the new ideology really was, whom it was intended to benefit and to what end. Many of the programmatic manifestations of the new ideology grew out of election promises which were proffered in the knowledge that a manipulation of the new instrumentalities by the advantaged (and the incapacity of the disadvantaged to develop the skills to take charge of the programmes intended for their benefit) would make a nonsense of these promises. There was also a great deal of doubt as to whether the system of State regulation — itself enmeshed in the economic realities of a mixed economy — would have the double effect of legitimating the demands of the emerging bourgeoisie as well as creating more processes for mutually advantageous double dealing between administrators, politicians, and their benefactors. It soon became apparent that the new State-backed 'co-operatives' and the much publicised redistribution of land would benefit the new *kulaks* rather than the poor peasantry. The new industrial policy would help business houses to the prejudice of others less fortunate and less enterprising.[54] There was a great deal of distrust about the new 'secularism'. The constitutional plea for a 'uniform civil code' was seen as a weapon to crowd the Muslim population into accepting a reform of their law. This would derive support from some sections of the Muslim community and alienate the rest. The new secularism did not aim to allow various communities to exist. It refashioned them in accordance with a new dispensation. The new secularism would affect the government's control of religious endowments, the growth of traditional centres of religious education and the overall policy of the Indian State towards a whole range of issues which are part and parcel of the day-to-day lives of contemporary Indians living in India's multi-structured pluralistic society.[55]

At the *level of pathology* everything seemed what it was not. The new instrumentalities did not work in the way in which they were supposed to. Empirical assessment and intuitive assessment both suggested that something quite different was happening 'out there'. The legal system as a social reality reflected and reproduced social realities not affected, or even moderated, by the proposed change. One way to assess this pathological practice was talk of the 'gap' between law in books and law in action. What followed were explanations and declarations about how this 'gap' is to be filled. It has been convincingly argued that this kind of analysis about the relationship between law and society is misconceived.[56] 'Gap' theory is really a political ploy to persuade the disadvantaged that the programmes which were designed for their benefit had, somehow, accidentally, rather than deliberately, misfired. 'Gaps' are created by powerful forces of change in society. Politicians make the promises of socialism and, then, shortchange these promises by maintaining 'gaps' to prevent their fulfilment. Legal techniques cannot by themselves be used either to comprehend these 'gaps' or cross them.

Many Indian lawyers and judges tried to develop broad legal approaches to match the ideological demands of Nehru's regulatory State socialism. Nehru's thinking was inspired by the Soviet experiment with planning rather than Roosevelt's New Deal. However, Indian planning had to be located within the political economy structured by India's new liberal Constitution. While Nehru was confident that the constitutional structure would assist the process of planning, he soon learnt that legal and constitutional intricacies intertwined with planning in a much more complex way than he had imagined. In the early fifties, the judges of the Supreme Court, unable to shake off the pervasive effect of the Anglo-American jurisprudence in which they had been so assiduously schooled, quickly found themselves at odds with the new programme; and, indeed, as we have indicated, with Nehru himself.[57] A lonely protest by a, then, recently retired Chief Justice of India in 1955, that there had been a big mistake and the approach of some of the judges had been misunderstood, was itself ignored and, perhaps, misunderstood.[58] The judges were called upon to support and develop a new 'planning commission jurisprudence'.

Some judges responded to this challenge with zest and creativity in various speeches, papers, and judgments.[59] Justice S.S. Dhavan of the Allahabad High Court explicated his understanding of the legal and constitutional implications of the new challenge. In one of his judgments, he described the problems:

> The Founders (of the Constitution) were fully conscious of the opposing forces of unity and disruption which battled for mastery in the endless corridors of the history of India. The disruptive forces were local traditions, regional royalties and a jungle of customs and laws which produced 'nations within a nation'.[60]

Dhavan took the view that, like English and American judges, Indian judges should take an instrumental view of law but remould it not for the support of vested interests but in a manner consistent with the redistributive justice championed by the new State-sponsored secularist socialism. He felt that all judges at all times have used various juristic techniques in order to achieve the ends that they believed in.[61] Thus, his interpretation of what Horwitz was later to call the age of formalism was that 'formalism' could itself be considered as a strategically devised instrumentalism designed to achieve certain purposes. Dhavan took the view that a proper instrumental secular socialist approach was not just enshrined in the Constitution but could be traced back in India's history as part of a continuous reforming impulse present in the *dharmasastra*.[62] While it is easy to prove that the *shastra* displayed both ingenuity and adaptiveness, it was somewhat more difficult to show that *dharmasastric* learning either contained a critique of distributive justice or asserted the claims of a State socialism which would transcend India's highly stratified hierarchical society. Reassessing the new legal system, Dhavan was the first judge to give primacy to the Directive Principles of State Policy (in the Constitution) as a guide to statutory and constitutional interpretation.[63] This was not just another version of the evolutionary progression of societies '*from* status, *to* contract, *to* plan', of the kind that had been suggested in Seidman's oversimplified prescriptions for African society.[64] Nor was this another planning commission' jurisprudence looking for new techniques to enforce existing regulatory laws. What was being emphasized was the need for a new ideological hegemony drawn from reform movements in previous tradition to give support to the new socialism. Dhavan called for a renaissance in Indian jurisprudence which would link the past and the future in one coherent whole.[65] In this, emphasis would be placed on the King's duty to his subjects to eliminate poverty and reduce social and economic differentials. Such an approach could easily be trivialized as rhetoric. But it was an important call for a new indigenous jurisprudence. The creativity that Justice Dhavan had in mind is exemplified in the vigour with which his own judgments sought new techniques, both to restrain the lawlessness of the State as well as advance the cause of the disadvantaged.

The search for a new judicial response was both advanced by, and suffered a setback in the hands of, P.B. Gajendragadkar — a one-time Chief Justice of India (1963–5) and Chairman of the Law Commission (1971–8). Through his judgments, Gajendragadkar stabilized the relationship between the courts and the contemporary government of the day. He settled many questions about the approach to labour law, the control of Hindu religious endowments, India's policy of compensatory discrimination and the implications of the new secularism.[66] The figure of Gajendragadkar strides across Indian jurisprudence in the sixties. However, his ideas were never critically assailed or evaluated. Gajendragadkar's status as the Chief Justice of India, his jurisprudence and his extra-judicial pronouncements acquired an *ex cathedra* status and character. Gajendragadkar gave respectability to the Planning Commission view of law without adding to its jurisprudence. Relying on the conservative American jurist, Roscoe Pound, Gajendragadkar's call for a new social engineering was purely hortatory in nature.[67] His own decision making is characterized by its capacity to project government's policy compromises in an elegant and dignified way.

His declarations cannot be taken as a contribution to legal theory but as an indication of a kind of badge of commitment that judges were supposed to wear on their political sleeves without losing their identity as judges. It is not surprising that, in the early seventies, the term 'committed' judge acquired a new and totally different meaning. A political attack on those judges who had resisted parliament's efforts to claim a plenary sovereign power to amend the Constitution was couched in terms of their being 'uncommitted' to the socialist developmental goals of the Constitution.[68] 'Committed' and 'commitment', in this context, soom came to mean a loyalty to a particular government or regime. The effect of this controversy was to politically discredit the whole 'socialist judge' approach to modern Indian post-independence law.

The cause of a 'socialist' jurisprudence gained some momentum at the hands of Justice Krishna Iyer and a handful of other judges in the Supreme Court. Before his appointment to the Supreme Court, Krishna Iyer had been a minister in the ill-fated Communist government in Kerala (1957–9), a judge in the Kerala High Court, and a member of the Law Commission of India. Krishna Iyer had delivered some significant judgments to reform Muslim law in the spirit of secularism and spoke the language of Nehru socialism.[69] At first he took a harsh and somewhat crude view of the law, called for better law enforcement and demanded harsher measures to secure the implementation of development programmes.[70] He supported a grandiose government dominated and sponsored legal-aid programme which was very rightly criticized by one of India's leading legal theoreticians.[71] In time, Krishna Iyer became acutely conscious of the fact that the pathology of the legal system was the reality; and that the points of suspicion that we have talked of earlier were, alas, validly taken.[72] In his work on the bench — which is too varied to admit to any consistent description — he paid considerable attention to redefining the manner in which the instrumentalities of welfare and justice should be disciplined. He inaugurated a new due process to arrest the continuing lawlessness of the Indian State. In time, his legal theory came to be elaborated into many strands.[73] Strongly committed to establishing a socialist jurisprudence, Krishna Iyer called for a more responsible attitude to law and the socialist programme emanating from the various institutions of State, while simultaneously attacking them for their corruption, susceptibility to manipulation, and their meaningless drift into entropy. Alongside, he called for and supported a non-State-based activism to gather the forces of socialism together. His piquant use of language, his rich — but not always easy to understand — metaphors, his undisciplined forays into various conflicting theoretical stances, his reliance on the writings of Maharishi Mahesh

Yogi to develop the 'integral yoga' of development tend to confuse the already considerably confused student of his writing.[74] In the end, it is never wholly clear whether Krishna Iyer is asking for a new sociology of law, a more populist use of the legal system, or simply continuing to adhere to his State-dominated theory of law and development.[75] Justice Bhagwati—Krishna Iyer's colleague on the Supreme Court—has gone a considerable way towards enlarging the theatre of legal institutions in which the poor—or, at least, those sponsoring the cause of the poor—can participate. While his work continues to emphasize the State-centred nature of law and development, he has paid considerable attention to the importance of voluntary groups in mobilizing the law.[77]

These judge-centred approaches to a socialist jurisprudence are matched by a vigorous scholarship sponsored by the Planning Commission on the effect and working of governmental programmes. A number of Planning Commission reports stressed the need to develop special administrative, bureaucratic, and other techniques which were, allegedly, required to convert planning into social reality. Theoretically, the Planning Commission-approach had a nice evangelical 'St. George and the Dragon' touch to it. It meant that India could hold on to its liberal institutions and develop huge bureaucracies to implement the allocative planning which would change its future. In a sense, it was a solution looking for a problem. The nature of the Indian State — variously described as 'soft' and 'over developed' — did much to damage its own plans. Undeterred, the Planning Commission has continued to espouse the integrity of its approach. It diverted its attention to law soon after the first phase of agrarian reform. In this phase, the various states had enacted legislations which took land away from traditional land-owners and gave them, ostensibly, to the poor, but in fact to a new class of emerging *kulaks*. The Planning Commission initiated a series of studies which analysed the working of this legislation.[78] The standard approach was to look at the formal objectives of the legislation, describe the instrumentalities through which it was to be achieved, and determine whether these formal objectives had been achieved. Unfortunately, these studies did not really examine the nature of the agrarian economy.[79] They were content to publicize the isolated successes of the plan. This standard approach to studying the effect and impact of legislation has become quite settled and stylized in India. It has been applied to various other programmes of the government.

Although the Planning Commission approach to law (and development) continues to gain adherents, it has not really helped to establish a radical sociology of law. It has also not really monitored the effect of legislative and other programmes. Its respectability hovers precariously between adhering to the undisguised rhetoric of planning and articulating meaningful but partial insights into the working of the State. Its support for the State leads as easily into the development of a privileged, *status quo*, law and order society as it does to facilitate a more just and equitable planned growth. In a sense, the Emergency represents the high point of this approach to law and development. During that phase all the problems and contradictions of the Indian State, and of this approach, were exposed. A new emergent class had corrupted their own institutions to a point where these institutions were incapable of securing any kind of substantive rationality to the political economy. Each part of the State was being privatized by various vested interests in various parts of the political economy. Each fragment pursued its own private aims with a ruthlessness which undermined both the pretence and the reality of Indian socialism. The 'official' jurisprudence of Indian socialism became suspect as the developmental aims of the regime came to be viewed with disbelief.

New Perspectives

Not surprisingly, a large part of the literature on law concentrates on the work and role of the judges. Both the Planning Commission-approach as well as the 'black letter law' tradition produced their own view of the judicial functions. We have already seen how the Planning Commission view — in the hands of crude instrumentalists — led logically to the concept of a 'committed' judge. The 'black letter law' tradition sought to narrow the perspectives of the judge. Seervai, a constitutional theorist, argued: 'By convention, judges are expected to abstain from public pronouncements on political, economic, and social issues which are likely to come before them in their judical capacity.'[80] In later years he citicized Justice Krishna Iyer personally for conflating legal and other considerations so as to coin the Constitution in his own image.[81] At the root of these controversies lies a political dispute between 'Statist' judges who support State planning and 'free market' judges who feel that the interests of the market and the corporate sector need protection. The latter are supported by the 'free market' bar whose political objective is to immunize the judicial process from the influence of the rest of the State so that it is more responsive to the intricate politics generated by litigation. In a sense, the controversy has become arid and overstylized. Judges cannot be straight-jacketed into a simple theory of judicial decision making. They must be responsive to the other agencies of the State as well as maintain their independence from these agencies. Not surprisingly, controversies about the role and function of the Indian judge have continued. In 1973 some allegedly 'free market' judges resigned because their claims to be appointed Chief Justice of India

were passed over.[82] Several years later, there was an acute controversy between Justice V.R. Krishna Iyer and one of his colleagues
about the role of the judge in India.[83] Today, the question, 'who is
a good judge' cannot even notionally be separated from the concept
of a politically acceptable judge. Successive regimes have tried to
discipline the judiciary into complicity. In the aftermath of the
Transfer of Judges cases[84] — permitting the executive to transfer
judges from one High Court to another — a great tussle is being
fought out — between the various constituencies which support
'State', 'committed,' 'socialist,' and 'free market' judges respectively
— about the future of the judiciary in India.

Academic writing has paid some attention to Oliver Wendell
Holmes's observation that what the judges did, in fact, was the law.
An American scholar — George Gadbois — showed that an analysis of decisional patterns demonstrated that judges decided variously
and inconsistently — even with themselves — in some classes of
cases.[85] Unfortunately, he did not examine the ideological sources
of their differences, their choice of juristic techniques, the reasons
for their choice, the nature of the social and political differences
between them, or the structural context in which these pressures
impinged on the judicial process. Some of these defects were remedied in a full-length study of the Supreme Court in my work,[86]
where I have stated that many of the stylized controversies about
the role and function of the judge were theoretical and had not
really looked at the work, predicament, and performance of the
Indian judge. The Indian judge's behaviour was a lot more complex
and confused than stereotyped models of judicial behaviour suggested. Apart from a few notable exceptions, most Indian judges
were quite baffled about what they could do and what was expected
of them.

> In the end, we must regard the attitude of Supreme Court judges as
> typical of the decision-making habits of middle class metropolitan Indians:
> technically unpredictable, not uninfluenced by imitative cosmopolitan
> habits, conditioned by a native instinct not yet predictable by the psycho-
> logist or documented even by the novelist, the dramatist or the fiction
> writer, and suffering from an over sensitive opinion of their lonely and
> unparalleled position.[87]

A different kind of analysis of judicial and legal institutions was
required. Judges courts, law, and legal institutions had other dimensions to them. Judges were part of a class. They were both the
custodians and creators of ideology. As conceptive ideologists, they
reconstituted the nature of Indian society and provided the basis
on which many social and public issues were thought through. In

my later works, I have examined the working of the courts from an
administrative and financial point of view, studied how they reacted
to political pressures, analysed problems of patronage in respect
of their appointments, described them as bureaucrats but also urged
an analysis of their work within not just the context of the State but
also after taking into account the private market economy of litigation and lawyers which controls and circumscribes the working of
the judiciary.[88] R. Sudarshan's fascinating paper exploring the
judiciary as a class institution and his research into the Supreme
Court's determination of questions of economic policy is an excellent
example of original work done in this field.[89]

A powerful plea has been made that we should consider the
judiciary as a political rather than a 'legal institution. Apart from
Gadbois's emphasis on gross data analysis and the institutional
maturity of the Indian judiciary, a left wing scholar has tried to
emphasize the 'class' nature of the judiciary.[90] Unfortunately, this
effort has pursued only a consequentialist analysis to show how some
of the *decisions* of the courts have favoured the well-to-do.[91] However, even within a sophisticated left wing tradition, the power of
a class is not just reproduced on the basis of a few coercive appeal
court decisions. The courts articulate and sustain an ideological
hegemony which elicits support for the system and determines the
manner in which it must be critiqued. Upendra Baxi's lectures on
the Supreme Court strike a different note.[92] Ignoring the distinctness
of 'law' as a *discrete* social and political reality, Baxi treats the Supreme
Court as an arena of politics and counsels judges to follow the fool-
hardy course of preaching a new distributive and 'populist' justice.
Many may query Baxi's refusal to understand the distinct nature
of judicial politics or question his weak analysis of the judiciary as
part of the State. However, Baxi's book urges us to assess the courts
as centres of social and political conflict rather than simply as producers of esoteric legal doctrine. Baxi remonstrates that analysts
and practical politicians alike should not ignore the ideological
dimensions of judicial decision making. But Baxi himself does not
pay enough attention to the nature of judicial ideology and the circumstances and constraints within which it is created and respected.
Judges are important creators of ideology. Some of this ideology is
simply rhetorical in nature. But, the more significant work of judges
is both constitutive and conceptive. As constitutive ideologists,
judges and others reconstitute our perceptions of society and develop
a theory about the distribution of entitlements and power. As conceptive ideologists, they determine the manner and framework
within which value preferences and public policy are discussed.
Much of the respect that is attached to their declarations stems
from the deceptive neutrality which cloaks what they do. Baxi's

prescriptive exhortation that they preach the new socialism from the pulpit compounds a righteous effervescence with naivety.

Despite the fact that there is so much discussion about the role and function of the Indian judiciary, very little is known about its work and everyday functioning. The Indian judiciary is plagued with a considerable backlog of cases. The 'arrears' of cases awaiting decision in virtually all of India's courts date back to many years. In many instances, the original parties to the cases are dead; and the parties who had taken their place have also passed away. In criminal cases, many prisoners have been kept in jail for terms longer than those permitted for the offence for which they were being tried. Each time, the problem of arrears is examined by a government committee or commission, it is done so on the basis of hopelessly inadequate information.[93] Routine information about the judiciary is no longer published or made available. There is even less information or analysis about the work of the judiciary and the nature of litigation in the context of the political economy.[94] Indian scholarship has given only partial attention to the study of the judiciary. It tends to jump into political controversies rather than study the judiciary with intellectual vigour. This is as true of the analysis in the 'black letter law' tradition as of socio-political analysis of the courts' decisional and ideological role. While the controversies are colourful and exciting, there are too many gaps in our knowledge and too few insights in our understanding of the judiciary and its relationship to the State (in which it is housed) and the private market of litigation and lawyering (which determines its uses). This is partly due to the shadow of the 'black letter law' tradition which has isolated the judiciary from trans-doctrinal criticism.

Just as the 'committed' judge approach came to be reassessed, fresh studies of legislation were also conducted. It was pointed out that much legislation is purely symbolic in nature. This was certainly true of some of the legislation on Untouchability.[95] It may also be true of the legislation that came to be collectively described as the Hindu Code.[96] In many instances — such as the enactment of the Forty-Second Amendment — the enactment did not do enough justice to the ideals it purported to serve. The same was also true of the Amendment which tried to amend the (Forty-Second) Amendment.[97] Many varieties of social and economic legislation did not just fail to achieve the planning purposes they were supposed to achieve.[98] They became the vehicle through which emergent groups and classes obtained social and economic power. Forest legislation and policy were shown not to have a lesser ecological purpose and used to harass the tribal poor and hill folk while advancing the interests of wood contractors.[99] It was also found that some legislation transformed over time, acquiring new beneficiaries and new purposes. It was shown that the Hindu Marriage Act was abused by parties to harass one another rather than to resolve their differences.[100] Although no clear lines of inquiry into legislation have emerged, the Planning Commission view that legislation would, in most circumstances, generate social change has been found to be naive and misleading.[101] In this alternative approach, legislation is just another declaration of government policy. The fact that such policy is consecrated as legislation does not mean that such initiatives are not spurious. Inevitably, a whole range of social forces appropriate both the legislation and the instrumentalities created by it to liberate them from their allegedly intended purposes. We are only one step away from the more radical view that most legislation is designed to mystify the poor by invoking grand social purposes to sustain false promises. Such legislation helps to legitimate the *status quo* by showing that society cares about its disadvantaged. Under the guise of social reform, some of this legislation really facilitates the power of the advantaged and the repression of the poor.

The central core of the Planning Commission approach to law and development makes a lot of assumptions about the relative autonomy and neutrality of the Indian State. The State is taken to be sincere in its commitment to the plight of the exploited and disadvantaged. It is also assumed that the centrality of purpose of the State will be maintained by the various instrumentalities and functionaries that it sets up. In a report on *Public Interest Litigation*, I argued that large parts of both the 'local' and 'higher' State had become virtually indistinguishable from the social groups and forces that seek to manipulate their use.[102] Over the years, the Indian State has been directly or indirectly 'privatized' to the use of some sectors of the political economy to the exclusion — or partial exclusion — of other sectors. Some parts of the State were privately auctioned to the highest bidder. Thus, all the assumptions that one could make about the integrity of planned socialism are not justified. A range of activists have argued that the real future of development lay in the hands of activist groups.[103] Some of these groups chose the path of armed 'liberation' in order to achieve their purposes. These groups were virtually annihilated by forces of the State. Other groups — such as the Gandhian groups — have continued their work in their own *ashramas*. The quest for an alternative development based on the disadvantaged finding their countervailing power, has been given much attention by a group of legal thinkers who are not just cynical about the State's planned development but who feel that the State's programmes repress and mystify more than they fulfill. The development debate will continue. Much depends on how the activists plan to use the legal and administrative system; and, to what effect.

The Sociology of Indian Law

These new alternatives went a long way towards developing, but did not establish a distinct sociology of Indian law. The growth of sociological studies in India has meant that more and more scholars began to take an interest in Indian society; and directly in Indian law.

Indian society has baffled and fascinated a vast number of Indian scholars. There have been major debates about the distinct nature of Indian society and whether it should be considered as a singularly unique society or whether it admits of comparisons with other societies. Dumont's powerful view of the Indian concept of man — *Homo Hierarchichus*[104] — asserts the existence of a paradigm notion on which Indian society's understanding of itself rests.[104] Alternative lines of inquiry stress the need for empirical data or view the struggles within Indian society from a distinct left-wing understanding of the political economy. The new scholarship has done much to dispel the more orthodox British view that Indian society, which was fissiparously divided into castes, clans, and factions,[105] rather than its foreign masters was basically resistant to change. According to this image, the task of scholarship was to understand a huge range of the complicated interconnections of this allegedly static society. An array of scholarship became interested in the basis of caste ranking and the importance of ritual.[106] All this was roughly equated with some kind of customary law which lay immanent and eternal beneath the ravages of history.

With the rise of anthropology of law, there was some kind of interest in the nature of dispute settlement. Many studies of Indian villages examined the social structure of the village and devoted part of their analysis to rule patterns followed in the village as well as the manner in which disputes within the village were resolved.[107] That this approach obtained the blessings of Srinivas[108] — whose influence on the growth and development of Indian sociology can hardly be exaggerated — did much to generate work in this field. Anthropologists and sociologists have done some increasing work on various aspects of dispute settlement amongst caste groups and tribals.[109] Less attention was paid to dispute settlement in less traditional groupings created by an emerging capitalist political economy. Towards the end of the sixties, the study of dispute settlement became more and more fashionable. Sociologists and lawyers began to examine various village-level dispute settlement mechanisms.[110] In addition, some work has also been done on the working of Nyaya Panchayats which have been created by the State in village areas — partly as an ill-conceived aborted revival of traditional dispute settlement bodies, partly to divert minor cases from

courts and partly as a means to displace existing or potential mechanisms for dispute settlement.[111] But these studies of dispute settlement have not produced new distinct methodology either about
the *form* in which legal relations are devised or about the political
economy pressures within which such dispute settlement takes place.

Ironically, there has been very little systematic work done on
Indian litigation.[112] Indian litigation is quite unique. More stories
can be told about Indian litigation than about litigation anywhere
in the world. Derrett has speculated about the nature of Indian
litigation through a sample of appeal court case law.[113] Galanter
has noticed how so much of Indian litigation is deep rooted in social
and political quarrels.[114] The press has used litigation to make a
nonsense of tough and repressive anti-press laws.[115] Historians have
an interest in certain aspects of litigation. Appadorai's work on litigation and religious endowments and Price's work on litigation
with regard to a large landholding, in the nineteenth century
are powerful examples of this interest.[116] Recently, there has been
some work on the work-flow of the courts. But all this does not constitute a mature examination of Indian litigation. Some work has
also been done on Indian lawyers. Kidder's work on image management amongst Bangalore lawyers offers many insights into Indian
lawyering as well as to a general understanding of the sociology of
the professions.[117] Khare's work on sea lawyers[118] and Morrison's
research on 'munshis and their masters'[119] re-examine the entre-
preneurial activity that is generated by India's legal system. Cohn's
use of the 'status/contract' metaphor as a tool of analysis[120] has to
be considered against Mendelsohn's view that land forms the basis
and clue to the understanding of Indian litigation.[121] Work on the
organization of legal services in India continues. Some of it is more
concerned with lawyers' attitudes and aspects of role theory rather
than set in the more relevant context of litigation or a wider structural analysis of Indian society.[122] Even at the simple level of fact,
we know very little about how lawyers and litigants bargain, negotiate, settle, or process a case, or the manner in which the case is
adjudicated. The attempts made to reform the system have concentrated on looking at formal processes rather than how the system
works. Nobody knows who uses the legal system and why. How has
the system been used over the years? There is virtually no research
in this area. Everyone seems to have answers to these questions based
on half-informed intuitions; but these answers are incomplete and
not always convincing.

The working of Indian law has not received systematic consideration. British administrators and others looked at the system through
Anglophonic eyes. Maine's broad generalizations about Indian law
formed the basis of reform but were never drawn together into a

coherent whole. There were a great number of books on legal history; but these concerned themselves with formal issues and described the jurisdiction of various legal institutions. Some historical studies have tried to explain law as part of the State and the political economy.

More recently, Baxi has put together a set of essays to examine what he calls 'the crisis of the Indian legal system'.[123] Baxi — to whom the increasing interest in the sociology of Indian law owes a great deal — tried to show that India's legal system is going through a crisis which is affecting both its capacity for self legitimation as well as its general sense of direction. From this he makes the somewhat dramatic jump to the prescriptive solution that the Indian legal system can be made to help the cause of the poor. It is not clear whether this advice stems from a wish fulfilment or is based on an analysis of what the legal system might actually offer the poor. At times, the general drift of his analysis seems to exude the parasitical interest which India's middle classes show in the predicament of the poor, without contextualizing the acquisitive and fratricidal struggle amongst the middle classes, no less in their legal scholarship as in any other sphere. Baxi's book represents more of a *tour de force* rather than an argument; more an exhortaion rather than an analysis. It is part of a growing social and political movement to develop a greater interest in the predicament of the poor and advance their cause through public advocacy and what Baxi calls, 'social action litigation'.[124] This has generated a great deal of interest and research on the plight of women, the repression of the tribals and forest dwellers, the importance of the development of a new poverty jurisprudence, the supply of legal services, patterns of legal mobilization, and a reassessment of the meaning of development. Developing an activist and participatory theory of development, the top down bureaucratic theory of development has been seriously questioned. These new initiatives have not created new systematic plans for the future. They have merely assembled together fleeting insights into the nature of Indian society and the latter's appropriation, use, and abuse of Indian law.

A serious attempt to look at Indian law and society was made by Professor Derrett of London University. Derrett's essays — including his *Religion, Law, and the State*, were one of the first to reassess Maine's over-generalizations of Indian law and society.[125] Derrett's work cuts across many subjects. He is wrongly identified as belonging to the administrative and judicial tradition of the Raj. Unlike one of his distinguished colleagues, Derrett was neither a judge nor an administrator of the Empire. Some of his work is written in a lucid, but somewhat abrupt and sharp, style which might tax the sensitivity of the reader who might be impelled to read some of his more acerbic comments as imperial condescension. Starting as a historian,[126] Derrett used his considerble linguistic skills to master several Indian and European languages in which the legal literature of, and on, India was to be found. At home both with intricate legal points and *dharmasastric* literature, Derrett sought to combine a multiplicity of talents to generate a new kind of approach to Indian — and, more particularly, Hindu — law. Beneath the analytical formulations of law, Derrett discovered attitudes, beliefs, and ideas which were central to the development of Indian law — as it took shape in the hands of both the judges and administrators of the Raj.[127] At the same time, Indian litigation threw up a pathological practice which constantly gave fresh and new, social meanings to the text of the law. Studying the development of personal laws, Derrett not only examined the various juridical forces which gave rise to Anglo-Hindu law but also implicitly — even if incompletely — examined the political and social tensions through which Anglo-Hindu Law was born.[128] He provides a unique — even if not always shared — insight about Indian law, how — and for what reasons — it was developed and whose interests it served. Even if we are not always happy with his answers, his writings take us well beyond the questions raised by students of legal doctrine and seek to explain the growth of modern Indian law in a socio-legal and historical context. There are, however, several problems with his approach. While he was right to dismiss some of the 'law and development' claims made by Indian judges and administrators he was overcautious in his approach to personal law reform. He assailed Hindu judges for trying to reform 'Muslim law and assumed that legal change must proceed on the basis of social consensus.[129] Derrett's writings display an imperfect understanding of the political economy of the Indian State and the extent to which the Indian State balanced and manipulated power, status, and authority on the basis of class, caste, and economic purpose. He was himself greatly disappointed to witness the collapse of one of his favourite theories about Indian law. Derrett had always believed in the ideological unity of Hindu — and Indian — law. Aware that the heyday of the *dharmasastra* was over, he, nevertheless, believed that the ideological hegemony of ideas that had seen Hinduism through many centuries would play a crucial role in the governance, control, and exploitation of Indian society. It is not surprising that he developed a 'critique of modern Hindu law' which looked at the tension between this ideological hegemony and practice.[130] By 1976, he reported on 'the death of a marriage law' and pronounced 'an epitaph for the rishis'.[131] Despite all this, his central concerns about the relationship between ideological hegemony and actual practice are as challenging for future studies as they were to studying the transformations to which Derrett had given so much considered attention.

The essays in this volume represent another attempt to look at law and society in India. Marc Galanter went to India in 1957 having studied philosophy and law, taught at a law school as a teaching fellow, and worked with the American Civil Liberties Union. The atmosphere of Indian law in which Galanter was introduced was both settled and changing. The Constitution had been passed. The Second Five Year Plan had been launched. Parliament and the Courts had had their first round of constitutional battles. A fresh spate of legislative enactments were being passed by the various legislatures, and various socio-economic programmes to help the poor — especially the Untouchables — were being launched. Funded by Fulbright to study the abolition of untouchability, he soon realized that it was not enough to study statutory declarations and analyse the meanings given by the legal fraternity and courts to these declarations. What existed was a far more intricate problem, enmeshed in Indian society. The real issues required an understanding of India's socially stratified society and its often contradictory patterns of social mobilization. It required a much deeper understanding of both the government's intentions as well as of Indian society's receptivity to change. A much more extensive study had to be undertaken of the legal system in order to determine its true social nature and the working, use, and abuse of legal and other social processes. Galanter's work on untouchability (which has been published elsewhere) and his work on compensatory discrimination (which has recently been published as a book under the title, *Competing Equalities*) demonstrate how this wider approach can be used to study various aspects of law and society in India.[134] Galanter later recounted that, as his research interests crystallized, he became increasingly interested in

> . . . the relationship of official regulation to non-governmental controls of social life — and more generally the relation of law to other normative orderings in society. I became intrigued by the emergence of distinctively modern legal institutions and their relation to 'traditional' social patterns, and in particular, the way in which modern law adapts to give expression to and trasforms the 'primordial' or ascriptive human affiliations (religion, caste, language, tribe, ethnic group, etc.). In policy terms (this meant examining) . . . how plural societies may combine a commitment to equality with responsiveness to the special problems of component groups. In studying these problems . . . (the point of interest lay) not so much in the development of legal doctrine, as tracing empirically the relationship between doctrine and practice and being concerned with the responsiveness of legal institutions to social needs and the effectiveness and limits of law as an agency of social change.[132]

But where was the method by which all these intricate questions were to be examined? Galanter had been a student of Max Rheinstein (and through him of the works of Max Weber) and Karl Llewellyn. Concerned both with the formal structure of law and argument and the workings of the legal and traditional systems, Galanter had been schooled in a tradition which was concerned with the development of rational doctrine as well as the need to develop an empirical understanding of the informal and formal system. This is not always apparent in some of his earlier writings— some of which is solely concerned with doctrinaire 'black letter law' questions.[133]

How did Galanter approach the question of law and society in India? One of the most difficult questions that engage the student of modern Indian law is to mark out a field of study. What exactly was modern Indian law? Indologists, sociologists, political scientists, and lawyers, each provided a different answer. Indian perceptions of law and *dharma* made the answer to this question even more difficult. We have already seen how Derrett sought to interpret the law emanating from Indian institutions in the light of his understanding of the ideology which bound the system together. Gandhians in India not only declared that Indian law was alive and well in the villages but called for a full and complete replacement of imposed British law by native law. Galanter, using a Weberian analysis, identified certain features as constituting the stuff that modern law was made of. He argued — see Chapters 2 and 3 below — that 'traditional law' was dead. It had been 'displaced' by modern law. The restoration of 'indigenous' law had been aborted. We had to take the modern Indian legal system as we found it. But the richness and variety that Galanter saw in the modern legal system must indubitably make us ask a whole range of questions about the legal system as we 'find it'. Was Galanter's sharp differentiation of the modern legal system from the traditional system helpful? Was not the field so hopelessly blurred that it made little sense to demarcate the contours of the 'modern' system and differentiate it from the 'traditional'? What exactly was the import of the Weberian analysis with which he had identified the cluster of features which characterized the modern legal system? While these — and other questions — are examined later, it is important to identify the manner in which Galanter marked out his field of study.

The marking out of this field of study led to an analysis of whether the manner in which the legal system perceived and presented Indian society was accurate. Galanter took the view that many of the modern legal system's assumptions about Indian society did not correspond with social reality. The law reconstituted Indian society so as to dilute certain aspects of social reality and give prominence to others. In an unpublished paper, Galanter showed how the doctrine of *varna* was virtually invented by British law in the nineteenth century to preserve traditional mores in an otherwise rapidly changing society. Post-independence India cultured its own image of the vastly complex plural structure of Indian society. Galanter was specially concerned with how modern Indian law viewed the question of group identification and membership. How did it view religious groups and their rights?[135] What were the 'changing legal conceptions of caste'?[136] Underlying all this lay the broad question of how a formally rational system would treat 'irrational' and asymmetrical social relations in a society. Oddly enough — as we shall see later — some of the answers which Galanter provides to these questions make us reassess the confidence with which he marked out modern Indian law as a distinct field of study.

No one can work on the modern law of any country without looking at the work of the judiciary. We have already seen the extensive range of debates and discussions over the role of India's judiciary. While much of Galanter's work in this area matured much later, his earlier pieces provide rich insights into his theory of judicial decision making.[137] Political scientists may well wish to quarrel with this approach and Galanter's understanding of the kind of pressures within which a modern judiciary operates. Even so, Galanter has given detailed attention to the predicament and ccontribution of the Indian judge.

Finally, Galanter looks at the way in which Indian law is mobilized. Reviewing the work of various social action groups, Galanter — using an analysis for which he is justly well known — spells out the kind of abilities, skills, and institutional formations required in order to use the legal system effectively.[138] His belief that the disadvantaged can also mobilize the law effectively and find 'turn around' solutions in and through the legal system might surprise many of the disadvantaged themselves. Others may question his optimism and the extent to which his theories would be borne out on a wider analysis of 'power' and political relations in society.

Galanter's essays attempt to examine the nature of the Indian legal system. We have identified four significant themes in his work. The *first* is the modernization thesis which helps us define the field of modern Indian law. The *second* is concerned with the manner in which a formally rational system perceives and adjusts to the demands of a pluralistic society. The *third* concerns the nature of judicial decision-making in a modern legal system. The *fourth* explores how the modern legal system is used by those who exploit it; and, can be used by those who are exploited.

MARC GALANTER ON INDIAN LAW

Galanter on Modern and Traditional Law

While Galanter's essays on the 'displacement' and 'aborted restoration' of indigenous law[139] can be seen as part of a narrow debate concerned with attempts to revive what is loosely called 'traditional' law, they also attempt to create a wider basis for the study of modern law. The narrow Indian debate was partly the product of a romantic vision of India's villages and partly the consequence of the Gandhian belief that the political and economic structure of India should be decentralized so as to establish the village as a pivotal administrative and social unit.[140] To some extent, the new Constitution nurtured the concept of a village-based economy. The Directive Principles of State Policy recognized the importance of the village *panchayat* (assemblies) and regarded them as a possible agency through which a viable local democracy and socio-economic development would take place.[141] We should not, however, make much of this constitutional blessing. It was really no more than a concession to Gandhi and his followers whose own blueprint for a Constitution has not really ever been considered as representing a viable alternative. B.R. Ambedkar, who piloted the Constitution through the Constituent Assembly, made his view of a village-based polity quite clear.

> No doubt the village communities have lasted where nothing else lasts. . . . But mere survival has no value. . . . What is a village but a sink of localism, a den of ignorance, narrow mindedness and communalism . . . ? I am glad that the Draft Constitution has discarded the village and adopted the individual as its unit. [142]

Apart from paying lip service to Gandhism and the romance of the Indian village, the *panchayat* has continued to be emphasized by various government plans because it was seen as a possible vehicle for developing co-operatives and co-operative farming. In time, the judicial *panchayats* became moribund and the co-operatives began to service the needs of the political influential and those who were already well-to-do.[143] But the debate has continued. It is not really clear what the debate is about, who the protagonists are, what they want, and how rigorous they are about their approach and methodology. 'Traditionalists' and 'modernists' mix with uneasy virtue, and the difference between 'tradition' and 'modernity' remains blurred which is why studying it is sometimes so exciting. Some foreign observers have been fascinated by the co-existence of these divergent tendencies and have romantically described the way in which modernity respects and absorbs 'tradition' and 'tradition', in turn, responds to modern demands.[144] Others have talked of the middle class problem of the 'intellectual caught between traditional and modernity'.[145] Since the terms 'tradition' and 'modernity' are not defined with any precision, the literature quibbles over definitions as it acquires a life of its own.

Galanter does not really answer these questions as part of the 'narrow' argument. His more limited case study on *nyaya panchayats* is about a specific attempt to revive traditional village dispute settlement agencies.[146] Torn from the context in which *panchayats* originally existed, the new *panchayats* represent moribund institutions. To the extent to which they are hardly used, they support Galanter's assertion that the old system cannot be revived. But the new *nyaya panchayats* are not indigenous institutions. They are, as Galanter 'hypothesizes', an 'extension of the State legal system in rural areas'.[147] If the new *panchayats* are not traditional institutions, they are part of the intervention of the modern system. And, if the modern system is here to stay, the *panchayats* are not really a good example of the power and influence of the modern State and modern law. Galanter himself accepts that

> ... (a)s members of the judiciary, the *nyaya panchayats* are effectively isolated from the power hierarchy of the village societies.[148]

Undoubtedly, every part of the 'modern system' cannot be expected to become a decisive and effective social influence. The government often enacts programmes which have no positive effect but exert a negative influence on the emergence of alternatives in civil society. Such programmes occupy a space — whether symbolic or real — and prevent the development of new mechanisms to resolve differences in the political economy. In a sense, the case study of *nyaya panchayats* is not a wholly convincing example of either the displacement of traditional law or the aborted restoration of indigenous law. Neither 'modern' nor 'traditional', they are just part of a badly devised plan which has had a negative rather than a positive effect.

Beyond this specific example, Galanter is participating in a political debate about the 'displacement of traditional law'. Galanter points out that, historically, the debate is over. A modern system (with the 'cluster of features' identified by Galanter[149]) is already there. The process of displacement has already taken place. Even if the political discussion about 'tradition' and 'modernity' continues to be inconclusive, history has already taken care of the result. But having established this, Galanter continues to emphasize that traditional *society* is still there, ready to interpret, adjust, take over, manipulate, change, and/or possess this new modern system. So, are we back where we started? Depending on how we assess the strength of the traditional structures, are we saying that the modern system is part of traditional society? Are we asserting that there is a 'gap' between the modern system and traditional society? These questions distinguish the formal legal system from pathological practice within or around it. However, these questions of shifts and adjustments between the formal and informal structures give rise to another set of questions concerning the relationship between law and the political economy. Are we saying that one part of the legal system belongs to an emerging part of the political economy more than another part? Does this take us back to all those debates about 'status' and 'contract' and to assertions about contract being the vehicle used by emergent capitalism for self-expression and legal domination? And, if we have come this far, are we not close to the Weberian dilemma as to whether a certain ethic or certain form of legal or substantive rationality are a pre-requisite for the rise of capitalism *or* the natural spin-off resultant from capitalist development?[150]

Galanter does not seem to answer all these questions. He seems to be content to assert the existence of the modern legal system without denying the existence of traditional society or the powerful influence of the latter on the former. But what kind of ontological reality is this new system? Clearly, it has not subsumed traditional society; and, indeed, may have been subsumed by it.[151] For Galanter, the modern system is a system of rules, institutions, professional activity, and social practice. It is a belief system. It has a culture. It seeks to comprehensively cover and recast all aspects of legal rights and duties that exist or struggle for existence in modern society. At one level, the modern legal system is a 'discipline' through which all legal demands are conceived, constituted, and articulated.[152] It gives a new linguistic expression to entitlements and powers in modern society and creates the method of reasoning by which they are to be discussed. At another level, the modern legal system constitutes a new set of 'power filters' through which social, political, and administrative powers are recognized and articulated. It is in this context that Galanter describes what happened to indigenous law:

> First the administration moved from 'informal' tribunals into the government's courts; second the applicability of indigenous law was curtailed; third, indigenous law was transformed in the course of being administered by the government's courts.[153]

The new 'discipline' and the new 'power filters' supplanted, evaded, and enforced traditional restrictions. While traditional attitudes and habits continued, they were under a considerable amount of pressure to express and sustain themselves in the 'discipline' and structure of the new system.

In a sense, Galanter himself is responsible for obscuring the clarity of his analysis. By talking of 'two political idioms' or a 'dualistic legal' system, he creates an image of two systems in which one is fighting back and not quite 'displaced' by the other. Again, by talking of the 'gap' between the 'higher law' and 'local practice', he concedes the greater strength of the latter in the formal presence of the former. And, if the 'higher' system was only a formal one, what was the point of arguing that the traditional system was 'displaced'?

This takes us to a deeper problem. Having argued that A has 'displaced' B as part of a dynamic process, is there any point in asserting that a legal system consists of A *plus* B — and, perhaps, a little more A than B? Is it not more crucial to ask why A has displaced B? What were the forces which sustained A and B respectively? How were these forces realigned as a result of the process of modernization? These are questions pertaining to the political economy. What is missing from Galanter's analysis is precisely the context of political economy within which the terms 'modernity' and 'tradition' must be understood. For 'modern law' is produced by the modern State, given certain conditions of economic development.[154] The modern State does not seek to produce crude empowering rules to accommodate emergent interests, but also creates a legal and administrative system and ideology which services these interests in both a direct and indirect way. In the new system, political expediency, ideological consistency, and the polity's capacity to sustain the turmoil created by change determine the nature, style, and extent of the displacement. Years later, Galanter turned to some of these questions of political economy:

> ... (D)o we have to look to some elite among the indigenes? If so, which elite do we look to? This is very much a question here because in India there has been a displacement of an older elite who were carriers of an earlier indigenous tradition.[155]

And, if by elite, he means just the custodians of the law (i.e. the *rishis* of old and the lawyers, legislators, and judges of today), why not go further and look at the political economy which nurtured and sustained the growth of these 'legal' custodians? The truth of the matter is that, when Galanter speaks the language of the co-existence of the two systems, he is speaking more as a liberal pluralist with a particular view of society rather than as a dispassionate social scientist concerned with looking at the process of modernization including, *inter alia*, the modernization of law.

Galanter's contributions to the specific debates about the challenge of modern law are significant and incisive. There is a 'no nonsense' quality about them, even if Galanter is able to preserve his passionate defence of liberal pluralism and avoid answering many important questions about the process of modernization. While we are left with nice mind settling images about the peaceful co-existence of a plurality of legal systems, these images do not detract from the central importance of the modern legal system and the kind of 'discipline' and 'power filters' created by, and under, it. Galanter has done a great deal in exposing how the apparatus of modern law, nurtured by an emergent capitalism, displaced and transformed — but did not destroy — traditional alternatives. Since his concern is with modern law, he does not really define traditional or indigenous law. The latter is sometimes referred to as local ar-rangements. Occasionally, the traditional system is to be found in the *dharmasastra* of the Hindus which are credited with creating both a general ideology and disciplinary framework as well as a set of institutional arrangements which have held Hindu society together through the ages. In the main, Galanter — following Rocher, Derrett and, ultimately, Nelson — feels that the British may have been wrong in giving so much prominence to the *dharmasastra*. Rocher depicts *dharmasastris* as a talking club of jurists — quite like jurisprudes in our day.[156] However, the ideological effect and influence of the *dharmasastris* can hardly be doubted, even though it was only discreetly manifest in the 'local' systems which Nelson wanted to profile more prominently.

Apart from these Indian controversies, Galanter's essays have general implications for the study of modern law. Ever since Bentham criticized Blackstone's approach to law and attacked rights as 'nonsense, rhetorical nonsense, nonsense upon stilts',[157] English jurisprudence has been preoccupied with its concerted attempt to try to define law.[158] While most analytical jurists have really been concerned with defining modern law, they have set themselves the more general aim to define *all* law. What has followed is a con-fusion of definitions depicting law as the expression of 'power' (whether *de facto* or *de jure*).[159] Later, law was the practice of law courts;[160] law was something articulated and accepted by certain institutions and officials; law was a hierarchically arranged and internally consistent reality;[161] law was what is normally accepted as the law — — in ordinary language — by officials and others;[162] and law was a method of reasoning concerned with the protection of rights.[163] Some jurists and social scientists refused to be drawn into controversies about the definition of law. These jurists felt the need to examine the whole phenomenon of what is happening 'out there' — whether it was identified as 'law', official behaviour, social inter-course, or customary interchange.[164] In particular, anthropologists, like Malinowski, argued for a more general look at social rules:

There must be in all societies a class of rules too practical to be backed by religious sanction, too burdensome to be left to mere good will, too personally vital to the individuals to be enforced by any abstract agency. This is the domain of legal rules.[165]

The literature on anthropology of law is full of controversies seeking to define law *either* as a specific type of rules (e.g. rules which call for an application of sanction by an authorized person),[166] *or* as a model of judicial decision making which could be distinguished from mere negotiation or political decision making.[167] The controversies are as inconclusive as the literature that produced them. They illumine methods of study rather than provide satisfactory definitions.

Galanter seeks a middle path between definitions which seek to present law as a distinct reality and those approaches which reduce law to patterns of regular social behaviour. Without seeking to involve himself in the definition of *all* law, he concerns himself with modern law. This recognition of modern law as a separate phenomena distinct from other pre-modern forms of law is significant. It takes us away from ambitious, and necessarily complex, attempt to define all law. Building on the Weberian notion of the increasing rationality of modern law, Galanter identifies the cluster of features that compose modern law. Thus, modern law consists of uniformly and universally applied rules, territorially administered by courts and supported by specialized professionals. The modern legal system enjoys a theoretical monopoly of existing law and emerging disputes and is identified in such a way that it can be differentiated from the rest of government.[168] This is an 'ideal type' and 'not a destination, but a focus or vector towards which societies move'.[169]

Is this just a *description* of certain feature of modern law or part of a deeper understanding of its nature? If it is just a description, then many systems, modern and ancient, possess these features. If these features acquire a uniqueness is it because of some distinct forces that mould these features into a modern system? What are these forces? Galanter does not really explain this. At places he talks of the political basis for modern law. In an otherwise un-noticeable footnote, he speaks of how

> . . . the very forces which support this movement (of modernization) and which are released by it deflect it from its apparent destination.[170]

But what exactly are these forces that give a distinctness to this cluster of features which one associates more with modern law than with pre-modern systems? The fact remains that it is not these features, though characteristic of the modern system, which in fact make it modern. The modern legal system is both an idea as well as a response to a historical process. Perhaps, Galanter's approach suffers because it provides an explanation (*erklaren*) on what we see rather than increases our understanding (*verstehen*) of what is happening. This takes us to the construction of 'ideal types' in social sciences. Are we merely concerned with grouping certain characteristics and putting them under a broad label? Or, are we concerned with fusing explanation with interpretation (i.e. *verstehende erklarung* or *erklarendes verstehen*)? Galanter's quest for discovering distinct entities (i.e. modern law) rather than providing interpretative explanation characterizes his approach to modern law. On the other hand, it cuts short a great number of jurisprudential controversies about the true nature of modern law. Apart from the identification of certain distinct characteristics — Galanter does not really treat his ideal type as part of a historical and social explanation. While he admits that the growth of modern law is connected with the growth of the State's declared monopoly over what can be called law and the emergence of complex and powerful adjudicating bureaucracies, he does not go further. He is concerned with identifying certain tendencies in modern, pluralist society and how modern society deals with the continuing tension between pluralist forces in a society and the empowering and assimilative effect of the forces of the modern State. After all this, if we are still left a little puzzled about the nature of modern law; it is because Galanter is more concerned with marking out a tentative domain of research rather than providing an understanding of modern law.

Modern law appears to elevate the forces of rationality over irrational forces in man, social relations, and society. Unable to assimilate society within its fold, modern law elaborates the compromise of allowing a significant amount of autonomy to various social orderings as long as this autonomy fits into and does not militate against the normative orderings of modern law. Thus, the institutions of the official legal system, send out messages to society without crowding the latter's plurality out of existence. Galanter detects a

> . . . most significant legal traffic (in) . . . the centrifugal flow of legal messages rather than a centripetal flow of cases from official forums. . . . These messages radiate into spaces that are not barren of normative ordering; instead the social landscape is covered by layers and centres of indigenous law. . . . Every legal system has to address the problem of autonomy and authority of the various other sorts of normative ordering with which it co-exists in society. The big legal system faces the question of how to recognise and supervise or suppress the little legal system. Legal centralism is one style of response to this generic question of legal ordering, and its exhaustion suggests the need for reflection of other models.[171]

All this should not be taken to emphasize the normative rather than the institutional spread of modern law, even if Galanter's emphasis on the 'centrifugal flow of messages rather than the centripetal flow of cases' might reinforce such a view. The real problem concerns what causes this tension between 'modern' and 'traditional' orderings and how and why it is resolved by different societies in different ways. What do we make of this tension? How do we study it? Galanter rightly argues that we must be sceptical about the chronic over-commitment of modern law and treat as 'illusory or partial' the deliberately projected images of courts as 'engineers of control' or as authoritative resolvers of disputes.[172] Nor should we be concerned with just the 'effectiveness' of modern law or with the impact of studies about the influence of its norms and institutions. Instead we should ask

> Under what conditions and in what locations does self regulation emerge? What are the features that it displays? Is there an explication of norms, formality of procedure, broad or narrow participation, etc.? The social settings in which such regulation takes place are not independent self-contained units, but interact with a larger complex legal order. How then is indigenous regulation related to the regulation projected by the big legal system? What is the relationship between official law and the indigenous regulatory activity? Does the latter rely upon or borrow from the norms, sanctions and style of the official law? Study of the spheres of indigenous ordering leads to exploration of their interface with official forms. How do courts (and other official agencies) attempt to supervise and control bargaining and regulation in various social settings?[173]

Galanter himself evaluates the limitations of such research by accepting that, although it will identify 'real problems' and 'real choices', it will not solve the 'intractable question of justice'.[174] But this may be precisely where our problems begin. Who can deny the importance of 'reading the landscape of disputes'?[175] But, how are we to read them? To what end? Is it just to know more so we can hypothesize about society more accurately? Or is it to evolve a more radical conception about society? Underlying these choices are very fundamental arguments about the overall aims of the sociology of law. Galanter's emphasis is on an empirical examination of what is going on, followed by an explanation (or, perhaps, interpretation) of what is happening. A great deal of his work exemplifies this approach; as for example, his very extensive research into positive discrimination in India,[176] his look at the crusading judge in American trial courts,[177] his incisive analysis of the nature of litigants and litigation,[178] his examination of the social implications of the litigation explosion,[179] and his reading of the tension between official and indigenous law.[180] Pitched at the level of explanation, they flow attractively into interpretation. But these descriptive concerns

do not even attempt to portray — leave alone explain and interpret — those factors in the political economy which reinforce the
existence of some orderings while forcing others out of extinction.
An alternative radical approach need not be less committed to the
need for empirical evidence. Its central concerns start with the
political economy, the State, the division of labour and groups in
society, an anlysis of law in terms of the 'productive forces' in society
and their relationship to the distribution of power and the manu-
facture of 'ideology' in a society.[181] The difference between these
approaches is one of degree; but it is fundamental. Galanter's own
inspiration for research finds sustenance in his pluralistic view of
society in which various social orderings lie alongside each other,
grappling and interacting with each other. There is a kind of optimism — though not complacency — about this view of society: the
more we read the landscape, the more we define problems and solve
them. The alternative more radical approach is sceptical about
beginning with such peaceful, pluralistic assumptions about society
or about the sensitivity of the legal system to absorb the various
demands that seek recognition on anything other than exploitative
differential terms. Set against the backdrop of the State, seen as a
powerful example of instrumental rationality, possessing a lesser
capacity for communicative rationality than is commonly supposed,
sophisticated in its manipulation of the modalities of 'coercion' and
'eliciting consent' to control society, impressively general and simultaneously specific in its reconstitution of society and more susceptible
to use and abuse by some classes in preference to others, modern
law is seen as proceeding from, rather than a casual imposition
on society. The pluralism which Galanter espouses allows 'oppression' and 'injustice' to lie alongside its selection of the kind of groups
which should receive greater protection at the expense of others.
All this should not be taken to mean that Galanter is not concerned
with problems of justice. Indeed, he seems to believe that modern
law can be made to find 'turn around' solutions for the poor and
oppressed. To this we shall return when we discuss Galanter's view
of law and development.

Galanter on Group Life and Group Membership

Contemporary India is an example *par excellence* of the kind of
society where a 'modern legal system' runs into close and extended
encounters with traditional society and indigenous methods of (self
and other) regulation. India is replete with all kinds of 'group'
orderings which are constantly subject to the strains of economic
development and change; and, which provoke new patterns of group
formation and social and occupational mobility. This 'modern legal
system' seeks to define, constitute, reconstitute, sustain, re-align,

transform, or destroy these groups.

In the papers presented here, Galanter is concerned with problems concerning 'group membership' and 'group preference'[182] as well as the 'changing legal concepts of caste'.[183] He has also examined the problems of 'backward classes' and 'untouchables' in greater detail.[184] If the latter research efforts are under-represented in this selections of papers, it is because they are readily accessible elsewhere. There is another important difference between the papers included here and the researches employed in his study of governmental plans for 'backward classes' and 'untouchables'. In the latter studies he is concerned with the whole government programme and not just — as in the papers included in this volume — about how courts reason about these things. In limiting himself to the judiciary, Galanter is not just concerned with the judiciary as an institution that resolves problems of government policy but with judges as 'conceptive ideologists' trying to recognize, define, change, transform, and eclipse complex understandings and ambitions in an extremely complex society.

Successive governments in independent India have sought to redefine the groups that co-exist in Indian society with a much publicized inflated rhetoric, a political guile, a moral sense of pride and embarrassment, a belief and disbelief in the efficacy of programmes ostensibly designed to better the life chances of depressed groups, and a pragmatic insight to ensure a balance between what India is actually prepared to do and what it claims it wants to do. Given the huge diversity of groups in India, their complexity and the almost unlimited range of socio-economic inequality, examining the nature of group membership in India and assessing the kinds of opportunities offered by each to its members remains one of the most difficult and challenging tasks that political — or, for that matter, legal — sociology faces. Since political exigencies rather than intellectual rigour determine the answers to these questions, implicit in some analysis is a tendency to falsify the nature of these groups as well as the nature of the government's plans in relation to them. Galanter's extensive work on backward classes and untouchables looks at some of the problems generated by the tension between government plans, the social expectations that these plans generate and their manipulation by civil society.

We are concerned here with how judges preside over dilemmas of group membership. One of the legal contexts in which this occurs generates a strange mixture of demands, assertions, and expectations. The government allows certain electoral and other privileges to members of certain 'Scheduled Tribes and Castes'. In order to take advantage of these privileges, the intended beneficiaries have to assert their 'Backwardness'. Their adversaries, on the other hand, while anxious to declare their commitment to help the 'Backward' classes, can only win their legal argument by showing that the alleged beneficiaries are no longer 'Backward' because they have been liberated from their caste status. Since discussion about peoples' liberation from their caste status is hopelessly complex at the best of times, what follows are all kinds of intricate arguments about the symbolic or real renunciation of a group's particular affiliation, conversion to another faith or caste, ex-communication, and so on. Willy nilly, relatively simple public law questions about the entitlements of the 'Backward' classes become complex social questions about who the 'Backward' classes are and the circumstances in which members of these groupings lose their identity, and, consequently, their entitlement to benefit under the government's programme. These questions can be answered in two distinctly different ways. The first approach is the social approach. Here, we are concerned with discovering the social truth about a person's group-identity. What follows is a difficult intellectual journey about the true nature of a particular group, the incidences of membership, the manner in which a person can leave his old group, his relation to the old group, the notions of 'intent', 'acceptance', and 'rejection', the theology of group membership, the social and ritualistic requirements attached to belonging, non-belonging, and exit; and so on. As judges undertake these complex journeys, they are exposed to the absurdity as well as diversity of the Indian society in which they live. They are under some social pressure to devise explanations so that the flavour of traditional society is preserved without the orthodoxy underlying it appearing unpalatable. This might explain Justice Kapur's celebrated protest in the *Dora* case about the flexibility of the *varna* system.[185] The second public law approach concerns itself with the purpose behind the government's programme. Accordingly, where the issue at hand is about special representation for certain disadvantaged groups we are concerned with the notion of an intended beneficiary (as a public law concept) rather than group membership (which examines aspects of traditional group life and the nature and significance of special social transactions). These two approaches overlap to a great extent but share different objectives and methods. Indian litigation has drawn the courts into examining and commenting on the social complexities of Indian life rather than limiting their concerns to defining public policy. What lies at the root of a majority of such cases is not a genuine dispute about a person's membership of a group but a post-election squabble in which the losing candidate's lawyer stretches the ingenuity of legal procedure to question the bonafides of the winner. Imagine a disgruntled loser sitting in a lawyer's office, offloading his final catalogue of facts: 'There is one more point, the winning candidate got converted to Buddhism three years ago at a formal ceremony which symbolically protested the plight of untouchables in Hindu society'. On this assertion is built a case which could lead the courts into asking about the relationship between Buddhism and Hinduism throughout the ages; or, the nature of *varna* in Hinduism itself. This is precisely the dilemma in which the Supreme Court was trapped in the *Dora* case,[186] where the Court was faced with a political skirmish dressed up as a deep traditional argument. Not wanting to invalidate an election simply because a person had left his group, the judges propounded a 'caste iron' theory that under true Hindu doctrines the voluntary act of 'leaving' a tribe or caste could not change a person's social status. Unfortunately, though effective in dealing with the case, such a view was also a comment on India, Hinduism, and the limited possibilities of social mobility afforded by Indian society.

Galanter does not approach these cases in this way. He distinguishes the 'pragmatic' from the 'fictional' view but does not really look at the 'pragmatism' behind the 'fictional' view.[187] Under the fictional view, what the judges of the Court seem to be saying to politicians is this: 'Look, do not come to court with arguments about conversion. For our purposes the act of conversion may not be significant. Nor do we wish to be drawn into arguments about conversion. We will assume that ascribed roles do not alter — indeed, cannot alter — merely as a result of these acts'. Galanter interprets the Court's doctrinaire arguments about conversion as 'concentrating on the theoretical consequences of certain acts'.[188] In reality, such an approach consists of asserting the practical and political non-consequences of certain acts. It is what Galanter calls the 'pragmatic' approach which is really laden with concern about traditional criteria. Here lie all the questions about 'intent', 'acceptance', and 'rejection' that we have outlined earlier. The effect of following the pragmatic line is not to foreclose any issue, but to concentrate on traditional factors and allow a plethora of assertions and counter assertions about social life while making a reformist gesture about the need for upward social mobility in only the most obvious cases. Given the government's commitment to respect traditional group orderings as well as obliterate the disadvantageous consequences arising out of a traditional ascription of status, the pragmatic approach — which Galanter prefers — makes it possible to ask a wider range of questions, find and declare a commitment for traditional life as well as keep alive a flexible support for the government's policies.

Why does Galanter express a preference for the pragmatic view with its emphasis on traditional acceptance and rejection? This preference must indubitably be traced back to Galanter's passionate belief in a voluntaristic pluralistic society in which all kinds of group life can co-exist alongside each other. He is not just concerned with an Indian problem, but with a wider and much more important problem. He treats this as an opportunity by courts

> ... to demonstrate how principles of equality and voluntarism can be implemented in a plural society with a social structure that is mainly traditional and where status is mainly ascribed. By a thoughtful and coherent treatment of these problems of group membership and group preferences, Indian jurisprudence can contribute much needed guidance both to new nations attempting to construct a viable plural society and to older nations who are unable to resolve the problems of diversity. By such guidance the Courts may play a crucial role in assuring the world wide transformation from traditional to modern takes place in conformity to the principles of freedom and equality.[189]

While we may find Galanter's notion of 'voluntariness' ambiguous, it is important to recognize his view of society where membership of groups is secured without inviting hostility, discrimination and disadvantage. In such a society there would be 'voluntarism and respect for group integrity' as well as 'equality and non-recognition of rank ordering amongst groups'.[190] People would be able to claim group rights, rights to denominational temples, freedom to exercise various social practices in as complete a way as possible whilst internally reforming the groups to move to a greater egalitarianism. In what appears to be a mixture of a wish fulfillment and confident prediction, he suggests that we may

> ... anticipate that the new legal view of caste will not only sanction but stimulate and encourage new forms of organisation, new self images and new values within caste groups; and the dis-establishment of the predominant organizing model of cultural unity may give vitality to lesser traditions and new scope for innovation.[191]

So, Galanter's notion of 'voluntariness' comes to acquire two meanings. In the first place, it means 'voluntariness' of membership — that is freedom from discrimination by virtue of being a member of any group. In the second place, 'voluntariness' also emphasizes the quest for new values within older dipensations. This suggests that, while the State is right to de-mystify the sacral view of society (of which caste is a part) and prevent discrimination, it is wrong to de-legitimize the claims of those traditional groups which are not discriminatory.

This approach is fundamentally opposed to the view of some modern regimes which seek to attack the identity of traditional groups themselves. Accordingly, it is not enough to attack some caste practices; what needs to be attacked is the notion of caste itself. If people are to find true freedom, they must abandon traditional groups for new associational relationships more suited to fighting for their rights in a modern world. One variation of this argument can be found in Karl Marx's article *On the Jewish Question*, that religious rights must be fought for on a wider canvass of political rights.[192] This must, perforce, affect both the content that we give to religious rights as well as importance we attach to the sentiments behind them. Some aspects of India's official policy go further to declare a preference for totally destroying traditional groups in favour of new 'secular' groups on a scientific basis which would educate and transform Indian society.[193] Against these alternative perspectives, Galanter argues for giving full recognition to non-discriminatory and 'voluntarily' acquired traditional rights.

Galanter's approach may also run counter to an analysis of group life in society, based on notions of 'class'. Such analysis emphasizes the need for new radical groups to secure political and civil rights and seeks to minimize the importance of traditional groups which are seen as an impediment to progress. In certain societies, traditional loyalties and ideas are used to oppress and suppress the poor and disadvantaged. The alternative analysis based on 'class' perceives and analyses the political economy in a different way, giving greater importance to new emerging groups and the social and economic context in which they operate. The reification of traditional groups is seen as part of the attempt to assert social, economic and political claims in modern society. From this perspective, Galanter's liberal pluralism obscures our understanding of modern society and of our perspectives of how social and economic change is to be achieved. Galanter's concern is not to prevent the growth of new political and economic groups. Much rather, it is to protect the vitality of practices and beliefs around which people have organized their lives. To Galanter, such freedoms—to be Jewish, Hindu, Christian, or Muslim—give the substance of what constitutes the good life. While we may be prepared to respect Galanter's view of the good life, we may question his incomplete analysis of group *and class* life in modern society.

Finding a position somewhere between description and prescription, Galanter forcefully argues that this vision of Indian society arises not just out of a sense of personal commitment to a particular concept of the good life but one which is thrown up by India itself. Quoting from Granville Austin's all too optimistic account of Constitution-making in India,[194] Galanter suggests that India has

> . . . this ability to reconcile, to harmonize, and to make work without changing their content, apparently incompatible concepts — at least, concepts that appear conflicting to the non-Indian and especially to the European or American observer. Indians can accommodate such apparently conflicting principles by seeing them at different levels of value, or,

if you will in compartments, not watertight but sufficiently separate so that a concept can operate freely within its own sphere and not conflict with another operating in a different sphere . . . . With accommodation, concepts and viewpoints, although seemingly incompatible, stand intact. They are not whittled out by compromise but are worked out simultaneously.[195]

India's capacity for accommodating and reconciling all kinds of claims without losing its essential character has been noticed by other foreign observers of Indian law. Professor Derrett makes much the same point when he analyses Indian concepts of ownership:

Indian jurists did not attribute to property a definite incidental content. There might be several owners of a thing, owning not merely shares, but extensive rights of different characters . . . (They seem to ask): what point is there in defining the owner of some rights over a thing as owner and the owner of other rights as something other than owner; particularly when the word 'owner' implies nothing more than 'belonging', 'mastery' and the like? It is relevant here to note that the fluid, syncretic, non-disjunctive approach to ideas is notoriously characteristic of Indian thought, gradually merging and broad identities being far more congenial even to their own category minded attitudes than the staccato separation of things which share a characteristic.[196]

But these modes of thought serve many social and economic purposes other than maintaining a flexibility to accommodate the plural demands and expectations in society. The example of ownership is an apposite one in this connection. Such an accommodating concept of ownership deliberately blurs the lines of entitlement, decision making and control. Thus, a Hindu son is not deprived of his entitlement to family property but may be allowed little participation in decision making and possesses virtually no control over what he owns. In a hierarchical society, based on holistic concepts, power, wealth, and status are distributed on the basis of both clear and blurred distinctions. Concepts describing entitlements in such a society have a multiplicity of meanings. Any 'theory of accommodation' in such a society cannot be understood without asking who is accommodating what, and to what end. In the end, everyone is accommodated but at different levels and in different respects. Sons accommodate the debts of their fathers in this world because of their obligations to him and the general order of things in the next. Yet, the same debt, though an absolute obligation to the father, is not an absolute obligation to the creditors.[197] Such complex arrangements baffled the judges of British India who preferred to re-conceptualize the relationship between bankers and moneylenders, on the one hand, and agrarian landed interests, on the other, on a much more clearly defined basis. This was a situation not visualized by the *dharmasastra* in which the money lender's interest was less crucial to the social apparatus as a whole. The theory of 'accommodation' in the *dharmasastra* derived from a holistic conception of society in which everyone was slotted into a particular status and could still be related to (or 'dis-related' from) others in some meaningful way. This reinforced a person's social position but diluted his claims to individuality and individual attention. It is in this sense that Derrett argues that an Indian never lives for himself but only for others.[198]

The Indian 'theory of accommodation' is a far cry from the kind of liberal pluralism which Galanter espouses and claims to have found in the intellectual fabric of Indian society. The Indian system has to be seen as a whole. It remains the most impressive example of how disadvantaged groups in Indian society did not just consent to their own low status (and consequential exploitation) but willingly accepted their position as part of the *dharmic* order of things. India's genius for accommodation can only be understood against the backdrop of this *dharmic* order which holistically encompassed all of society. Given the complexity of this *dharmic* understanding and the hierarchical, stratified society that this order of things produced, it took all the 'wisdom' of the ancients to stress both the need for general accommodation as well as decree a tolerance for deviant practices, local practice, idiosyncratic demands and proposals for change. All this was possible as long as the basis of the *dharmic* order of things was not questioned. Galanter's work forces us to ask whether we can treat this sense of 'toleration' and 'accommodation' as independent qualities divorced from the context which circumscribed and contained them. Are these qualities strategic devices to deal with interstitial situations or are they a universal variable in Indian thought which have an independent life of their own?

Indian society preaches toleration while maintaining an otherwise intolerantly cruel society. Who can leave the group and claim the benefit of toleration? Who can vertically challenge a higher group and expect accommodation? Even where such efforts are successful, they can be legitimized only on the basis of an ingenious attempt to reconstruct the archaeology of the 'order of things'. Articulating sentiments similar to those underlying Renner's approach to the 'institutes of law', Galanter extends that kind of analysis to predict a 'controlled transformation of central social and cultural arrangements' in a way that will make the 'turn away from the older hierarchical model to a pluralistic participatory society . . . prov(e) vigorous and enduring'.[199] The characteristics of 'accommodation' to be found in a holistically conceived hierarchical society are not the same as those which will bring about a liberal egalitarian society. While we may not wish to quarrel with Galanter's attempts to examine the Indian 'theory of accommodation' as a more general

theory in order to support his plea of indulgence to various indigenous groups, it is difficult to accept his more general jurisprudential
analysis of the relationship between the form and content of social
and legal concepts. In pursuing that analysis, he also over-simplifies
the problems of pursuing equality in a land of hierarchy. Galanter's
appeal to the general spirit or *volkgeist* of Indian thought to gather
support for his modern liberalism is ingenious, but obscure and not
always convincing. Implicit in what he is saying is the hope that a
de-reified version of India's tradition might help us to grapple with
injustices created and sustained by the social structure. Such an
approach does not obviate the need for an alternative analysis of
Indian society. This alternative analysis may be less obscurantist
in its approach than Galanter. It may refuse to see a future (just)
society, as part of the hidden agenda of either past practice or,
virtually, any version of previous tradition.

Galanter on Law and Morality

Starting from the distinct viewpoint which we have described,
Galanter makes an incisive contribution to arguments about the
relationship between 'law' and 'morality'.

In analytical jurisprudence, the central concerns about the relationship between law and morality vary from well trodden philosophical arguments about the 'is/ought' distinction to descriptive accounts about the high incidence of congruence (or overlap) between
law and morality.[200] The other stand of the debate concerns the
right of individuals to pursue their own morality, free from constraints
imposed by the State.[201] Without putting anyone's back up, Galanter
gently steers his audience away from these controversies by suggesting
that the latter debate rests on a false picture of the individual in
society. Not surprisingly, he starts with the group to show that the
individual is already part of a network of moralities. In that sense
the tensions to be resolved are not the tensions between an individual
with a self-generating capacity for creating his own highly individualistic normative environment and the State which might inhibit
this capacity. Instead, we are concerned with the individual caught
in overlapping networks of morality by a process which, although
seemingly voluntary, is fraught with all kinds of pressures, tensions
and compromises. The choice of moralities is the product of a complex set of social factors rather than the product of an exercise of
free will. While it is only right to respect the ultimate personal and
psychological autonomy of the individual, the task of the State and
law—as Galanter sees it—is much less straightforward.

Lawyers have tended to view the relationship of law and morals as a
problem of the source of legal norms and the scope of legal regulation,

a problem, that is, whether law should express or enforce 'morals' (usually taken to mean some generally accepted rules . . .). There is another side to the relationship — the problem of autonomy and authority of various traditions of normative learning with which the law co-exists in society. The law must face the question of the mode in which it should recognise and / or supervise them. The alternatives emerge vividly in the law-religion relationship because religions are systems of control with complex learning doctrines, expounded by their own specialists and even their own doctrines concerning their relationship to law. But the same general problems of relative authoritativeness and competence are present in the relation of law to various simpler and unlettered traditions (for example, the custom of trade, the usage of a caste) as well as to complex and learned traditions like Hinduism. The 'law and morals' problem reappears in the relation of law to every body of normative learning. Our analysis of Indian secularism as presenting alternative possibilities for managing the relationship of law to Hindu tradition suggests the possibility of a more generally applicable typology of the relations between law and other modes of normative learning.[202]

So, it is not just a question of keeping the individual free from (State) imposed moralities. He is already subject to a whole range of seemingly voluntary but equally — if not more — onerous, imposed moralities. The State may have a role in protecting the individual from the diverse pressures which emanate from these 'other' moralities. But, for Galanter, the major problem for the State is the management of multiple, overlapping, co-existing and often conflicting moralities. His concern is not just for the autonomy of the individual (who might require the intervention of the State) but the autonomy of a plurality of moral systems which vary in the intensity with which they influence, manage or control the life and thoughts of individuals and groups. Some of these moral systems seek to control every aspect of the life of an individual or group; others are more forebearing. How are we to deal with an 'array of normative traditions which . . . coexist in society?'[203] In other writings, Galanter decisively rejects one American interpretation of the strict neutrality theory whereby religion can never be a basis for any kind of State activity.[204] This would lead to the absurd result that the State could not do anything about religious discrimination. Nor does he opt for the simple liberal solution that the law cannot, and should not, enforce morality. Morality is being enforced by various groups in various ways all the time. Membership of these groups is often as involuntary as membership of the State. The advice that the State should not impose what it cannot control is not a principle about individual autonomy but a strategy concerning the limits of effective legal action. To protect the autonomy of the individual, the State would have to play a highly interventionist role in liberating the individual from a multitude of moral regimes. There are two distinct questions

here. First, there is the question arising from liberalism: is it part
of the duties of the State to enforce its moral regime over the individual? Second: is it part of the State's duty to liberate the individual from the existing array of normative regimes to which he
may be voluntarily or involuntarily subject? Further, apart from
protecting the individual, are there any other reasons — or, any
other basis—on which the State should regulate or reform existing
normative regimes? For Galanter, there are no clear cut answers to
these questions. The State may have a duty to liberate the individual
from other normative regimes. The State may allow use of or lend
the State machinery to uphold (or enforce) certain aspects of other
normative regimes (e.g. marriage customs, right to group membership, right to establish religious and other institutions). The normative orderings of the State themselves constitute a normative
regime. The State may, to some extent and for well recognized
reasons, be forced to reform existing normative regimes to bring the
latter within the framework of the State's normative regimes. In
the Indian context, Galanter's quarrel is as much with the unholy
alliance of liberals and traditionalists who want to protect the
absolute autonomy of the existing normative regimes as it is with
reformers who wish to reform these existing normative regimes out
of existence to be replaced by a new rational, scientific approach
to individual and group life. And, since it is the reformers who are
on top, his major quarrel is with them. The reformers are, allegedly,
trying to do too much; assuming unconvincing stances, attempting
to denude society of its pluralistic variety and failing to recognize
the limitations of State and official action. The reformers, on their
part, are not always convinced that this 'array of normative regimes' (even after the reform of any oppressive practices which may
be an inherent part of them) does not constitute a welter of political
and social regimes which imprison the lives and perspectives of those
which fall within their catchment. They argue that the direct and
indirect, actual and hegemonic influence of these regimes could
deepen so as to stratify further an already over-stratified hierarchical
society. To such reformers, the State must assist in de-mystifying
both the influence of these normative regimes (and the power
structures they sustain and generate) as well as the pluralistic belief
that preserving such an 'array of normative orderings' is, itself,
necessarily a good thing. While arguments about the true nature
of the State's role and responsibility will continue, Galanter rightly
takes us well beyond the usual sterile controversies about the relationship of law and morals to confront a much more real and
challenging problem about the co-existence of the various normative
regimes to which individuals, groups and various segments of society
are subject.

Galanter's discussion raises several questions about the manner in which the management of this 'array of normative regimes' will take place. It must, perforce, take us to discussing the nature of the State that will undertake this management. Galanter has given a great deal of considered attention to one part of the Indian State which has housed some of this discussion, namely the judiciary. While the implications of his theory of State and the latter's place in 'development' are discussed later, it is necessary to make some prefatory remarks about his concept of the State; and, of the judiciary as part of it. The picture that emerges is that of a State consisting of a range of proactive and reactive agencies which can be systematically used by civil society to prevent and fight exploitation. *Prima facie*, Galanter does not enter into doctrinaire controversies about the legitimate functions of State. Just as in studying 'modern law', Galanter is concerned with examining a 'cluster of features', in studying the agencies of State, Galanter looks at the State as an aggregate of institutions and processes. He starts empirically by examining the working of these institutions and processes. Thus, he looks at the various arenas of operation which lawyers generate around themselves and the kinds of skills—institutional or otherwise—required to use these arenas effectively.[205] For Galanter, the State is not just an authoritative source of power and rules. It constitutes a cluster of bargaining and negotiating arenas which co-exist with other such arenas constituted by society. Galanter's sociology of law examines the nature of these various arenas and conclaves of power and influence and concludes that they can co-exist and be made to co-ordinate their work so as to produce 'justice in many rooms'.[206] His concept of State is consistent with his pluralistic view of society. For him the State is neither a conspiracy on behalf of one class against another nor a vanguard regulatory instrument for reform, welfare and development. In his later strictures on the literature on law and development, he attacked 'liberal legalist' assumptions about the State and its instrumental capabilities.[207] He does not deny that the State may have these instrumental tasks but suggests deeper and more incisive journeys into identifying each one of those tasks and the manner in which they are planned, executed and received. This emphasis on empirical examination has the dual effect of asking for more facts as well as disaggregating the work of the State into separate compartments. Such an approach invariably takes us deeper into Galanter's pluralistic view of society in which various groups and institutions—new, old and those about to be created — interact with other groups, institutions and processes — State, non-State, and quasi-State — to explore various versions of the good life. In this way, Galanter perceives the relationship between public and private forces to maintain the 'array of nor-

mative regimes' which go to make up the pluralistic society to which
he espouses such a passionate commitment. Many may feel that
Galanter gives much too low a profile to the responsibilities of the
modern Indian State to create a more equitable and less exploitative
society. They may argue that the defence of the co-existence of this
array of normative regimes is following the path of least resistance
which enjoins India to entrench its traditionalism and pursue incremental change. Galanter, on the other hand, seems to explicate
the caution of pragmatism (i.e. people will not give up their birth-
right of beliefs for a mess of potage) as well as assert his view that
individual and group preferences about the good life should be
maximized.

Galanter on the Judiciary

Our discussion about Galanter's concept of a pluralistic society —
and the role of the State in preserving this plurality — is essential
to understanding his analysis of the judiciary. The judiciary —
though established by the State — is featured as a bargaining arena.
Different kinds of bargaining take place in different kinds of adjudicatory processes. Galanter rejects the somewhat staid 'triadic' model
of the judiciary depicting the court as a neutral apex body presiding
over disputes in which the judge applies law between two or more
parties.[208] Under the 'triadic' model, the judge applies 'law' to the
dispute. Those who accept the 'triadic' model as a basis for analysis,
concentrate their research concerns on evaluating the work of the
judiciary against this 'triadic' ideal or examining the basis and
values on which judicial decision making is based. All this results
in prescriptive propositions about the importance of 'neutrality' in
constitutional decision making, typologies about the 'indiscriminate'
judge, the 'policy' judge, and the 'principled' judge or suggestions
that the politics of the judiciary is quite like the politics of any other
institution.[209]

In rejecting the 'triadic' model as a tool for analysis, Galanter
does not rule out the possibility that empirical investigation may
discover exactly the same typologies as those envisaged by the 'triadic'
model. Galanter does not trivialize the importance of legal discourse
as do those who see no difference between the discourse used in
judicial institutions and the discourse used in other administrative
and political institutions. 'Law' is an important symbolic and
persuasive entity. But all decisions are not made according to law
but in the light of various factors after taking law into account. Law
is depicted as a 'bargaining endowment' used by various actors who
participate in the judicial process. In his understanding of the
judicial process, Galanter has substituted a theory of 'bargaining in
the shadow of the law'[210] for a theory of 'deciding according to the

law'. This forces us back to the problems of defining law which we thought we might have got rid of in earlier discussions in this essay. We have seen how Galanter claimed that modern law possessed a cluster of features whose ontological existence was unquestioned. In this seemingly revised analysis, law emerges, far less securely, as a bargaining endowment. These two views do not necessarily pose a contradiction. The 'cluster of features' which Galanter identified as part of modern law are an important part of the modern legal system. But the rules of law are less imposing than the fanfare and mystique which accompany them. In the social world — no less the legal world subjected to social analysis — they do not loom as large, or appear as important, as they appear to analytical jurists. The declarations of law in the form of rules, principles, and policies are, at one level, only impressive declarations. The mere act of their declaration does not immediately give them social substance, either generally or as part of the decision-making process. Galanter argues that these declarations are most influential in the litigation and adjudicatory system as bargaining endowments. While he is aware that these symbolic declarations are part of a 'constitutive' intellectual discourse and mark out an ideological domain, he is less preoccupied with this dimension of legal rules than those whose central concern is to understand law as an ideology. Galanter's views are directed to reinforcing an image of law, legal institutions and the State as forces which are malleable at the hands of society and social forces. This takes us away from analysing the central role of law to maintain patterns of power in the political economy and to focus on a much more disaggregated view of law and how it is used by a differentiated pluralistic society.

Galanter's work on the judiciary is an application of his approach to treating law and legal institutions as a malleable bargaining arena constituted by the State. In the American context, he has looked at the role of the judge in lower criminal court proceedings. From this emerges his picture of the crusading judge.[211] The crusading judge is more of a third party to the dispute than an independent natural apex. He may possess the ultimate power to make a decision, but the nature of the proceedings is better understood as existing on a horizontal rather than a vertical plane. And, if they take a vertical form on the day of judgment (and, there is not always a day of judgment), the final metamorphosis is one only of form. There is, according to Galanter, no reason for us to abandon our view of the judicial process as a horizontal inter-party negotiation in which the judge participates, negotiates and presents a view. It is, however, not clear whose view the judge represents, other than his own.

A lot of Galanter's work on India has been on the higher judiciary. At this level, decision making is not just directed to determining

specific decisions and practical outcomes. The higher judiciary is
also an instrument of policy declaration. It creates new laws, pro-
nounces new policies and re-draws the ideological framework of law
and society. Judges in such courts are often regarded as much more
credible, trusted and effective makers of ideology than other parts of
the State. They carry a certain mystique about their role. They provide detailed and 'balanced' explanations about their ideological formulations on the basis of allegedly universalized principles. Galanter
pays a considerable amount of attention to the role of these judges as
conceptive ideologists.

We have already referred to the role of the Indian judiciary in
shaping the views of group membership and the 'changing conceptions of caste' in India. In this area, Galanter finds the work of
the Indian judiciary pragmatic and, in the main, sustaining an
egalitarian pluralistic society even though he has some misgivings
about its support for orthodox justifications of caste and its,
occasionally exaggerated notions of social reform. Galanter is
sceptical about the reformist stance that many judges in India have
assumed. He gives detailed attention to a judgment of Justice
Gajendragadkar who sought to reform Hinduism according to some
kind of pro-government secular vision of society.[212] He is also critical
of Krishna Iyer's symbolic activism in treating positive discrimination as part of — and not an exception to — the principle of
equality while simultaneously creating a super classification in which
the gains of such an approach are limited to people whose entitlement to such benefits was never really in question anyway.[213]
Galanter seeks to argue that the judiciary could be put to far better
use than simply obtaining such symbolic declarations.

By way of research, Galanter suggests two courses of action. The
first course of action is to analyse the kind of intervention that
judges make as third parties to the 'negotiation'. He sees, in this, a
far greater freedom on the part of judges to take an interventionist
stance. He also sees far greater need for social analysis beyond the
simple application of legal doctrine or the explication of official
policy. As a third party intervener, the judge is not just the creature
of legal doctrine or an instrument of mainstream politics. The second
course of action is to assess the kind of reform that the judges promulgate. He surmises that

> . . . when court decisions are influential it is not through doctrinal pro-
> nouncements but through the re-channeling of major institutional oppor-
> tunities and controls and by their liberating effect. The court may provide
> dissident and progressive (groups) institutional support and with ration-
> alization for their non-traditional behaviour and beliefs.[214]

By this time we should not be surprised to read Galanter making a plea that judges should not rush in to reform what they do not understand and cannot effect. Apart from protecting the integrity of groups, Galanter is concerned to emphasize that judges do not just declare doctrine but simultaneously create enabling social and legal processes. They create fresh bargaining arenas and help to sustain and change existing ones.

This view of the judiciary leaves many questions unanswered. Especially disturbing are those questions which arise from the premises on which this view of the judiciary rests. Can we really view judges as third party interveners without asking a basketful of questions about their relationship with the State of which they are a part? Can we treat law as a mere bargaining endowment when we can see that it does create a sense of boundedness and is not a mere will-o'-the-wisp? Of course, judicial institutions are bargaining arenas, but are they really that open-ended? Is not the symbolic activism of the judiciary an important part of their role as the conceptive ideologists of the State and their class? These questions lead to exploring the role, position and function of the judiciary as part of the State and the political economy within a broader concept of power. To some, the judiciary is a unique bureaucracy within the State. Neither wholly dependent or independent, neither an open-ended bargaining process nor a wholly closed one, the judiciary, its work, its declarations and the use and abuse of its processes is, perhaps, best understood after looking at the whole cluster of institutions and processes which form part of the State. For we must not overlook Galanter's own declaration that modern law is a political creation. Even if we look at the areas Galanter writes about (group rights, religious groups, and positive discrimination in favour of backward classes) the declarations of the judges correspond very greatly to the general value orientations presented by other agencies of the government.[215] And, if we sense such a congruence, is it not crucial to divert our research to examining the nature of such an ideology and the role that such an ideology plays in sustaining patterns of power in Indian society? This is not just a crude suggestion that judges — like their administrative counterparts — directly and indirectly sustain class interests. It is to examine the nature of modern law as part of the State. In this more complicated vision, the judges are a unique bureaucracy as well as subtle conceptive ideologists in a society much too complex and much too sensitive to accept only direct methods of oppression, control and influence. For Galanter, such questions are important but speculative. His commitment to empiricism and pluralism (and the two are very closely inter-related) suggests we must start by looking at the work of these institutions and processes and discover who uses them and why. But, are these more general questions about the State and power just an added bonus or are they essential to our understanding of what is going on? The methodological dispute between these alternative approaches is not just about *where* we initiate our inquiry. It is really a dispute about *how we view* society; how we view the groups, processes, and institutions within society and the interconnections between institutions and processes. The approach of the empiricists has, implicitly, been assailed by a group of 'critical' theorists who examine the ideological dimensions of law and State and seek to liberate the individual and the community from the structural, institutional and ideological constructs in which law has trapped civil society. The domain assumptions of this 'critical' sociology of law and the empirical approach are quite different, as is their view of society and their political objectives. This alternative 'critical' analysis of law concentrates on how a particular exploitative power structure is maintained in society. Galanter seems to take the view that such 'critical' explorations border too greatly on the speculative. They initiate an 'argument without end'216 in which everyone is his own master. He, therefore, argues for deeper and more incisive journeys into the pathology and contours of the 'law' we encounter every day.

Galanter on Law and Development

All this takes us to the crucial area of law and development. So far we have discussed Galanter's commitment to pluralism but only declared, without analysing, his commitment to preventing exploitation. Galanter and Trubek have argued that law and development was the major American export industry which, till the seventies, was much too enthused by a proselyting motivation to subject itself to criticism. This led to an absurd and somewhat acerbic personal debate within American scholarship.217 In a sense, that debate is irrelevant to India and the Third World except to the extent to which they are recipients of advice and resources on the basis of what is suggested by the protagonists and antagonists in the debate. If the *sahibs* must agonize amongst themselves, let them do so. But the debate is not just an American debate. The debate also has Indian parallels which are not inspired, or influenced, by the American discussion. The American debate concerns disputes about the core paradigm on which American notions of law and development are based. This is a 'liberal legalist' paradigm. In a sense, this paradigm draws on the kinds of perceptions precipitated by Roosevelt's New Deal. The State promulgates norms, regulates various sectors of the economy and society, creates bureaucracies to sustain change, uses courts and the rule of law to define the rules and processes of change and tries to cross the 'gap' between prescriptive rules and existing social reality. Within this paradigm there are many variations. One variation—which we can call the 'strong instrumental' view—makes government and bureaucrcy central to the whole process of spearheading change. The second variation—which may be termed the weak instrumental view—gives greater primacy to law, the rule of law and a liberal theory of rights as the means of development. Courts are very important agencies to preserve the essence of this liberal approach. In both these variations, the State plays a very central role, in determining the kind of corrective that can be administered to ensure the direction of change. The 'liberal legalist' view sees society as something 'out there'—an amorphous mass, confused and confusing and which, in varying degrees, is receptive to some changes and resistant to others. Attention is, therefore, diverted to the 'gap' between normative purpose and social reality. Optimism—or pessimism—is expressed about whether the 'gap' can be crossed. This entails the need to analyse 'society' which seems to lie so inconveniently 'out there' thwarting the government's plans to achieve change. And since the 'out there' seemed to vary greatly from society to society, a compendium term called 'legal culture' was devised to encompass studies about the nature of the 'out there'.[218] Unfortunately, legal culture is not an analysis of specific forms of resistance to, or support of, change but a discussion of general attitudes. Law and development is supposed to take into account legal culture. Equally, there is emphasis on the need to create a new legal culture which would be more sensitive to change. And so, the *sahibs* talk amongst themselves. 'Strong' or 'weak' instrumentalism? Small gaps or big gaps? Respecting legal culture or changing it? Since the discussions are hopelessly inconclusive, some have withdrawn from the discussion, some have left the field altogether and some continue to pursue these controversies.

But these are not just controversies that exist amongst American scholars. They also take place in the Third World. Strong instrumentalism has a very credible following in India. It is not drawn from American New Deal conceptions but from Nehru's Soviet-inspired system of Five Year Plans. The demand for changing 'legal culture' is explicit in attempts to re-define the nature of group membership in India or changing the legal conception of religious rights. The great controversies about the importance of courts and law in India—which has led to many constitutional battles and the 'Emergency'—notionally represents quarrels between the 'strong' and the 'weak' instrumentalists. Even if American debate has ended on the self-indulgent note that the scholars themselves are in a state of self estrangement, the Indian debate continues. But the Indian debate is not just concerned with academic explorations about competing alternatives. Each labelled position in the debate is backed by powerful interest groups. The 'strong instrumentalists' argue for more powers, less court intervention, more attacks on traditional culture, and a more ruthless political apparatus to achieve their declared instrumental aims. The 'weak instrumentalists' seek less government power, more court intervention, some emphasis on traditionalism and the further perpetuation of a liberal theory of rights. As we penetrate beyond the doctrinaire discussions in which these demands are expressed, we may encounter paradoxical lines of support for these approaches. The 'strong instrumentalists' may not be interested in reform at all. The 'weak instrumentalists' may be less interested in supporting a general liberal theory of rights in society and mainly concerned with the advancement of a narrow range of interests. The motives and reasons for supporting a particular view of development may be varied, the 'gap' between plans and their implementation may vary according to whom it is intended to harm or benefit and the entire apparatus for development may be designed to preserve the *status quo*.

These preliminary observations are enough to cause scepticism about the 'top down' model of development. If governments and planning commissions continue to adhere to this model, it is because they must. If people are persuaded to lend credence to this model, it is as much a tribute to their optimism as a reflection of the naivete of their social analysis. To that extent, Galanter and Trubek's plea that this model should at least be re-examined is much too modest a proposal to provoke the sense of outrage that their sometimes indiscreet use of language has provoked.

So, what happens next? The estrangement of American scholarship may, perhaps, be discussed as part and parcel of American foreign policy. And, if American scholarship rushed in where American Presidents no longer fear to tread, such scholarship will surely re-group in some other more alluring form. From India's point of view the quest for fresh analysis and fresh solutions must continue even if the new insights have their origins in finding public solutions for private vested interests.

One serious alternative to the 'top down' model is to look at development from below. The starting point of development perspectives in this line of analysis is not an increase in Gross National Product or Productive Capacity but in distributive justice for the disadvantaged. One strand of such analysis sees the State as an oppressor linked very closely to class interests. Another strand of this analysis sees the State as an important agency which creates a cluster of institutions to repress the disadvantaged, maintain the *status quo* and, paradoxically, provide, as a result of its own rhetoric, some undefined opportunities for obtaining distributive justice. The first strand of argument calls for an extreme distrust of the State, a wide ranging social and political struggle and an emphasis on creating new groups and alliances on a class basis to fight oppression. The other strand, although suspicious of the State, regards it as providing a range of 'turn around' solutions which can be exploited by the disadvantaged. Such solutions may be possible within the existing political economy and social structure. These are two distinct position, even if there are agreed areas of overlap whilst vast areas of disagreement are void for uncertainty.

Galanter belongs very clearly to the second strand of argument. With his commitment to a tolerant pluralist society, it is not surprising that he views development from below. Given his analysis of the State as providing a range of bargaining arenas, he is not overwhelmed by the notion that the State is class conspiracy. Since modern law represents a 'cluster of features' which, in turn, precipitate bargaining, there is room for negotiation and, therefore, development. But before we commit Galanter to unbridled optimism, it is necessary to emphasize that, although he believes that the dice are loaded against the disadvantaged, he feels that they have no other choice but to identify and learn to exploit the opportunities provided by the institutions and processes constituted by modern law.

In order to understand Galanter's quest for 'turn around' solutions in a modern 'State sponsored', liberal, pluralist society, it is necessary to return to a celebrated article in which Galanter explains how the 'haves' come out ahead under the Anglo-American system of litigation.[219] The legal system is not an arena for the crude exercise of power. It can only be effectively used by those possessing the institutional skills to exploit it. The 'haves' come out ahead because they are able to create these skills on a continuous basis. They are 'repeat' players; not just 'one shotters'. If the 'have nots' wish the legal system to work in their favour, they must find similar or comparable institutional skills. Applying this to India, Galanter argues that the record of the 'Untouchables' and other Backward classes represents a series of 'missed opportunities'. They have never really used the law; and, where they have, they have done so mainly as unskilled 'one shotters' rather than institutionally efficient 'repeat' players. Given the right kind of skills, the oppressed can find 'turn around' solutions in the law.

There are various problems with this view which has been oversimplified in explanation. Some of these problems are underlined by Galanter himself. At the practical level, there is the problem of resourceless disadvantaged groups finding the resources to acquire the institutional skills of 'repeat' players. Galanter believes that the Indian landscape provides instances of groups who have found such resources. In any event, the disadvantaged have no other

choice but to face the legal system. It follows them everywhere, defining their rights, controlling their oppression, whittling down their
entitlements and, where necessary, hunting them down. It is possible
that what Galanter suggests may be valid for the American legal
system where the range of resources available to the disadvantaged
is much greater and where the political economy has been forced
to design greater 'turn around' solutions for them. Galanter's
arguments are built on a series of assumptions about the receptivity
of all (or some) modern law to this kind of induced change. It is
not part of Galanter's argument that modern law must, by its very
nature, provide these 'turn around' solutions. He does not seek to
argue that the universalistic, egalitarian rhetoric of modern law
necessitates providing opportunities for the disadvantaged. He
assumes that some opportunities must exist and can be exploited.
There is no analysis, structural or otherwise, of the kind of opportunities provided by any particular system, for whom, why and in
what measure. Indeed, there is every reason to believe that the so-
called opportunities which modern law designs for the disadvantaged
are, in fact, cleverly designed 'traps' which offer marginal relief
in a sedulously designed system which obtains social and political
legitimacy in exchange for its overtly expressed concern for the poor.
Galanter does not argue that the 'legal route is the only route to
change'. Nor does he deny that the ultimate strength of the disadvantaged lies in their capacity to find the social and political
counter-vailing power which will help them to bargain into a better
strategic position. He does not counsel against using other routes to
change, including, perhaps, violent revolution. Independent of these
insights, Galanter concentrates on devising the ways and means in
which the various institutions and processes of modern law can be
used to advantage by the poor. However, his prescriptions seem to
rest on the implied assumptions that they will work to fruition only
in a modern liberal State blessed with the kind of pluralist society
which Galanter believes in so passionately. Adherence to these
prescriptions would, indubitably have the effect of a loss of opportunity for competing political alternatives.

At the end of the day, we cannot divorce Galanter's approach to
change from his distinct views about State and society. For him
society is enriched by the co-existence of a mosaic of horizontally
and vertically arranged groups. For him such groups are so super-
abundant, so luxurious in their texture, so complex in their commitments, so flexible and inflexible in their range, so resilient in their
capacity for accommodation and, yet, so hostile to change that they
cannot constitute a threat but must be seen as providing a vista of
possibilities. Of course, some groups conspire, some oppress, some
manipulate. Equally, others expose conspiracy, fight oppression and

resist manipulation. The various sides are not evenly matched. But 'out there' lies this pluralist society which offers protection to all and which, in turn, must be protected—both as practical necessity and as an act of faith. Perhaps, if he were persuaded to a different view of power and society, his insights might differ. His view of society is matched by his view of the State. Too big to be just a conspiracy, too varied to be subject to unified control, performing too many diverse functions to admit of possession by one or the other group, it simply sprawls. Not necessarily a Leviathan, it stretches in many directions and for many purposes. And, to borrow Walt Whitman's phrase

> If it contradicts itself,
> Well, then, it contradicts itself,
> For it is large and can contain multitudes.

Galanter's Indian Construction of Modern Law

Galanter's image of law is derived very greatly from his understanding of Indian society. In India, one can see a vast variety of group life. One can also see a State that is both concerted and disconcerted in its aims. Part of the State manufactures what Galanter calls modern law. Other parts of the State are privatized by Indian society in measures that make it impossible to distinguish private from public power. Parts of the State simply drift into entropy. Although Galanter begins by establishing the ontological existence of modern law, he dilutes the importance of this existence by showing how it is, inevitably, sustained and undermined by social processes. Traditionalist to the extent of recognizing both primordial, traditional and emergent groups, romantic about preserving the rich contradictions of heritage and uncompromising about fighting exploitation, Galanter tries to approach the legal system from below to draw pictures of socio-legal reality as they appear to those who encounter and participate in the legal system. He looks for solutions which preserve group life while simultaneously fighting oppression. His insights into the nature of society, its groupings and the nature of modern law as a neatly expressed, but malleable, reality provide rich insights.

At the end, there are some clear messages. Modern law is not a self fulfilling prophecy. It has both conspiratorial and non-conspiratorial elements. Faced with the power asymmetry of modern life, its task is to manage the co-existence of various normative and group orderings without losing its commitment to prevent oppression. Much of Galanter's analysis is pitched at the level of explanation. This is reflected in his research agenda which begins by examining pluralistic group orderings and demands an exacting amount of

data before it proceeds beyond what he considers speculative interpretation. And, if after all this, we have reason to suspect Galanter's
emotional attachment to a particular view of the good life, who can
say that we are wrong? Least of all, Galanter himself.

END NOTES

1. For some of the problems of chronology in dealing with the *dharmashastra*, see R. Lingat: *The Classical Law of India* (Berkeley, 1973) 123 ff., esp. at 132. 'The authenticity of the attribution and, consequently, the very value of the extract as a rule of *dharma* rests entirely and definitively upon the authority enjoyed by commentators or digest writers who selected and fixed it. From then onwards it becomes part of a mass of texts whose dates, relative to each other, were of no interest to the interpreter.' Contextualizing the work of the *sastris* has not been easy. For example, K. P. Jayaswal's *Manu and Yajnavalkya: A Basic History of Hindu Law* (Calcutta, 1930), which presents Manu as defending a Brahmin usurpation to the throne in the second century BC, has rightly been described as standing 'at the boundary between brilliance and guesswork' (See J.D.M. Derrett: *Religion, Law and State in India* (London, 1968) 567).

2. There are enormous difficulties in concentrating on just the intellectual inspirations that motivated the Raj when looking at normative legal change (e.g. E. Stokes: *The English Utilitarians in India* (Oxford, 1959)). We need to look more closely at the actual effect of changes in the structure of judicial administration on the political economy. For an interesting case study, see R. Kumar: *Western India in the Nineteenth Century* (London, 1968) 74–83, 153–60, 209–28.

3. Quite apart from 'the British as patrons of the *sastra*' (on which see *infra* n. 6), British administrators themselves began to compile their version of the native law. Beginning with Halhed's *Code of Gentoos Laws* and Hamilton's *Hedaya*, a vigorous scholarship — associated with the names of Sir William Jones, Colebroke, Macnaghten, Strange, Sutherland, Ellis, J.D. Mayne and, in our times, Derrett in Hindu law; and Baillie, Macnaghten, Wilson, and others in Muslim law — produced British versions of Hindu and Muslim law. For detailed accounts of the lives and work of some of the key figures, see Lord Teignmouth: *Memories of the Life and Writings and Correspondence of Sir William Jones* (London, 1804); Garland Cannon: *Oriental Jones* (New York, 1954); S.N. Mukherjee: *Sir William Jones: A Study in Eighteenth Century British Attitudes to India* (Cambridge, 1968); R. Rocher: *Alexander Hamilton 1762–1824: A Chapter of History of Sanskrit Philosophy* (New Haven, 1968); B.S. Cohn: 'The Command of Language and the Language of Command', in R. Guha (ed.), *Subaltern Studies: Writings on South Asian History and Society* (Delhi, 1985) IV, 276–329; R. Rocher: *Orientalism, Poetry and the Millennium: The Checkered Life of Nathaniel Brassey Halhed 1751–1830* (Delhi, 1985), esp. 48–72. For brief surveys of this transformation, see M.P. Jain: *Outlines of Indian Legal History* (Bombay, 1971 edn.) 702–12; J.K. Mittal: *An Introduction to Indian Legal History* (Allahabad, 1980 edn.) 362–8. This was a period of considerable turmoil when English adventurers made a lot of money and did not desist from directly corrupting the judicial system to their use. There is a considerable literature on this early period of development: e.g. J.F. Stephen: *The Story of Nuncoomar and the Impeachment of Sir Elijah Impey* (London, 1885); Derrett: 'Nandkumar's Forgery' (1960) *English Historical Review* 223–38; Jain (*supra*) 105–65; B.N. Pandey: *The Introduction of English Law into India* (New York, 1967); N. Dhar: *The Administrative System of the East*

India Company in Bengal 1774–1784 (Calcutta, 1966); A.C. Patra: *The Administration of Justice under the East India Company in Bengal, Bihar and Orissa* (Bombay, 1962); B.B. Misra: *The Judicial Administration of the East India Company* (Delhi, 1961); N. Majumdar: *Justice and Police in Bengal 1765–1793: A Study of the Nizamat in Bengal* (Calcutta, 1960). For accounts of seventeenth-century justice, see C. Fawcett: *First Century of British Justice in India* (London, 1934); B.S. Jain: *Administration of Justice in Seventeenth Century India: A Study of Salient Concepts of Mughal Justice* (Delhi, 1970). Meanwhile, an enduring Indian scholarship in these subjects has continued culminating in practitioners' works on Anglicized versions of these laws. However, scholarship on these areas has not abated.

4. For the Roman and English law origins of the clausula, see J.D.M. Derrett: 'Justice, Equity and Good Conscience' in J.N.D. Anderson: *Changing Law in Developing Countries* (London, 1963) 114–53.

5. J.D.M. Derrett: 'Justice, Equity and Good Conscience in India' (1962) 64 *Bom.L.R.* (Jnl.) 129, 145; R. Dhavan (*infra* n. 86) 95–101; Cf. M.P. Jain (*supra* n. 3) 576–90; and Mittal (*supra* n. 3) 282–9.

6. J.D.M. Derrett: 'Sanskrit Treatises Compiled at the Instance of the British' (1961) 63 Z.V.R. 72–117 (Z.V.R. = *Zeitschrift fuer vergleischende Recthswissenschaft*); reprinted with amendments in Derrett (*supra* n. 1) 225–73.

7. On the role of *pandits* and *moulvis* during the British period, see J.D.M. Derrett: *History of Indian Law* (Leiden/Koln, E.J. Brill, 1973) 22–5; ibid.: (*supra* n.1) 296–303; Kane: *History of the Dharmashastra* (5 volumes in 7 parts, Poona, 1932–68) III. 969–73, on the Anglo-Indian attitude to custom (cf. 856–84) For summary accounts of their use in British courts from 1772 to 1864, see M.P. Jain (*supra* n.3) 698–701; J.K. Mittal (*supra* n.3) 368–2. Act IX of 1864 obviated the need for recourse to pundits and *moulvis* whose assistance was, subsequently, adjudged as unnecessary. (See *Masjid Shahid Ganj Committee v. Shiromani Gurdwara Prabandhak Committee* (1940) 67 *I.A.* 251 at 260) and misleading (*Hori Dasi v. Secretary of State for India* (1880) 5 *Cal.* 228 at 242). This prejudice against the pundits was voiced soon after the induction into the judicial administration of the company. (See Derrett, *supra* n. 1, 243). It gave a greater leeway to the judges to find the law from the *sastra* itself. Whether the judges themselves became the new *dharmasastris* is doubtful. The modern judge, as, *inter alia*, a servant of the State, is not in an analogous position in respect of the political economy.

8. For Gandhi's statement, see *Hindustan Times*, 7 August 1926, quoted in G. Gadbois Jr, 'Evolution of the Federal Court in India' (1963) 5 *J.I.L.I.* 19 at 26.

9. See *A Digest of Hindu Law on Contracts and Successions with a Commentary by Jaggan' Tha Tercapanchanana*, Translated from the Original Sanskrit by H.T. Colebroke (Madras, 1874, fourth edn., first published 1796). For an account of how the *Digest* was received, see Derrett (*supra* n. 1, 254–9). For an example of the fusion between the *sastra* and equity principles, see Derrett: *Critique of Modern Hindu Law* (Bombay, 1970) 74 ff. about the proportionate allotment of gains at the expense of the family; see further his 'Acquisition of Joint Family Property through a Coparcener — Let Shastric and Equity Principles Join Hands' (1969) 71 *Bom.L.R.* (Jnl.) 75–81.

10. For Nelson's writing, see his *The Madura Manual: A Manual Compiled by the Order of the Madras Government* (Madras, 1868); *Hindu Law at Madras* (London, 1881); *A Prospectus for the Scientific Study of Hindu Law* (1881); 'Hindu law in

Madras in 1714' (1880) *Madras Journal of Literature and Science* 1–20. Note the reply to his approach by Justice Innes: *Examination of Mr. Nelson's view of Hindu Law in a letter to the Right Hon. Monstuart Elphinstone Grant Duff, Governor of Madras* (Madras, 1882). For a record of the work and contribution of Nelson, see J.D.M. Derrett: 'J.H. Nelson—A Forgotten Administrator-Historian of India', in C.H. Phillips (ed.): *Historians of India, Pakistan and Ceylon* (London, 1961) 354–72.

11. For the transformation of Hindu law, see Derrett: 'The Administration of Hindu law by the British' (1961) 4 *C.S.S.H.* 10–52. However, to gauge the social and political realities underlying this transformation, deeper and more incisive journeys have to be taken into specific areas of this transformation; as, for example, the Hindu joint family (on which see Derrett: 'A History of the Juridical Framework of the Joint Family' (1962) 6 *Contributions to Indian Sociology* 17–47; G. Sontheimer: *The Joint Hindu Family: Its Evolution as a Legal Institution* (Delhi, 1977)) or religious endowments (see G. Sontheimer: 'The Juristic Personality of Hindu Deities' (1965) 67 *Z.V.R.* 45; also Derrett: 'The Reform of Hindu Religious Endowments', in D.E. Smith (ed.): *South Asian Politics and Religion* (Princeton, 1966)). However, even these more specific studies do not explore the relationship between the transformation of doctrine and the socio-political context. For a general attempt to formally link *sastric* learning with modern 'transformed' Hindu law, see J.D. Mayne: *Treatise on Hindu Law and Usage* (Madras, 1950, eleventh edn.). The general transformation of Muslim law is considered by Derrett (*supra* n. 1) 513; A.A.A. Fyzee: 'Muhammadan law in India (1963) 5 *C.S.S.H.* 401; 'The Impact of the English Law on the Shariat in India' (1904) 66 *Bom.L.R.* (Jnl.) 107–16. See also T. Mahmood: *Islamic Law in Modern India* (Delhi, 1971); *Muslim Personal Law: Role of the State in the Sub-Continent* (Delhi, 1977). Anglo-Muhammadan law produced its own fund of textbooks e.g. R.K. Wilson: *Anglo-Muhammadan Law* (London, 1930); Abdur Rahim: *Principles of Muhammadan Jurisprudence According to the Hanafi, Maliki, Shafi and Hanbali Schools* (Madras, 1911); S. Amieer Ali: *Mahommeddan Law* (Calcutta, 1912); F.B. Tyabji: *Muhammadan Law: The Personal Laws of Muslims* (Bombay, 1940); A.A.A. Fyzee: *Outlines of Muhammadan Law* (London, 1964); *Mulla: Principles of Mahommedan Law* (Bombay, 1977 edn. by M. Hidayatullah and A. Hidayatullah) is the current practitioner's text. For the exchange between Hindu law and Roman and Continental law, see Derrett: 'Hindu Law in Goa: A Contact between Natural, Roman, and Hindu Laws' (1965) 67 *Z.V.R.* 131; 'The Role of Roman and Continental Laws in India' (1959) 24 *Z.f. auslandisches und internationales Privatrecht* 657–85. For the contribution of Italian and Portuguese jurists, see Derrett: 'Juridical Ethnology: The Life and Work of Guiseppe Mazzarella, 1868–1958' (1969) 71 *Z.V.R.* 137; 'Luis da Cunha' Gonsalves (1975–1956) Jurist, Comparative Lawyer and Orientalist' (1974) 74 *Z.V.R.* 137–62. Once again, these studies, even though enthused with considerable social insights, do not attempt to analyse developments in the context of the political economy.

12. For a review and anthology of this legislation, see Whitley Stokes: *The Anglo Indian Codes*: Volume 1: *Substantive Law;* Volume II: *Procedural Law* (Oxford, 1887); see further Jain (*supra* n. 3) 600–96; Mittal (*supra* n. 3) 312–53; H.P. Dubey: *A Short History of the Judicial Systems of India and Some Foreign Countries* (Bombay, 1968) 233–41; S.V. Desika Char: *Centralized Legislation: A*

History of the Legislative System of British India from 1834 to 1861 (London, 1963).
167–225 (on Law Commissions) and 270–321 (on the legislation); R.C. Mau-
jamdar and K. Datta: 'Legislation and Justice' in R.C. Maujamdar, *et. al.* (ed.):
*The History and Culture of the Indian People: Volume 9: British Paramountcy and Indian
Renaissance* (Bombay, 1963) 339–53; B.K. Acharyya: *Codification in British India*
(Calcutta, 1914). For earlier accounts, see W. Hunter: *Seven Years of Legislation*
(London, 1870); C. Ilbert: 'Indian Codification' (1889) 5 *L.Q.R.* 347–69;
'Sir James Stephen as a Legislator' (1894) 10 *L.Q.R.* 222. T.B. Macaulay's
speech in the House of Commons (Third series, Hansard XIX, 531–3, 10 July
1833) displays the concern of the British before this legislative period. For comments on the utilitarian basis of the then impetus for reform, see E. Stokes:
supra n. 2. For an overview of the legislation, see C. Ilbert: *The Government of
India: Being a Digest of the Statute Law Relating Thereto* (London, 1967); G. Rankin: *Background to Indian Law* (Bombay, 1946).

13. This is quite crucial to the understanding of the *dharmasastra*. Thus, my unpublished paper 'Two concepts of Law' (mimeo., 1983) contrasts the western
   'institutional' approach of law from the *sastric* ideological approach. To try and
   understand the *sastra* within the institutional framework is to misunderstand
   its purpose. Kane (*suptra* n. 7) I, 466. stresses how the *sastra* '. . . created great
   solidarity and cohesion among the several classes of Aryan society in India
   despite conflicting interest and inclinations and enabled Hindu society to hold
   its own against successive invaders.' The ideological and hegemonic effect of
   the *sastra* and, indeed, of all law, is a profitable and important area of research.

14. For Kane's writing, see n. 7 (*supra*).

15. As Derrett remarks in his Preface to Lingat (*supra* n. 1) ix: 'P.V. Kane('s) . . .
   gigantic and solid encyclopedia still daunts even the best intentioned enquirer.
   (His) . . . objectivity and accuracy were phenomenal; but he too had a motive,
   namely to show that India's *own* jurisprudence shown in its historical setting
   and that when judged by the appropriate criteria it could stand comparison
   with any system likely to be compared with it.' For Kane's own account of his
   life, see his *History* (*supra* n. 7) V, 1712 ff. At the end of his survey of the *dharmasastra* (*supra* n. 7) Kane (at I, 466) committed himself to the view that '(a)ll
   these numberless authors were actuated by the most laudable motives of regulation in Aryan society in all matters, civil, religious and moral and of serving for
   the members of that society happiness in this world and the next.' This basic
   paradigm about the intent of the *sastra* serves as the basis for later students.
   Accounts of the *dharmasastra* concentrate on internal controversies amongst the
   sastris rather than on the social, political and economic circumstances which
   occasioned the controversies.

16. The phrase is taken from A.L. Basham's *The Wonder that Was India* (London,
   1971 edn.) which (at 112–22) presents a very rosy picture of Indian law.

17. e.g. S. Varadachariar: *The Hindu Judicial System* (Lucknow, 1946); Dhavan
   (*infra* 62 and 65) and a whole host of writers wishing to prove the modernity of
   the *sastra*. Reference to the *sastra* as the basis of the rule of law is part of a normal
   rhetoric (see S. Varadarajulu Naidu: 'The Rule of Law as Dharma' (1961) 2
   M.L.J. (Jnl.) 11–12; 'The Rule of Law as Karma' (1960) 1 *M.L.J.* (Jnl.) 42;
   B.N. Chobe: 'Judiciary in Ancient India: (1954) *S.C.J.* 85; 'The Art of Governance in Ancient India' (1956) *S.C.J.* (Jnl.) 19). For accounts of the judicial
   period, see M. Akhbar: *The Administration of Justice by the Mughals* (Lahore,

1948); W. Husain: *Administration of Justice During the Muslim Rule in India* (Calcutta, 1934); and, above all, M.B. Ahmad: *The Administration of Justice in Mughal India* (Aligarh, 1941). It is doubtful that courts during either era consciously articulated either *sastric* or Koranic textual learning. For a review of the literature, see R. Dhavan: 'Judge and Jurist in India: Notes Towards an Understanding of "Judicial Elites" in India' (paper to the International Political Science Association's Comparative Judicial Studies Group in June 1985 at Bellagio).

18. Much of modern *sastric* learning was systematized at the end of the last century. The early scholars did a great deal to map out the evolution of the *sastra* resulting in Kane's monumental *History (supra* n. 7). For other significant works, see N.C. Sen-Gupta: *Evolution of Ancient Indian Law* (London, 1953); K.P. Jayaswal (*supra* n. 1); R.B. Pal: *The History of Hindu Law in the Vedic Age* (Calcutta, *no date*); S.C. Banerjee: *A Study of the Dharma Sutras: A Story of their Origin and Development* (Calcutta, 1962); U.C. Sarkar: *Epochs of Hindu Legal History* (Hoshiarpur, 1958); Derrett (*supra* n. 7). For social aspects of Hindu law, see R.V. Ranga-swami Aiyangar: *Aspects of Social and Political System of Manusmriti* (Lucknow, 1949); R.V. Rangaswami Aiyangar: *Some Aspects of the Hindu View of Life According to Dharmasastra* (Baroda, 1952); *Rajadharma* (Adyar, 1941). For collections of the *dharmasastra*, see G.N. Jha: *Hindu Law and its Sources* (Allahabad, 1933, in two volumes). Concurrently scholars studied Hindu jurisprudence (for the best of which see P.N. Sen: *The General Principles of Hindu Jurisprudence* (Calcutta, 1918); Lingat (*supra* n. 1); K.R.R. Sastry: *Hindu Jurisprudence* (Calcutta, 1961); A.S. Natraja Aiyar: *Mimamsa Jurisprudence* (Allahabad, 1952). Throughout this period there have been more incisive excursions into specific aspects of Hindu law (e.g. J.Jolly: *Outline of a History of the Law of Partition Inheritance and Adoption* (Calcutta, 1885); R. Sarvadhikari: *The Principles of the Hindu Law of Inheritance* (Madras); B.K. Mukherjee: *Hindu Law of Religious and Charitable Endowments* (Calcutta, 1954, 1962, 1970 and 1985 ends.).) Such specific studies are characteristic of the extensive and unparalleled work of Derrett (*infra* n. 125). Such learning continues (e.g., Sontheimer (*supra* n. 11); R. Lariviere: *The Divyattatva* (Delhi, 1982); G.D. Sontheimer and P.K. Aithal (ed.): *Indology and Law: Studies in Honour of J. Duncan M. Derrett* (Weisbaden, 1982)). Another strand of the literature concerns itself with statecraft — see especially Beni Prasad: *Theory of Government in Ancient India (Post Vedic)* (Allahabad, 1927); A.S. Altekar: *State and Government in Ancient India* (Delhi, 1958); B.A. Sale-tore: *Ancient Indian Political Thought and Institutions* (Bombay, 1963); J.W. Spell-man: *Political Theory in Ancient India* (Oxford, 1964); Drekheimer: *Kingship and Community in Early India* (Stanford, 1962).

19. See G.W. Keeton: 'How Ancient is Ancient Law?' (1971) *Question* 78; also S.S. Dhavan: 'Indian Judicial System' (*infra* n. 65).

20. See H. Maine: *Ancient Law* (London, 1930, F. Pollock edn., first published 1871); *Village Communities in the East and West* (London, 1895); *Dissertations on Early Law and Custom* (London, 1901); *Lectures on the Early History of Institutions* (London, 1905). Galanter's library contains the otherwise unavailable *Minutes of Sir Henry Maine 1862–9* (Calcutta, 1869).

21. For an interesting review of his work, see Derrett: 'Sir Henry Maine and Law in India 1858–1958' (1959) *Juridical Review* 40–55; more generally, see W.A. Robson: 'Sir Henry Maine Today' in Jennings (ed.): *Modern Theories*

of Law (Oxford, 1933) Chapter 9; P. Vinogradoff: *The Teachings of Sir Henry Maine* (London 1904); M.E. Grant-Duff: *Sir Henry Maine* (London 1892); Sir F. Pollock on 'Sir Henry Maine' in *Essays in the Law* (London, 1922); L. Stephen: 'Maine' in the *Dictionary of National Bibliography* (London 1893); W.S. Holdsworth: *Some Makers of English Law* (Cambridge, 1938) 266–73; R.H. Graveson: 'The Movement from Status to Contract' (1941) 4 *Modern L.R.* 261; R. Redfield: 'Ancient Law in the Light of Primitive Societies' (1950) *Western P.Q.* 574; J.H. Landman: 'Primitive Law, Evolution and Sir Henry Sumner Maine' (1928) 28 *Michigan L.R.* 404; J. Stone: *Social Dimensions of Law and Justice* (1966) 119 ff. Note also obituary notices (1884) I.L.Q.R. 129 ff. (A.C. Lyall, E. Glasson, Franz Von Holtzendorff, P. Cogliolo).

22. e.g., B. Cohn: 'From Indian Status to British Contract' (1961) 21 *Journal of Economic History* 613; cf. B.S. Cohn: 'Some Notes on Law and Change in North India' (1959) 8 *Economic Development and Social Change* 69–83; 'Anthropological Notes on Dispute and Law in India' (1965) 7 *American Anthropologist* 82–122.

23. Even the incisive D.A. Washbrook ('Law, State and Agrarian Society in Colonial India' (1981) 15 *Modern Asian Studies* 649–721) seems to lay great emphasis on the 'form' in which the law is expressed. Thus, he states (at 653): 'From its very beginnings, then, the Anglo-Indian legal system was distinctly Janus-faced and rested on two contradictory principles with different social implications. If the public side of the law sought to subordinate the rule of 'Indian status' to that of 'British Contract' and to free the individual in a world of amoral market relations, the personal side entrenched ascriptive (caste, religious and familial) status as the basis of individual right. Strangely the paradox seems not to have been grasped by the official mind of the Raj. There appears to be awareness of a contradiction between two parts of the law and no concern that rigid Hindu social tradition might stand in the way of free market economic enterprize.' Apart from the ideological implications of the two traditions — never discussed with any conviction by Washbrook — both systems were consciously and selectively adapted by the law — processually and substantively — to balance competing and emerging interests. The real effect in the law arose out of the intrusion — and consequently extended use — of new legal procedures to capture all the more important disputes of the political economy connected with land and credit within the state system. Washbrook, whose other work stresses the importance of local factions and networks to create their own dynamic (see Washbrook: *The Emergence of Provincial Politics: The Madras Presidency 1870–1920* (Cambridge, 1976)) is generally sensitive to the adaptation of traditional and modern social apparati to emerging needs. Indeed, his admirable essay on law (*supra*) does much to link 'law', in all its forms, to this dynamic.

24. Unfortunately, despite the excellent work done on the Hindu joint family by Sontheimer and Derrett (*supra* n. 11), no work has been done on how the law of the joint family was actually adapted to give a greater power to alienate rights in land and meet the demands of creditors. Clearly, there was some adaptation (e.g. the doctrine of the pious obligation of the Hindu son to pay his father's debts on which see Dhavan (*infra* n. 86) 325–43 and the literature cited there). But despite Derrett's many insights, the exact social and political measure of this transformative adaptation has not been worked through systematically.

25. The more famous statutes in this code were the Indian Penal Code, 1860; Indian

Contract Act 1972; Transfer of Property Act 1984; Indian Easements Act 1888. Procedurally, there was the Indian Evidence Act 1882, Code of Criminal Procedure 1883 (re-enacted 1973), Civil Procedure Code 1859 (re-enacted 1908), Specific Relief Act 1877; Insolvency Act 1909 (and 1920). Commercial law was codified by the Sale of Goods Act 1893; Partnership Act 1930; Companies Act 1908; see further *supra* n. 12.

26. For a detailed analysis of the contribution of Justice Holloway, see J.D.M. Derrett: 'The Role of Roman and Continental Laws in India' (*supra* n. 11) at 179–82.

27. Until the late fifties, Roman law was actually taught in various Indian law schools as a basis to understand both modern legal concepts as well as to compare the *dharmasastra*. Efforts to compare the *sastra* with Roman law are an important feature of earlier *sastric* studies (e.g. P.N. Sen (*supra* n. 18) 251–69 on the son's debts).

28. For an exhaustive and interesting account of the legal literature that nurtured these developments, see Derrett: 'Legal Science During the last Century: India', in Rotondi (ed.): *Inchieste di dirrito Comparato* (Padua, Cedam, 1976) 413–35. Much of the literature on the legal system seems to have commented on the formal aspects of the constitutional and legal system. The pattern for this was set in H. Cowell: *The History and Constitution of the Courts and Legislative Authorities in India* (London, 1905; fifth edn.; Tagore Law Lectures of 1872); A.B. Keith: *A Constitutional History of India 1600–1936* (London, 1967); J.P. Eddy and F.H. Lawton: *India's New Constitution: A Survey of the Government of India Act 1935* (London, 1935). Indian commentators also followed this approach e.g B. Prasad: *A Few Suggestions on the Problem of the Indian Constitution* (Allahabad, 1928); M. Ruthnaswamy: *The Political Theory of the Government of India* (Madras, 1939); P.N. Murty and K.V. Padmanabhan: *The Constitution of the Dominion of India* (Delhi, 1947); C.H. Alexandrowicz: *Constitutional Development in India* (Oxford, 1957). This became a paradigm approach for the study of Indian law as well as Indian legal history. Accordingly, scholars concentrated on the 'power' and 'jurisdiction' of the courts. (See Jain, *supra* n. 3 generally; B.S. Chowdhury: *Studies in Judicial History of British India* (Calcutta, 1972).

29. Characteristic is K.T. Telang's famous speech asking 'Must social reform precede political reform' rep. in K.T. Telang: *Selected Writings and Speeches* (Bombay, 1914) I, 288. Similar attitudes can be seen in the writing of Justice Ranade: *Religious and Social Reform: A Collection of Essays and Speeches* (Bombay, 1902; M.B. Kolaskar edn); *Miscellaneous Writings* of *Mr. Justice M.G. Ranade* (Bombay, 1915); N.G. Chandavarkar: *Speeches and Writings of Sir Narayan G. Chandavarkar* (Bombay, 1916); see generally C.H. Heimsath: *Indian Nationalism and Hindu Social Reform* (Princeton, 1964). For an excellent account of the social background of twenty-one famous Indian judges in British India, see G.A Natesan (ed.): *Indian Judges: Biographical and Critical Sketches with Portraits* (Madras, 1932). For an account of the social background of Bombay's judges, see P.B. Vacha: *Famous Judges, Lawyers and Cases of Bombay: A Judicial History of Bombay During the British Period* (Bombay, 1962) 63–104. Earlier accounts of High Court judges seem to stress this socio-political aspects of the lives of High Court judges. See Sanyal: *Life of the Honourable Justice Dwarakanatha Mitter: One of the Judges of Her Majesty's High Court of Calcutta* (Calcutta, 1883); Sir Raymond West: *Telang's Legislative Council Speeches with an Essay on his Life*; with

Notes by D. W. Pilgramker (Bombay, 1885); N. G. Chandavarkar: *The Study of Law* (Bombay, 1910); Husain B. Tyabji: *Badruddin Tyabji—A Biography* (Bombay, 1952); K.A. Nilkantha Sastri (ed.): *Speeches and Writings of Sir P.S. Sivaswami Aiyar* (Bombay, 1965); J.R.B. Jee Jee Bhoy (ed.): *Dr Sir Chiman Lal Setalvad— A Biography* (Bombay, 1939); K.P.S. Menon: *C. Sankaran Nair, Builders of Modern India* (Delhi, 1967); P.C. Sinha: *Sir Ashutosh Mukherjee: A Study* (Calcutta, 1928). Judges and lawyers writing autobiographies devote more attention to their socio-political rather than 'legal' life (e.g. Judges Chimanlal Setalvad: *Recollections and Reflections: An Autobiography* (Bombay, 1946); M.C. Mahajan: *Looking Back: The Autobiography of Mehr Chand Mahajan, Chief Justice of India* (London, 1963); M.C. Chagla: *Roses in December: An Autobiography* (Bombay, 1974); M. Hidayatullah: *My Own Boswell* (London, 1980); and G.D. Khosla: *The Murder of the Mahatma and Other Essays* (London, 1963); B.P. Sinha: *Reminiscences and Reflections of a Chief Justice* (Delhi, 1985); H.R. Khanna: *Neither Roses nor Thorns* (Lucknow, 1986); on lawyers, K.N. Katju: *Reminiscences and Experiments with Advocacy* (Calcutta, 1952); G.V. Mavalankar: *My Life at the Bar* (Delhi, 1953); and, more significantly, M.C. Setalvad: *My Life: Law and Other Things* (Bombay, 1971). Recently judges have used public platforms to comment on the legal and judicial system (e.g. the works on and by V.R. Krishna Iyer *infra* n. 72–3; and the piquant G.M. Lodha *infra* n. 94). Some judges have been profiled as innovators for their work on the bench and their extra curricular writing and work (See V.D. Mahajan: *Chief Justice Gajendragadkar* (Delhi, 1966); *K. Subbarao: Defender of Civil Liberties* (Delhi, 1967); K.K. Mathew: *Democracy, Equality and Freedom* (Lucknow, 1978, ed. by U. Baxi). For a review of the other literature, see R. Dhavan *infra* n. 86), Chapter I.

30. Indian lawyers got drawn into preparing the documents of constitutional change. Lawyers like Motilal Nehru, Sir H.S. Gour, M.A. Jinnah, A.K. Ayyar, K.M. Munshi, B.R. Ambedkar, B.N. Rao and others played a crucial role in constitutional discussions (see B. Shiva Rao (ed.): *The Framing of India's Constitution* (Delhi, 1968) in 5 volumes including 4 volumes of documents. Years later Nehru ruefully remarked: 'Somehow we have found that this magnificent Constitution that we have framed was later kidnapped and purloined by lawyers.' (1951) XII-XIII *Parl. Deb.* (Pt.II) col. 8832 (17 May 1951).

31. While Gandhi's attitude may have been tongue in cheek, his declared fidelity to law cannot be dounted, see F. Watson: *The Trial of Mr Gandhi* (London, 1969) 73 (quoting Gandhi on how civil disobedience showed the 'ultimate sovereignty of British justice') and 115–21 (replies to the Hunter Commission on the Jallianwalla Bagh massacre). On the view that Gandhi was not just acting out these campaigns as political expediency, see F.R. Frankel *infra* n. 46, 32–54 stressing Gandhi's insistence that ''political' and 'social' grievances be handled separately'' (35).cf. Dhanagare: *Agrarian Movements and Gandhian Politics* (Agra, 1975); J. Brown: *Gandhi's Rise to Power* (Cambridge, 1972); see also D. Conrad: 'Gandhi's Egalitarianism and the Indian Tradition' in G. Sontheimer and P.K. Aithal, *supra* n. 18, 359–416.

32. On Champaran specifically, see S. Henningham: 'The Social Setting of the Champaran Satyagraha' (1976) 13 *Indian and Economic Social History Review*. For an analysis of the politics of other campaigns, see D. Hardiman: *Peasant Nationalist of Gujarat: Kheda District 1917–1934* (Delhi, 1981). For a discussion of 'peasant revolts' in this context see G. Pandey: 'Peasant Revolt and Indian

Nationalism. The Peasant Movement in Awadh, 1919–22' in R. Guha (ed.):
Subaltern Studies: Writings on South Asian History and Society (Delhi, 1982) I, 143–97.

33. There is an impressive literature on the Constitution. G. Austin's *The Indian Constitution: Cornerstone of the Nation* (Oxford, 1966) is a masterly but optimistically misleading account of constitution making. See also Shiva Rao *supra* n. 30. Apart from case digests on Constitutional law, legal accounts of Constitution tend to be analytical and isolate legal jurisprudence from social context (e.g. M.C. Setalvad: *The Indian Constitution 1950–65* (Bombay, 1966); Seervai: *Constitutional Law of India* (Bombay, 1967; 1975; 1978; 1983). Academic reflection on the Constitution tends to be legalistic. (Note several anthologies: Jacob, *et al.* (ed.): *Constitutional Developments Since Independence* (Bombay, 1975); S.N. Jain, *et al.* (ed.): *The Union and the States* (Delhi 1972). R. Dhavan and A. Jacob (ed.): *The Indian Constitution: Trends and Issues* (Bombay, 1978). See also M.P. Jain: *Indian Constitutional Law* (Bombay, 1970); S.N. Jain and M.P. Jain: *Indian Administrative Law* (Bombay, 1986); M. Kagzi: *Administrative Law of India* (Delhi, 1976). This is also true of accounts of Constitutional law by judges (e.g. D.D. Basu: *Limited Government and Judicial Review* (Calcutta, 1972)). Political scientists have shown a greater eye for detail when analysing defections (e.g. S.C. Kashyap: *The Politics of Power: Defections and State Politics in India* (Delhi, 1974)), political federalism (e.g. S. Maheshwari: *President's Rule in India* (Delhi, 1977); R. Dhavan: *President's Rule in the States* (Bombay, 1978)), Constitutional crisis (R. Dhavan: *The Amendment: Conspiracy or Revolution* (Allahabad, 1978), M.V. Pylee. *Crisis, Conscience and the Nation* (Delhi, 1982)), or the abuse of political power (e.g' D.C. Wadhwa: *Re-promulgation of Ordinances: A Fraud on the Constitution of India* (Delhi, 1982); L.N. Sharma: *The Indian Prime Minister: Office and Powers* (Delhi, 1976)). However, a well-rounded literature on the Constitution does not exist, even though there is a considerable literature on political development, principally from a typically American pluralist point of view; see R. Kothari: *Politics in India* (Boston, 1970); R. Hardgrave: *India: Government and Politics in a Developing Nation* (New York, 1981 edn.); cf. W.H. Morris Jones: *The Government and Politics of India* (London, 1964).

34. V.S. Naipaul: *An Area of Darkness* (London, 1964) 90.

35. While there has been a considerable amount of writing on Constitutional and legal matters in various legal journals, there has been very little serious writing on jurisprudence. Sethna's 'synthetic jurisprudence' (see his *Progress in Law* (Bombay, 1962); *Contributions to Synthetic Jurisprudence* (Bombay, 1962); *The Essentials of an Ideal Legal System* (Bombay, 1968); *Jurisprudence* (Bombay, 1969)) has been rightly described as more of 'an aggregate than a synthesis' (Paton: *Jurisprudence* (Oxford, 1964), 3 f.n.3). For other works on jurisprudence, see R.B. Mitchell: *Jurisprudence with Special Reference to the Law of India* (Madras, 1881); K. Basu: *Modern Theories of Jurisprudence* (Calcutta, 1921); T.B. Sapru: *Modern Jurisprudence* (Allahabad, 1954): G.C. Venata Subbarao: *Analytical and Historical Jurisprudence* (Guntur, 1956 edn.); *Jurisprudence and Legal Theory* (Madras, 1960); G.S. Biswas (ed.): *Metrics of Legal Philosophy* (Calcutta, 1972); K.C. Agarwal: *Legal Thought and Comparative Law* (Lucknow, 1969); Bakhasi-sasimha: *Lectures on Jurisprudence — Legal Theory* (Meerut, 1967); A.R. Biswas: *Legal Theory* (Calcutta, 1970); M.P. Tandon: *Legal Theory and Jurisprudence* (Allahabad, 1970); Monica David: *Ancient Law* (Madras, 1961); *Jurisprudence or Legal Theory* (Madras, 1967). Indian academics display at least a competent

Western jurisprudence (e.g. G.S. Sharma (ed.): *Essays on Jurisprudence* (Lucknow, 1964)) and occasionally very insightful excursions into theory (e.g., Gobind Das: *Justice in India* (Cuttack, 1967); V. Dhagamwar: *Law, Order and Power* (Bombay, 1979); and, for an earlier example, I.S. Pawate: *Res Nullius: an Essay on Property* (Dharwar, 1938); *Contract and the Freedom of the Debtor in Common Law* (Bombay, 1953)). Derrett (*supra* n. 28) rightly comments: 'By and large we are forced to acknowledge the scarcity of juridical originality in any field in which judicial authority is conclusive. The same may, alas, be said of the field technically known as jurisprudence . . . in which contributors are much better at putting together the opinions of others than finding some novel solution on their own'. Indian Academics funded by the University Grants Commission have organized various workshops which critically and constructively examined the state and future of Indian jurisprudence. C. Singh's work (*infra* n. 201) adds further impetus to these explorations. In recent years, there have been four major reviews of the world of Indian legal writing. The explorations of Derrett (*supra* n. 28) and Veena Das (*infra* n. 107) are joined by U. Baxi: *Towards a Sociology of Indian Law* (Delhi, 1986) and R. Dhavan: 'Means, Motives and Expectations: Reflecting on Legal Research in India' (1987) 50 *Modern Law Review* 725–49.

36. On the importance of English law, see Setalvad: *The Common Law in India* (London, 1960); ibid.: *The Role of English Law in India* (Jerusalem, 1966) and note the book review by Derrett (1981) 10 *I.C.L.Q.* 206; Galanter (1961) 10 *A.J.C.L.* 292. Although English law has been manipulated imaginatively by Indian judges (see generally Dhavan *infra* n. 86') it remains the basic framework for discussion and development.

37. The bulk of Indian students who went to England went to the Universities of Oxford, Cambridge, or London. Most of them were interested in being called to the Bar because of the status of an England returned Barrister. After some time, serious work on Indian law was done at the School of Oriental and African Studies by S.V. Vesey Fitzgerald, J.N.D. Anderson and, more significantly, by A. Gledhill (see his *The Indian Constitution* (Oxford, 1964)) and J.D.M. Derrett (see *infra* n. 125). Gledhill and Derrett supervised a large number of dissertations on aspects of public law, Hindu law and jurisprudence. Few of these scholarly treatises have been published (e.g. M. Imam: *The Supreme Court and the Constitution*; T.K.K. Iyer: *The Concept of Reasonableness in the Indian Constitution* (Madras, 1978); Dhagamwar: (*supra* n. 35); Dhavan (*infra* n. 86). (Such research has not really had any real impact on Indian law outside academic circles. On Indian legal education, see the *Gajendragadkar Report of the Committee in the Organization of Legal Education in the University of Delhi* (Delhi, 1964). Yet the suggestions for reform are not incisive (e.g. S.K. Agarwal (ed.): *Legal Education in India: Problems and Prospects* (Bombay, 1973), N.R. Madhava Menon: *Legal Education in India: States and Problems* (Delhi, 1983)).

38. A recent survey of the publication of law books (see R. Dhavan: *Law Publishing in India: a Report on a Preliminary Investigation* (mimeo. 1986) suggests that the black letter law tradition is further entrenched by unenterprising law publishers comfortable with a captive practitioners' market.

39. See generally R. Dhavan: 'Borrowed Ideas: On the impact of American Scholarship in Indian Law' (1985) 33 *A.J.C.L.* 505. These essays look at various kinds of traffic from America: of precedent, of funds, of scholars and of research. Given the research funds available to American scholars, American scholarship

has acquired a dominant position and trapped Indian law into a love for citation and the 'black letter' law tradition. Not surprisingly, American scholarship has also exported American prototypes and solutions for Indian development. On this see further, Trubek and Galanter *infra* n. 207; and on specific legal programmes. J. Gardiner, *infra* n. 216.

40. Most of the Indian academics who dominate Law Faculties in India — U. Baxi, Alice Jacob, S.N. Jain, A.T. Markose, P.K. Tripathi, S.P. Sathe and others — went through crucial periods of their own education in American law schools. This has a profound impact on their writing. They reflected — with notable exceptions — the public law concerns of American academics of the fifties. Ironically — but not surprisingly — American scholarship in America has moved on to new analysis and insights whilst Indian scholarship remains trapped in some of these earlier insights.

41. The initial concern was in public law (see L. Ebb: *Public Law Problems in India: A Survey Report* (Stanford, 1981); H.C.L. Merrillat: *Law and the Constitution* (Bombay, 1970); 'Chief Justice S.R. Das — a Decade of Decisions on the Right to Property' (1960) 2 *J.I.L.I.* 83; 'Compensation for Taking Property: A Historical Footnote to Bella Banerje's Case' (1960) 1 *J.I.L.I.* 345; see also his 'The Soundproof Room: A Matter for Interpretation (1967) 9 *J.I.L.I.* 521; G.O. Koppell: 'The Emergency Courts and the Indian Judiciary' (1966) 8 *J.I.L.I.* 287; D. Bayley: *Preventive Detention in India* (Calcutta, 1961). However, later efforts have tended to more complex examination of Indian society to include work on lawyers (see *supra* n. 117, 119, esp. Kidder and Morrison), anthropology (*infra* n. 109 esp. Hayden), and history (*infra* n. 116 esp. Cohn, Price and Appadorai). There was also some interest in legal education and legal aid in India (e.g. Von Mehren: 'Law and Legal Education in India' (1963) 70 *Harvard L. Rev.* 1180; B. Metzger: 'Legal Aid and Law Students in the Developing Nations' (1974) 2 *Jnl. Bar Council of India* 322–29). While some of the material has been published, some American scholarship is secreted in not, or never to be, published papers and monographs. For a contrasting view of the American influence, see L. Beer: *Constitutionalism in Asia: Asian View of the American Influence* (Berkeley, 1979), esp. P.K. Tripathi: 'Perspective on the American Constitutional Influence on the Constitution of India' at 56–98.

42. For background material see Comment: 'Conference on the Indian Law Institute' (1958) 7 *A.J.C.L.* 519–24; S.N. Jain: 'Some Reflections on the Research Programme of the Indian Law Institute' (1982) 24 *J.I.L.I.* 450; R. Dhavan: 'Legal Research in India: The Role of the Indian Law Institute' (1985) 27 *J.I.L.I.* 223–52.

43. *infra* n. 216–18.

44. Seervai: Preface to *Constitutional Law in India* (Bombay, 1967) ix.

45. In practically all their reports the Law Commission of India has concentrated on narrow 'black letter' law issues. For a critical analysis see U. Baxi: *Crisis in the Indian Legal System* (Delhi, 1982). Although many of the reports of the Law Commission have been implemented (see Baxi 244–94) the Law Commission is not an influential body except to the extent to which it occupies a space and reinforces the black letter law tradition. Judges continue to think of legal development in purely legal terms. Recently, Vice President Hidayatullah in his C. Daphtary Memorial Lecture (14 August 1984) castigated judges and lawyers who tried to deviate from this narrow tradition. Even those connected

with the uplift of poverty seem to stay within narrow legal traditions, looking to the western legal systems for sustenance and inspiration e.g. L.M. Singhvi: *Law and Poverty: Cases and Materials* (Bombay, 1973).

46. On Nehru's socialism see F.R. Frankel: *India's Political Economy, 1947–77— The Gradual Revolution* (Princeton, 1978).

47. There is a vast amount of literature on 'Backward' Classes. The major Central Government reports include: (Kalelkar) *Report of the Backward Classes Commission* (Delhi, 1955); *Ministry of Home Affairs: Memorandum on the Report of the Backward Classes Commission* (Delhi, 1956); (Renuka Ray) *Report of the Study Team on Social Welfare and Welfare of Backward Classes* (Delhi, 1959); (U.N. Dhebar) *Report of Scheduled Areas and Scheduled Tribes Commission* (Delhi, 1961); (Bhargava): *Report of Special Working Group on Cooperation for Backward Classes* (Delhi, 1962); (Mandal) *Report of the Backward Classes Commission* (Delhi, 1981). Indian law academics have concentrated on narrow 'black letter' issues (e.g. M.C.J. Kagzi: *Segregation and Untouchability Abolition* (Delhi, 1976); Paramanand Singh: *Equality, Reservation and Discrimination* (Delhi, 1982), even though an effort is made to contextualize the law in the context of an incompletely worked out socio-political perspective (P.T. Boralie: *Segregation and De-Segregation in India: A Socio-Legal Study* (Bombay, 1968); cf. B.A.V. Sharma and K. Madhusudan Reddy: *Reservation Policy in India* (New Delhi, 1982); Ratna G. Revankar: *The Indian Constitution: A Case Study of Backward Classes* (Rutherford, 1971).) For a full review of the literature see M. Galanter: *Competing Equalities . . . infra* n. 134.

48. The Directive Principles were initially not well received by the Courts (see R. Dhavan, *infra* n. 86, 87–95). A distinct change of heart was discerned in the Fundamental Rights case (*Kesavananda Bharati v. State of Kerala A.I.R. 1973 S.C.* 1461) and affirmed in later cases. On the background to the Directive Principles see K. Markandam: *Directive Principles in the Indian Constitution* (Delhi, 1960). The new importance given to Directive Principles seems to have generated its own jurisprudence (see P. Diwan and V. Kumar (ed.): *Directive Principles Jurisprudence* (Delhi, 1982)) which is, in the main, characterized more by its spirited sense of socialist euphoria rather than its capacity for social and legal analysis.

49. T.T. Krishnamachari: VIII *C.A.D.* 583 (24 November 1983).

50. The shadow of Dicey's *Introduction of the Law of the Constitution* (London, Macmillan, 10th. edn. 1962) thwarts independent thinking esp. his view on the 'rule of law' (see 202–3) which resists even the administration's 'wide discretionary authority', to the law. After the Emergency the Indian courts are trying to develop their own theory of 'due process' in India. See R. Dhavan: *Due Process in India . . .* (Delhi, Indian Law Institute, mimeo, 1981).

51. A very thorough study of the administration was made in the *Reports of the Administrative Reforms Commission* (Delhi, 1969), on which see S. Maheshwari: *Administrative Reforms Commission* (Delhi, 1972); see generally C.N. Bhalerao: *Administration, Politics and Development in India* (Bombay, 1972). Modern India's administrative problems dwarf the perceptions of older Indian Civil Service officers (for their accounts see K.L. Panjabi (ed.): *The Civil Service in India* (Bombay, 1965)). For a full survey of research into the administration see Indian Council of Social Science Research: *A Survey of Public Administration in India* (Delhi, 1973).

52. See A. Jacob: 'Nehru and the Judiciary' (1974) 19 *J.I.L.I.* 169–81.

53. Nehru's intervention in the debates on the first and fourth Constitutional Amendments was particularly sharp. He felt (see *supra* n. 30) that lawyers had 'purloined' the Constitution.

54. On the state of these socialist plans, see generally F. Frankel (*supra* n. 46).

55. The new secularism was enshrined in the Constitution by the Constitution (Forty-Second Amendment) Act 1976; note a summary of the discussions in R. Dhavan: *The Amendment: Conspiracy or Revolution* (Allahabad, 1978). Some of the literature (*infra* n. 193) on secularism is discussed in Galanter (*infra*, Chapter 10, L. and R. Dhavan, *infra* n. 193).

56. R. Abel: 'Law Books a.c. Books about Law' (1972) 26 *Stanford L.R.* 175 at 184–9.

57. See especially *Kameshwar Rao v. State of Bihar AIR 1952 S.C.* 252; *Dwarkadass. v. Sholapur Spg. and Wvg. Mills AIR 1954 S.C.* 119; *State of West Bengal v. Subodh Gopal AIR 1954 S.C.* 92; *State of West Bengal v. Bella Banerji AIR 1954 S.C.* 170 – discussed in R. Dhavan (*infra* n. 86) Chapter 3; and generally R. Dhavan: *The Supreme Court and Parliamentary Sovereignty* (*infra* n. 88).

58. Justice Patanjali Shastri: 'Speech at the Madras Lawyers' Conference' *A.I.R.* 1955 Jnl. 25.

59. S.S. Dhavan, P.B. Gajendragadkar and Krishna Iyer are singled out for special attention because during the period under discussion they were the main coherent proponents of the new 'secularism' and 'socialism'.

60. *Union of India v. Firm Ram Gopal A.I.R. 1960 All* 672 at 681, see S.S. Dhavan: *Judicial Process and Judge Made Law* (Mussorie, National Academy of Administration, Printed Lectures, 1962) where he reviews his own judgments, especially those on Muslim law (*Itwari v. Asghari A.I.R. 1960 All.* 684 (on the duty to maintain a second wife); *Basai v. Hasan Raza* App. n. 1230 of 1955 (right of *purdah* on modern times), transport nationalization (*Kashi Prashad v. State* (Sp. App. No. 8 of 1960), the plight of the poor (e.g., Unreported Case W.P. 2623–30 about the rickshawwallahs of Kanpur; also *Balwant Raj v. Union of India A.I.R. 1968 All.* 14 (on the notion of a living wage)). Not just interested in the kind of crude instrumentalism described by Horwitz (see his *The Transformation of American Law 1780–1860* (Harvard, 1980). Justice Dhavan concentrates on the symbolic and conceptual needs of developments as well as the decisional aspects of the judiciary's work.

61. Justice S.S. Dhavan in *R. T. Authority v. Kashi Prasad A.I.R. 1962 All* 551 at 567.

62. S.S. Dhavan: 'Secularism in Indian Jurisprudence' in G.S. Sharma (ed.): *Secularism: Its Implications for Law and Life in India* (Bombay, 1966) 102–38.

63. *Balwant Raj* . . . *supra* n. 60; note the comments of P.K. Tripathi: *Spotlights on Constitutional Interpretation* (Bombay) Preface; U. Baxi: 'Directive Principles of State Policy' (1969) 11 *J.I.L.I.* 245 at 260–1.

65. See also S.S. Dhavan: *Indian Jurisprudence and the Theory of State in Ancient India* (Mussorie, National Academy of Administration, Printed Lectures, 1962); 'Indian Judicial System—A Historical Survey' in *Allahabad High Court Centenary Celebrations Volumes* (Allahabad High Court, 1967) I, 53; S.S. Dhavan: 'The Indian Judicial System' in M.G. Gupta (ed.): *Aspects of the Indian Constitution* (Allahabad, 1964) 337–55. The new jurisprudence was concerned with the future. S.S. Dhavan had strong views on the organization of the Bar (see his 'The Challenge of Communism and the Legal Profession' (1960) 58 *All.L.J.* (Jnl) 1; 'The Role of the Bar and the Judiciary in the Democratic State' in

Allahabad High Court Centenary Celebrations Volumes (Allahabad High Court, 1967) II, 303). When he reiterated his views on law, lawyers and the judiciary in a speech as the Governor of Bengal, he appears to have provoked some disquiet (see 'Dhavan Provokes Vested Interests' and the editorial 'Anti-Dhavan Mania' in the *National Herald*, 21 December 1970).

66. For his contribution on various aspects of the law see S.N. Dhyani: 'Justice Gajendragadkar and Labour Law' (1967) 7 *Jai L.J.*69; P.K. Tripathi: 'Mr. Justice Gajendragadkar and Constitutional Interpretation' (1966) 8 *J.I.L.I.* 479; P.W. Rege: 'Contributions of Mr Justice Gajendragadkar to Hindu law' (1966) 8 *J.I.L.I.* 588. The leading judgment on positive discrimination is *Balaji v. State of Madras A.I.R. 1963 S.C.* 649; on religious endowments *Tilkayat v. State of Rajasthan A.I.R. 1963 S.C.* 1638 esp. at pr. 61 p. 1661 and *Durgah Committee v. Hussain Ali A.I.R. 1961* S.C. 1402 at pr. 33 p.1415. On Gajendragadkar's attempt to reform Hindu law in *Yagnapurushdasji v. Muldas A.I.R. 1966 S.C.* 119 see Galanter, Chapter 10, *infra*, and Derrett: 'The Definition of a Hindu' (1966) 2 *S.C.C.Jnl.* 67; 'Hindu—a Definition Wanted for the Purpose of Applying a Personal Law' (1968) 70 *Z.V.R.* 110.

67. See esp. his *Law, Liberty and Justice* (London, 1965); *Constitution of India* (London, 1970); *Secularism and the Constitution* of India (Bombay, 1972); *Indian Democracy: Its Major Imperatives* (Delhi, 1975). On other cognate matters see his *The Indian Parliament and the Fundamental Rights* (Calcutta, 1972); *The Hindu Code Bill* (Karnataka University, 1951); *Kashmir: Prospect and Retrospect* (University of Bombay, 1967); *Tradition and Social Change* (Bombay, 1967); *Jawaharlal Nehru — Glimpses of the Man and his Teaching* (Nagpur University, 1967); *Imperatives of Indian Federalism* (Bombay, Indian Institute of Science, 1968). The running theme in all these lectures is the emphasis on State socialism and the new secularism.

68. The concept was created by Mohan Kumaramangalam: *Judicial Appointments* (Delhi, 1973). See further, *infra* n. 82.

69. *Shahulameedun v. Subaida* (1970) K.L.T. 4 (on polygamy); *Khader v. Kunhamina* (1970) K.L.T. 237 (on the doctrine of *mushaa*); note J.D.M. Derrett: 'A Hindu Judge's Animadversions to Muslim Polygamy'. (1970) 70 *Bom. L.R.* (Jnl.) 61 3.

70. See especially V.R. Krishna Iyer: *Law, Freedom and Change* (New Delhi, 1973) esp. on sentencing (p. 57) tough punishment (89–92). For critical review see R. Dhavan (1976) 5 *Anglo-American Law Review*, 174.

71. *Report of the Expert Committee on Legal Aid: Processual Justice to the People* (Delhi, Government of India, 1973). For an excellent and spirited evaluation see U. Baxi: 'Legal Assistance to the Poor: A Critique of the Export Committee Report' (1975) 10 (27) *E.P.W.* 1005–13.

72. V.R. Krishna Iyer: 'The Judicial System: Has it a functional future in our Constitutional Order (1979) 3 *S.C.C.* (Jnl.)1. For other writings by Justice V. R. Krishna Iyer see *Law and the People: A Collection of Essays* (Delhi, 1972); *Law, Freedom and Change (supra* n. 66); *Indian Secularism: Proclamation Versus Performance and a Viable Philosophy* (Delhi, 1975); *India's Wobbling Voyage to Secularism: The Nehru Facet Plus a New Gloss* (Ahemadabad, 1976); *Jurisprudence and Jurisconscience a la Gandhi* (Delhi, 1976); *Social Mission of Law* (Bombay, 1976); *Law and Social Thought: An Indian Overview* (Chandigarh, 1978); *Some Half Hidden Aspects of Indian Justice* (Lucknow, 1979); *The Integral Yoga of Public Law and Development in the Context of Development* (Delhi, 1979); *Minorities, Civil Liberties,*

and Criminal Justice (Delhi, 1980); *Law, Society and Collective Consciousness* (Delhi, 1982).

73. For an attempt to do just that see K.M. Sharma: 'The Judicial Universe of Mr Justice Krishna Iyer' in R. Dhavan, R. Sudarshan and S. Khurshid (ed.): *Judges and the Judicial Power: Essays in Honour of Justice Krishna Iyer* (Bombay, 1985) 316–36; Hari Swarup: *For Whom the Law is Made. Mind and Faith of Justice V.R. Krishna Iyer* (Seelisburg, 1984); also N.C. Beohan: 'Justice V.R. Krishna Iyer – A Statesman Judge', (1981) 1 *Jabalpur Law Quarterly*, 24.

74. For reference to Maharishi Mahesh Yogi and his teaching see *Mohammad Giasuddin v. State of A.P. A.I.R. 1977 S.C.* 1926 at 1934; *Hiralal Malik v. State of Bihar A.I.R. 2977 S.C.* 2236 at 2243–4, with some implied support from his colleague, Justice Goswami (at 2245). On his language note comments by H.M. Seervai: *Constitutional Law of India* (Bombay, 1979) III, 1880 ff; A.R. Blackshield: 'Capital Punishment in India' (1979) 21 *J.I.L.I.* 139; see further R. Jethmalani: 'Judicial Gobbledygook' (1973) 2 *J.B.C.I.*20; K.M. Sharma: 'Judicial Grandiloquence in India: Would Fewer Words and Shorter Judgments Do?' (1973) 4 *Lawasia* 192; K.M. Nambyar: 'Mr. Jethmalani and Judicial Gobbledygook' (1974) 1 *S.C.* (Jnl.) 68. For the view that it is wrong to concentrate on these linguistic excesses see R. Dhavan: 'Judicial Law Making and "Committeed" Judges: Doctrinal Differences in the Supreme Court and a Recent Case' (1978–9) 6 & 7 *Delhi L.R.* 24. For an interesting analysis of Justice Tulzapurkar who 'criticised' Judge Krishna Iyer, see U. Baxi: 'Judicial Terrorism: some Thoughts on Mr Justice Tulzapurkal's Pune Speech' (1977) 19 *Jaipur L.J.*, 1. For critical accounts of his judgments see R. Dhavan: 'The Morality of Trade: A Socialist Supreme Court's Animadversions to Rural Debt Collectors' (1978) 1 *S.C.C.* (Jnl.) 10. For a similar evaluation of the symbolic rather than substantial gains in some of his judgments see Galanter, Chapter 11 *infra*.

75. In a personal interview, Justice Krishna Iyer stressed his faith in the State Planning method as a means of bringing about social change.

76. See, in particular the (Bhagwati) *Report on National Juridicare: Equal Justice-Social Justice* (New Delhi, Government of India, 1971); As Chairman of the (National) Committee for the Implementation of Legal Aid Schemes (from 1980), Justice Bhagwati has tried to create a structure to mobilize the poor into a more strategic use of the legal system. For developments in this area see R. Dhavan: *Public Interest Litigation in India* (*infra* n. 102); see also M. Galanter, Chapter 12, *infra*.

77. This should not obscure Justice Bhagwati's work on the bench which has extended the Krishna Iyer technique of allowing letters by interested individuals and groups about the plight of the poor to be listed as petitions. For a representative example see *People's Union of Democratic Rights v. Union of India A.I.R. 1982 S.C.* 1473. For a preliminary review see U. Baxi: 'Taking Suffering Seriously: Social Action Litigation before the Supreme Court' in Dhavan, Sudarshan and Khurshid (*supra* n. 73) 289–315.

78. e.g., A.M. Khusro: *Economic and Social Effects of Jagirdari Abolition and Land Reforms of Hyderabad* (Hyderabad, 1958); B. Misra: *Enquiry into the Working of Orissa Tenants Protection Act 1946 and Orissa Tenants Relief Act 1955 (1948–49, 1960–61)* (Delhi, 1963); B. Singh and S.D. Misra: *A Study of Land Reforms in Uttar Pradesh* (Delhi, 1964); G. Parthasarathy and B.P. Rao: *Implementation*

of Land Reforms in Andra Pradesh (Calcutta, 1969); B.S. Rao: *The Economic and Social Effects of Zamindari Abolition in Andhra* (Delhi, 1963); K.S. Sonachalam: *Land Reforms in Tamil Nadu* (Delhi, 1970); V.M. Dandlekar and G.C. Khudanpur: *Working of Bombay Tenancy Act 1948* (Poona, 1957); R.R. Misra: *Effects of Land Reforms in Saurashtra* (Bombay, 1963); Kolhatkar and Mahabal: *An Enquiry into Effects of the Tenancy Legislation in the Baroda District of the Bombay State* (Baroda, 1958); N.C. Datta: *Land Problems and Land Reforms in Assam* (Delhi, 1968). See generally the Planning Commission's own evaluation: *Progress of Land Reform* (Delhi, 1963); *Report of the Committee on the Panel on Law Reforms* (Delhi, 1959); cf. *Report of the Task Force on Agrarian Relations* (Delhi, 1973 mimeo); and M.L. Dantwala and C.H. Shah: *Evolution of Land Reforms with Special Reference to the Western Region in India* (Bombay, 1971); cf. T. Jannuzi: *Agrarian Crisis: The Case of Bihar* (Delhi, 1974). This literature to law and social change remains uncritical about its own methodology.

79. For a comprehensive survey of the literature on land reforms see P.C. Joshi: *Land Reforms in India* (Delhi, 1976). For deeper and more incisive studies into land see the varied, even if uneven, contributions in R. Frykenburg (ed.): *Land Control and Social Structure in Indian History* (Madison, 1969); *Land Tenure and Peasants in India* (Delhi, 1975), there is a considerable emerging literature into land and other questions from the point of peasant movement. For a survey of the literature see S. Sen: *Peasant Movements in India: Mid Nineteenth and Twentieth Century* (Delhi, 1982); and on social movement M.S.A. Rao (ed.): *Social Movements in India* (Delhi 1984 edn.).

80. See Seervai (*supra* n. 44). For a spirited attack see U. Baxi: 'On How Not to Judge the Judges: Notes Toward Evaluation of the Judicial Role' (1982) 25 *J.I.L.I.* 211.

81. Seervai: *Constitutional Law of India* (Bombay, 1979) III, 1880ff.

82. See the discussion in the Lok Sabha (1973) XXVII (No. 46) *L.S.D.* (Vth. Series) cols. 311–402 (2 May 1973); ibid.: No. 45 cols 136–59 (26 April 1973); ibid.: No. 48 cols 228–32 (27 April 1973). For an account of the ensuing controversy see Mr Kumaramangalam: *supra* n. 68; N.A. Palkhiwala: A *Judiciary Made to Measure* (Bombay, 1975); K.S. Hegde: *Crisis in the Judiciary* (Delhi, 1973); A.R. Antulay: *Appointment of a Chief Justice* (Delhi, 1972); Seyid Muhammad: *The Constitution: For 'Haves' or 'Have Nots'* (Delhi, 1973); K. Nayar (ed.): *Supersession of Judges* (Delhi, 1973); H.M. Seervai: *Constitutional Law of India* (Bombay, 1975) II, 1415–20; R. Dhavan (*infra* n. 86) 19–31.

83. See Justice Tulzapurkar's comments in *Manohar v. Marotrao A.I.R. 1979 S.C.* 1084 and Justice Krishna Iyer's reply in *Oregano Chemical Industries v. Union of India A.I.R. 1979 S.C.* 1803; for comments on this exchange see R. Dhavan: 'Judicial Law Making (*supra* n. 74); Seervai (*supra* n. 74).

84. See *S.P. Gupta v. Union of India A.I.R. 1982 S.C.*149; see the comments of A. Jacob: 'Constitutional Law-II' (1982) *A.S.C.L.*325. The judgment was attacked by a sitting judge of the Court, Mr Justice Tulzapurkar—*Statesman*, 27 February 1983—on which see also S. Sahay: 'Mr Justice Tulzapurkar Speaks Again' (*Statesman*, 29 March 1983); note also the views expressed at a seminar (*Statesman*, 5 February 1983); L.M. Singhvi: 'Judicial Appointments and Transfer of Judges'; note also the articles on the Transfer Case in (1982) 9, Number 2 *J.B.C.I.*, esp. S. Khurshid: 'How Not to Judge a Judge,' 345–55.

85. See generally George Gadbois: 'Selection, Background Characteristics and

Voting Behaviour of Indian Supreme Court Judges' in G. Schubert and D. Danelsky (ed.): *Comparative Judicial Behaviour* (Oxford, 1969) 221–56; 'Indian Supreme Court Judges—a Portrait' (1968–9) 3 *Law and Society Review*, 317–36; 'Indian Judicial Behaviour' (1970) 5 *E.P.W.* (Annual Number) 153; 'The Supreme Court of India—A Preliminary Report on an Empirical Study' (1970) 4 *J.C.P.S.* 33–54; 'Supreme Court Decision Making' (1974) 10 *Banaras* L.J. 1; 'The Decline of Dissent in the Supreme Court' in R. Sharma (ed.): *Justice and Social Power in India* (Delhi, 1984) 235–59; see also, in the same mould, M. Chakraborty: 'Chief Justices of the Supreme Court of India' (1978) 17 *Pol. Science Review* (No. 3 and 4); M. Chakraborty and S.N. Ray: 'Supreme Court Justices in India and the U.S.A.: A Comparative Study of Background Characteristics' (1979) 13 *J.C.P.S* 35–45. This scholarship totally transcends a previous scholarship which looked at the contribution of the court with a generalist perspective looking at the Western legal tradition from which the Constitution was derived (see M. Imam: *The Indian Constitution and the Supreme Court* (Delhi, 1970)) or merely recounting doctrine (e.g. S.R. Sharma: *The Supreme Court in the Indian Constitution* (Delhi, 1959)). For a survey of the political science literature on the judiciary see R.V. Chandrasekhara Rao: 'Studies in the Judicial Behaviour and Process' in Indian Council of Social Science Research: *A Survey of Research in Political Science: Volume I: Political System* (Delhi, 1979) 131–60; and in Iqbal Narain and Suresh Rahtore: 'Studies of the Judicial System in India', ibid., 161–86.

86. R. Dhavan: *The Supreme Court of India: A Socio-legal Analysis of Juristic Techniques* (Bombay, 1977).

87. Ibid.: 461.

88. R. Dhavan: *The Supreme Court under Strain: The Challenge of Arrears* (Bombay, 1979); *Justice on Trial: The Supreme Court and Parliamentary Sovereignty* (Delhi, 1976); on the patronage of appointment see R. Dhavan and A. Jacob: *Selection and Appointment of Supreme Court Judges: A Case Study* (Bombay, 1978); on analysing the judiciary as a bureaucracy within the State see R. Dhavan: *Judicial Decision Making: The Challenge of Democracy* (Faculty of Law, Delhi University, mimeo, 1979); on the proliferation of cases in the higher judiciary see R. Dhavan: *Litigation Explosion . . . infra* n. 94.

89. See R. Sudarshan: 'Judges, State and Society' in Dhavan, Sudarshand and Khurshid (*supra* n. 73) 268–88.

90. S. Datta Gupta, 'The Supreme Court and Indian Capitalism' in K. Mukhopadhyay (ed.), *Society and Politics in Contemporary India* (Calcutta, 1979) 167–87; S. Datta Gupta, *Justice and Political Order in India* (Calcutta, 1978).

91. e.g. J.A.G. Griffiths: *The Politics of the Judiciary* (London, 1978).

92. U. Baxi: *The Supreme Court and Politics* (Lucknow, 1980).

93. After the Rankin Committee *Report on Civil Justice* (Government of India, 1924), the major official impetus for study has come from the Law Commission of India. See *Fourteenth Report: Reform of Judicial Administration* (Delhi, 1958); *Forty-Fourth Report: On the Appellate Jurisdiction of the Supreme Court* (Delhi, 1971); *Forty-Fifth Report: On Civil Appeals to the Supreme Court on a Certificate of Fitness* (Delhi, 1971); *Fifty-Eighth Report: Structure and Jurisdiction of the Higher Judiciary* (Delhi, 1978); *Seventy-Seventh Report: Delay and Arrears in Trial Courts* (Delhi, 1978); *Seventy-Ninth Report: Delay and Arrears in High Courts and Other Appellate Courts* (Delhi, 1979). The massive backlog of cases before High Courts inspired

the Shah *Report of the High Court Arrears Committee* (Delhi, 1972) which also also reviews (pp. 4–14) the work of earlier committees.

94. There appear to be few academics who have worked in this area; see R. Dhavan: *The Supreme Court Under Strain* (*supra* n. 88); *Litigation Explosion in India* (Delhi, Indian Law Institute, 1986); U. Baxi: *Crisis in the Indian Legal System* (Delhi, 1982) 56–83. For a provocative—but otherwise undigested—account see G.M. Lodha: *Judiciary: Fumes, Flames and Fire* (Delhi, 1983) 268 ff. For representative comments see the following contributions in P. Diwan and V. Kumar (*supra* n. 48); P.H. Madhusudan Rao: 'Laws Delay' (II 275); H. Mehta: 'Delay in the Dispensation of Justice . . . (II 295); K.T.S. Tulsi: 'Cancer of Delays in Law Courts' (II, 299). Much of the knowledge about the working of courts was based on intuitive informal exchanges and impressionistic accounts of those associated with the operation of the law. For an early account, under a *nom de plume*, see Panchkouree Khan: *The Revelations of an Orderly: Being an Attempt to Explore an Abuse of Administration by the Relation of Everyday Occurrence in the Mofussil Courts* (London, 1849); also Anon: *The Necessity of Criminal Appeals Demonstrated by the Working of the Magisterial Courts* (Calcutta, 1855); F. Laslelles: *Reminiscences of an Indian Judge* (London, 1880); Om-Tom-Oie: *Judicial Portraits* (Allahabad, 1935); and the delightful *Letters of an Indian Judge to an English Gentlewoman* (London, 1934). Criminal law and Indian idiosyncrasies have attracted attention (Sir Cecil Walsh: *Indian Village Crimes* (London, 1919); *Crime in India* (London, 1930); Sir Thomas Strangman: *Indian Courts and Characters* (London, 1931); E.L. Iyer: *Crimes, Criminals and Courts: Extracts from My Scrap Book* (Madras, 1940); M. Collins: *Trials in Burma* (London, 1938); S. Barkataki: *The Escapades of a Magistrate* (Bombay, 1961); also, the forthright, even audacious K.L. Gauba: *Famous Trials: For Love and Murder* (Lahore, 1945)). Criminal law and the police have received some detailed attention, both from a limited legal point of view (e.g. Durga Charan Singh: *A Guide to the Administration of Justice in British India* (Allahabad, 1930); S.K. Ghosh: *Law Breakers and Keepers of Peace* (Calcutta, 1970) and from wider historical perspective, Inspector General of the Police: *The History of the Madras Police Centenary 1859–1959* (Madras, 1960); J.C. Curry: *The Indian Police* (London, 1934); S. Venugopala Rao: *Facets of Crime in India* (Bombay, 1965)) and sociological (e.g. Perin C. Kerawalla: *A Study of Indian Crime* (Bombay, 1959); V. Dhagamwar: *Law, Power and Justice: Protection of Personal Rights under the Indian Penal Code* (Bombay, 1974); L. Panigrahi: *British Social Policy and Female Infanticide in India* (Delhi, 1972); and more recently on sentencing and probation see S. Chabbra: *Quantum of Punishment in Criminal Law* (Chandigarh, 1970); R.K. Raizada: 'Trends in Sentencing: A Study of Important Penal Statutes and Judicial Pronouncements of the High Courts and the Supreme Court 1970–1975' (Jaipur, Unpublished Ph.D. thesis); J.H. Shah: *Studies in Criminology: Probation Services in India* (Bombay, 1973); and, above all, David H. Bayley: *The Police and Political Development in India* (Princeton, 1969).

95. See M. Galanter: 'The Abolition of Disabilities: Untouchability and the Law' in J.M. Mahar (ed.): *The Untouchables in Contemporary India* (Tuscon, Arizona, 1972) 227. Note also the work of D. Khosla: 'Social Structure and Human Rights Legislation: Studies in Fourteen Indian Villages (Yale Doctoral Dissertation, 1981) on the 'communicative' effect of the legislation against untouchability. For the government approach see *Report of the Committee on Untouchability, Economic*

and Educational Developments of the Scheduled Castes (Delhi, 1969). On the limited
use of this legislation see Galanter, Chapter 12.

96. e.g. J.D.M. Derrett: *Hindu Law — Past and Present* (Calcutta, 1957).

97. e.g. R. Dhavan: *The Amendment: Conspiracy or Revolution* (Allahabad, 1978);
Amending the Amendment (Allahabad, 1979).

98. B.B. Chatterjee: *Impact of Social Legislation and Social Change* (Calcutta, 1971).
See Dhagamwar (*infra* nn. 94 and 100); Panigrahi (*supra* n. 94) and studies of contemporary legislation R.V. Kelkar: 'Impact of Medical Termination of Pregnan-
cy Act 1971: A Case Study' (1974) 16 *J.I.L.I.* 593–8; N.R. Madhava Menon:
'Population Policy: Law Enforcement and the Liberalization of Abortion:
A Socio-legal Enquiry into the Implementation of Abortion Law in India' (1974)'
16. *J.I.L.I.* 626; R. Dhavan: *Contempt of Court . . . infra* n. 115.

99. On the recent forest legislation see D. D'Abreo: *People and Forests: The Forest
Bill and a New Forests Policy* (Delhi, 1983); W. Fernandez and S. Kulkarni (ed.):
Towards a New Forest Policy: People's Rights and Environmental Needs (Delhi,
1983); see further R. Guha: 'Forestry in British and Post-British India: A
Historical Analysis' (1983) XVIII *E.P.W.* Nos: 44 and 45–66.

100. See Derrett: *Critique . . supra* n. 9; *The Death of a Marriage Law: An Epitaph for
the Rishis* (Delhi, 1976). On the impact of marriage laws see Derrett: 'Aspects
of Matrimonial Causes in Modern Hindu Law' (1964) *Reveu de Sud-est. Asiatique*,
203–41; M.S. Lutchinsky: 'The Impact of Some Recent Government Legislation on the Women of an Indian Village' (1961) 3 *Asian Survey* 73–83; Y.B.
Damle: 'Divorce in Poona District' in A. Aiyappan and Bala Ratnam (ed.):
Society in India (Madras, 1966); R.K. Agarwal; 'Attitudes of Social Groups to-
wards Uniform Civil Code with Special Reference to Marriage' (Poona,
Department of Law, 1978, mimeo); V. Dhagamwar: *Women and Divorce*
(Poona, 1971 ICSSR, now published Delhi, 1987). On the impact of the law re-
lating to the rights of women see A.J. Almenas-Lipowsky: *The Position of Indian
Women in the Light of Legal Reform: A Socio-Legal Study of the Legal Position of
Indian Women as Interpreted and Enforced by Law Courts Compared and Related to their
Position in the Family and at Work* (Wiesbaden, Franz Steiner Verlag, 1975);
Government of India: *Report of the Committee on the Status of Women in India* (Delhi,
1974), esp. Chapter IV, 102–47.

101. The 'planning commission' approach is another example of what has been called
'naive instrumentalism' (see R. Summers: 'Naive Instrumentalism and the
Law' in P. Hacker and J. Raz (ed.): *Law, Morality and Society* (Oxford, 1977)
119–31; even though Summers himself got trapped in such instrumentalism
by concentrating on a narrow range of legal and bureaucratic techniques as
the means of making law effective; see R. Summers: 'The Technique Element
in Law' (1971) 59 *California L.R.* 733–51. On the Law Commission of India's
adoption of a 'naive instrumental approach' see, e.g., *Forty-Seventh Report: Trial
and Punishment of Social and Economic Offences* (Delhi, 1972). There is a considerable amount of literature on societal perspectives on legal change and the
'symbolic' and 'instrumental' aspects of legal change. See generally J. Gusfield:
Symbolic Crusade: Status Politics and the American Temperance Movement (University
of Illinois, 1963); W. Carson: 'Symbolic and Instrumental Dimensions of Early
Factory Legislation: A Case Study of the Social Origins of Criminal Law' in
R. Hood (ed.): *Crime, Criminology and Public Policy* (London, 1974) 142–73. The
'planning commission' view has also reflected in a number of 'impact' studies

e.g. Khosla (*supra* n. 95); Lutchinsky (*supra* n. 100); M.A. Khan: *Social Legislation and the Rural Poor* (Delhi, 1981 – on debt relief legislation: Mr Khan is the author of several other works of a similar nature). For an interesting Planning Commission and other examples see *Planning Commission, Report of the Prohibition Inquiry 1954* (Delhi, 1955); *Report of the Study Team on Prohibition* (New Delhi, 1963). See also *Report of the Expert Committee on Consumption Credit* (Delhi, 1976).

102. R. Dhavan: *Public Interest Litigation in India: An Investigative Report* (Delhi, National Committee for the Implementation of Legal Schemes, 1981), 10–40.

103. For an alternative theory of development see A. Das and V. Nilakant (ed.): *Agricultural Relations in India* (Delhi, 1983); and more generally K. Bhasin and Vimala R. (ed.): *Readings on Poverty, Politics and Development* (Delhi, FAO Office, 1980). For an interesting structural analysis see P. Chatterjee: 'More Modes of Power and Peasantry' in R. Guha (ed.): *Subaltern Studies: Writing on South Asian History and Society* (Delhi, 1983) II, 311–50. For a review of the new legal activism in India see R. Dhavan: 'Managing Legal Activism: Reflecting on India's Legal Aid Programme' (1987) 15 *Anglo-American Law Review* (281–309); and from a liberal point of view Galanter, Chapter 12. An interesting collection of papers was presented to an International Workshop on the Effective Uses of Law for Disadvantaged Groups in Ahmedabad in 1982–3 (available from the Consumer Education and Research Centre, Ahmedabad); note also International Commission of Jurists: *Conclusion and Recommendations of the Seminar on Rural Development and Human Rights in South Asia* (Geneva, 1983 – reporting on another seminar in December 1982). For an alternative theory of 'legal development' see Baxi: 'People's Law, Development and Justice' (1979) 12 *Verfassung and Rechts in Ubersee* 97; and more generally his *Crisis* ... (*supra* n. 94).

104. L. Dumont: *Homo Hierarchicus* (London, 1971).

105. While the traditional view of India was adopted by Indian sociologists (e.g. Ghurye: *Caste and Race in India* (London, 1932); Srinivas: *Caste in Modern India and Other Essays* (Berkeley, 1962); *Social Change in Modern India* (Berkeley, 1966); Beteille: *Caste, Class and Power* (Berkeley, 1965)) and British writers (e.g. A. Mayer: *Caste and Kinship in Central India* (Berkeley, 1980); cf. Bailey: *Caste and the Economic Frontier* (Berkeley, 1963)), it also had a profound influence on the analysis of history (see generally J. Gallagher, G. Johnson, A. Seal (ed.): *Locality, Province and Nation* (Cambridge, 1973); D.A. Washbrook: *Emergence* ... *supra* n. 23) and politics (e.g. P. Brass: *Factional Politics in an Indian State: The Congress Party in India* (Berkeley, 1965)). For a good critical review, see D. Hardiman: 'The Indian "Faction": A Political Theory Examined' in R. Guha (ed.): *Subaltern Studies* ... *supra* n. 32 (Delhi, 1982) I, 198–231.

106. Apart from theories about caste and ritual (e.g. Hocart: *Caste* (London, 1952)) there is some controversy about discussing caste in terms of attributional (e.g. H.N.C. Stevenson: 'Status Evaluation in the Hindu Caste System' (1954) *Journal of the Royal Anthropological Society* 45–65.) or interactional characteristics (e.g. M. Morris: 'International and Attributional Theories of Caste Rank' (1950) 39 *Man in India* 94–6); and other less holistic socio-economic analysis of caste (e.g. E. Kathleen Gough: 'Criteria for Caste Ranking in India' (1959) 30 *Man in India* 115–26). For a recent review of the literature see M. Davis: *Rank and Rivalry: The Politics of Inequality in Rural West Bengal* (Cambridge, 1983) 1–19.

107. Following Nelson (*supra* n. 10), there were a considerable number of studies into customary law which sought to codify the local law of specific areas. The more famous of these 'customary codes' include W.H. Rattigan: *A Digest of the Civil Laws for the Punjab Based on the Customary Law as at Present Ascertained* (Allahabad, 1953 edn.); P. Lall: *Kumaon Local Customs* (Allahabad, 1921); L.P. Joshi: *The Khalsa Family Law in Himalayan Districts of the United Provinces in India* (Allahabad, 1929); see also L.S.S. O'Malley: *Indian Caste Customs* (Cambridge, 1922); A. Steele: *Law and Custom of Hindu Castes* (Bombay, 1968); S. Ray: *Customs and Customary Law in British India* (Calcutta, 1971); Kilkani: *Caste in Courts* (Rajkot, 1912); P.V. Kane: *Hindu Customs and Modern Law* (Bombay, 1980). Veena Das in her 'Sociology of Law: A Trend Report' in *ICSSR: A Survey of Research in Sociology and Social Anthropology* (Bombay, 1974) 367–400, examines how a new crop of village studies have generated a new interest in dispute settlement. Reflecting on the work of Kane (see *supra* n. 7) and the earlier sociologists (e.g. I. Karve: *Hindu Society — an Interpretation* (Poona, 1961; Ghurye *supra* n. 105)), she observes that such works 'do not give any understanding of the hierarchy of norms within the texts or on the general principles which explain the perscriptive rules . . . . A simple listing of rules does not give any fresh insight into the structure of legal rules.' She argues that this is also true of other studies on tribals by British administrators or anthropologists cited by her (e.g. W.V. Grigson: *The Maria Gonds of Bastar* (London, Oxford, 1949); J.H. Hutton: *The Angama Nagas* (London, 1921); *The Sema Nagas* (London, 1921); J.P. Mills: *The Lohota Nagas* (London, 1921); *The Ao Nagas* (London, 1926); *The Rengma Nagas* (London, 1937)). There is no doubt that new studies on villages are more rewarding. But it is impossible to club in the same category works which approach villages from an economic point of view (e.g. F.G. Bailey: *supra* n. 105) or new notions of caste (e.g. Beteille: *supra* n. 105) with more traditional analysis in terms of caste (e.g. C. V. Furer-Haimendorf: *The Raj Gonds of Adilabad* (London, 1948)) or reapprisal of how the social structure houses networks of relationships (e.g. M. Davis, *supra* n. 106). Although there are separate accounts of the law (e.g. Ishwaran: 'Customary Law in Village India' (1964) 4 *International Journal of Sociology* 228; Furer-Haimendorf: 'Notes on the Tribal Justice of Among Apa Tanis' (1946) 26 *Man in India*), a proper analysis of how the form, substance and process of village law relates to the political economy needs fresh and separate attention.

108. M.N. Srinivas: 'The Study of Disputes in an Indian Village' in his *Caste in Modern India and Other Essays* (Bombay, 1962); also, 'A Joint Family Dispute in a Mysore Village' (1952) 1 *Journal of the M.S. University of Baroda* 7–31; 'A Caste Dispute of the Washermen of Mysore' (1954) 7 *The Eastern Anthropologists* 148–60; 'The Caste of the Potter and Priest' (1959) 39 *Man in India* 190–209.

109. On anthropological studies see R. Hayden: *No one is stronger than Caste: Arguing Cases in a Caste Panchayat* (Buffalo, State University of New York, Ph.D. 1981). See also 'Excommunication as Everyday Event and Ultimate Sanction: The Nature of Suspension from an Indian Caste' (1983) 42 *Journal of Asian Studies* 291; and his 'Fact Discretion and Normative Implications: the Nature of Argument in a Caste Panchayat' (Paper to a conference on Indian law, Madison, 1982). I am indebted to Hayden for directing me to the following research: see further S.V. Biswas: Harere Kumkhars and their Caste Panchayat' (1962) 11. *Bull of the Anthropological Survey of India* (hereafter, *Bull of A.S.I.*) 41–6; R. Freed: 'The Legal Process in a Village in North India:

the Case of Maya' (1971) 33 *Transactions of the New York Academy of Sciences*
423–435; B.B. Goswami: 'Caste Panchayat among the Hindu Nai (Barber)
of Patna City' (1963) 12 *Bull. of A.S.I.* 153–63; Working of a Caste Panchayat in Bihar' (1963) 12 *Bull. of A.S.I.* 129–141; S.G. Morah: 'Caste Council
of the Bhandari of Dapoli' (1966) 46 *Man in India* 154–63; 'The Bhandari
Caste Council' (1965) 45 *Man in India* 152–8; 'Caste Panchayats of the Bhan-
daris of Ratnagiri and North Canara Districts' (1964) 13 *Bull. of A.S.I.* 77–84;
R. Nicholas and T. Mukhopadhyay: 'Politics and Law in Two West Bengal
Villages' (1962) 11 *Bull. of the A.S.I.* 15–40; N. Patnaik: 'Assembly of Barbers
in Dimiria Pari District' (1961) 42 *Man in India* 194–209; B. Saraswah: 'A
Note on Rabbari Caste Panchayats' (1962) 42 *Man in India* 195–205. Such
studies are, however, less concerned with the difficult questions relating to the
political economy and more with problems of social cohesion.

110. U. Baxi: 'From Takrar to Karar: The Lok Adalat at Rangpur' (1976) 10
    J.C.P.S. 53; cf. 'Popular Justice, Participatory Development and Power Politics'
    (mimeo, 1979). I am grateful to Baxi for a copy.

111. For a review of the literature see Galanter Chapter 4. This piece was
    written jointly by Baxi and Galanter. A plea for more rigorous studies into the
    operation and functioning of nyaya panchayat was made by U. Baxi: 'Access,
    Development and Justice: Access Problems of the Rural Population' (1976) 18
    J.I.L.I. 376–430.

112. For a survey of the literature in India and elsewhere 1987 see R. Dhavan:
    Litigation Explosion . . . (supra n. 94).

113. Derrett's explorations on modern Hindu law can be found in his *Critique . . .*
    supra n. 9 (Bombay, 1970). His more specific reflections on marriage can be
    seen in his *Death of a Marriage Law . . . supra* n. 100.

114. Some examples may be found in Galanter, Chapter 7.

115. R. Dhavan: *Contempt of Court and the Press* (Bombay, 1981); *Only the Good News:
    On the Law of the Press in India* (Delhi, 1987).

116. e.g. A. Appadorai: *Worship and Conflict under Colonial Rule* (Cambridge,
    1981), esp. Chapter 5, 'Litigation and the Politics of Sectarian Control,
    1878–1925', 165–209; C.A. Breckenbridge: 'From Protector to Litigant:
    Changing Relations Between Hindu Temples and the Raja of Ramnad' in
    B. Stein (ed.): *South Indian Temples: An Analytical Consideration* (Delhi, 1978)
    75–106; Pam Price: 'Raja Dharma in the Nineteenth Century India: Law,
    Litigation and Largesse in Ramnad Zamindari' (1978) 13 *Contributions to
    Indian Sociology* 207–39; N.B. Dirks: 'From Little King to Landlord: Property,
    Law and Gift under the Madras Permanent Settlement' (1986) 28 *C.S.S.H.* 307.

117. R. Kidder: 'Formal Litigation and Professional Insecurity: Legal Entrepre-
    neurship in India' (1974) 9 *Law and Society Review* 1; 'Court and Conflict
    in an Indian Society: A Study of Legal Impact' (1973) 11 *J.C.P.S.* 121; 'Liti-
    gation as a Strategy for Personal Mobility: The Case of Urban Caste Associ-
    ation Leaders' (1974) 33 *Journal of Asian Studies* 177; see also M. Galanter:
    'The Study of the Indian Legal Profession' (1963) 3 *Law and Society Review*
    201; S. Schmitthener. 'A Sketch of the Development of the Legal Profession in
    India.' (1969) 3 *Law and Society Review.* Typically, lawyers have concentrated
    on giving accounts of the formal rules about the etiquette and behaviour of
    lawyers. See K.V. Krishnaswami Aiyar: *Professional Conduct and Advocacy* (Oxford,
    1960 edn.); N. Dutt-Majumdar: *Conduct of Advocates and the Legal Profession:
    Short History* (Calcutta, 1974).

118. R.S. Khare: 'Indigenous Culture and Lawyer's Law in India' (1972) 14 *C.S.S.H.* 71.

119. C. Morrison: 'Munshis and their Masters: The Organization of an Occupational Relationship in the Indian Legal System' (1971) 21 *Journal of Asian Studies* 309; 'Clerks and Clients: Paraprofessional Roles and Cultural Identities in Indian Litigation' (1974) 9 *Law and Society Review* 39; 'Social Organization in the District Courts: Colleague Relationships among Indian Lawyers' (1968–9) 3 *Law and Society Review*; see also P. Rowe: 'Indian Lawyers and Political Modernization: Observations in four District Towns' (1969) 3 *Law and Society Review*; S.P. Sathe, S. Kunchur and S. Kashikar: 'Legal Profession: Its Contribution to Social Change—A Survey of the Poona City Bar (Delhi, ICSSR, 1982, mimeo); S.W. McKintry: *The Brokerage Role of the Indian Lawyer: A Law and Society Approach* (1974, University of Mussorie, Ph.D.).

120. B. Cohn: *supra* n. 22.

121. O. Mendelsohn: 'The Pathology of the Indian Legal System' (1981) 15 *Modern Asian Studies* 823.

122. e.g. J.S. Gandhi: *Lawyers and Touts: A Study of the Sociology of the Legal Profession* (Delhi, 1982); *Sociology of Legal Profession, Law and Legal System* (Delhi, 1987): T.K.K. Oomen: 'The Legal Profession in India: Some Sociological Perspectives' in N.R. Madhava Menon (ed.): *The Legal Profession: A Preliminary Study of the Tamil Nadu Bar* (Delhi, 1984). The latter anthology also contains other articles on lawyers, legal education and legal aid in Tamil Nadu.

123. U. Baxi: *Crisis . . . (supra* n. 94). Baxi's latest exploration *Towards a Sociology of Indian Law, supra* n. 35 celebrates the advent of inter-disciplinary scholarship on Indian law without examining the forces that moulded it or identifying the intense middle-class struggle for the mastery over law, legal processes and institutions and legal ideology.

124. Baxi (*supra* n. 77).

125. See Derrett: (*supra* n. 1). His later essays are collected in J.D.M. Derrett: *Essays in Classical and Modern Hindu Law* (Leiden, E.J. Brill, 1976–8 in four volumes viz., I. *Dharmasastra and Related Ideas* (1976); II. *Consequences of the Intellectual Exchange with the Foreign Powers* (1977); III. *Anglo-Hindu Legal Problems* (1977); IV. *Current Problems and the Legacy of the Past* (1978)). Strangely enough, no critical evaluation of Derrett's work has been written (cf. T.C. Phadkare: 'Derrett on Hindu Law' (1984) XI *Indian Bar Review*, 322–36).

126. e.g. Derrett: *The Hoysalas* (Madras, 1957).

127. See Derrett: 'Tradition and Law in India' in R.J. Moore (ed.): *Tradition and Politics in South Asia* (Delhi, 1979) 32–59; 'The Predicament of Law in Indian Traditional Culture' in Franke-Kohler (ed.): *Entsehung und Wadel Rechtlicher Traditionen* (1980). I am grateful to Professor Derrett for a copy of the latter article.

128. For his collected writings on this area see his *Essays . . . (supra* n. 125) Volume III, generally. His *Critique . . . supra* n. 9 tries to connect basic *sastric* principles to modern learning.

129. e.g. Derrett: (*supra* n. 69).

130. Derrett: *Critique . . . (supra* n. 9).

131. Derrett: *Death of a Marriage Law . . . supra* n. 100; see in this connection L. Rocher: 'The Theory of Matrimonial Causes According to the Dharmasastra' in J.N.D. Anderson (ed.): *Family Law in Asia and Africa* (London, 1968) 90–117.

132. Galanter's personal files 'Narrative Account' (1966).

133. e.g. Galanter: 'Protective Discrimination for Backward Classes' (1961) 3 *J.I.L.I.* 39–69; 'Equality and Preferential Treatment: Constitutional Limits and Judicial Control' (1965) 14 *Indian Yearbook of International Affairs* 257–80.

134. His work on Backward Classes, which assimilates a great deal of his earlier writing, can be found in *Competing Equalities: Law and the Backward Classes in India* (Delhi, 1984). His work on untouchability can be found in 'The Abolition of Disabilities . . .' *supra* n. 95 227–314.

135. *infra*, Chapter 10.

136. *infra*, Chapter 7; also Chapter 6.

137. See the articles cited *supra* n. 129; also 'Caste Disabilities and Indian Federalism' (1961) 3 *J.I.L.I.* 205–304; 'The Problem of Group Membership: Some Reflections on the Judicial View of Indian Society' (1962) 4 *J.I.L.I.* 331–58; above all his 'A Dissent on Brother Daniel' (1963) 36 *Commentary* 10–17 (No. 1).

138. Galanter: 'Why the "Haves" Come out Ahead: Speculations on the Limits of Legal Change' (1974) 9 *Law and Society Review* 95.

139. *infra*, Chapter 2 and 3.

140. For a blueprint of the Gandhian Constitution see S.N. Agarwal: *Gandhian Constitution for Free India* (Allahabad, 1948). For a Gandhian perspective on rural development see J.C. Kavoori and B.N. Singh: *A History of Rural Development in India* (Delhi, 1967) Vol. I; I.P. Desai and B. Chaudhary (ed.): *A History of Rural Development* (Delhi, 1977) Vol. II. Village studies have gone through transformation beyond the somewhat simplified Gandhian debates (e.g. M. Marriot (ed.): *Village India* (Chicago 1955); M.N. Srinivas (ed.): *Indian Villages* (Calcutta, 1955)). For an account of village administration see R. Ratzlaff: *Village Government in India* (New York, 1962).

141. Article 40, Constitution of India states: 'The State shall take steps to organize village panchayats and endow them with such powers and authority as may be necessary to enable them to function as units of self government.'

142. B.R. Ambedkar, Speech to the Constituent Assembly VII C.A.D. 38–9 (4 November 1948).

143. The commitment to developing *panchayat* systems appears to have continued (see Balwant Rai Mehta: *Report of the Study Team for Community Developments and National Extension Service* (New Delhi, 1957) and Ashok Mehta Committee; *Report on Panchayati Raj Institutions* (New Delhi, 1978). Despite the fact that such institutions were creating asymmetrical patterns of development in the political economy (see Frankel *supra* n. 46, 195–200, 366–40).

144. L. Rudolph and S. Rudolph: *The Modernity of Tradition: Political Development in India* (Chicago, 1967). For an incisive review, see Derrett: 'Rudolph and Rudolph: Modernity of Tradition' (1969) 71 *Z. V. R.* 89–94; also U. Baxi: 'The Ultra-modernity of Ttradition: some Comments on the Modernity of Tradition of Lloyd and Suzanne Rudolph' (1972) *South Asia* 101–8.

145. e.g., E. Shils: *The Intellectual Between Tradition and Modernity* (Cambridge, 1971).

146. Galanter, *infra*, Chapter 4; see also C. Meschievitz and Galanter: 'In Search for Nyaya Panchayats: the Politics of a Moribund Institution' in R. Abel (ed.): *The Politics of Informal Justice* (New York, 1982) II, 47.

147. Galanter, *infra*, Chapter 4.

148. Galanter, *infra*, Chapter 4.

149. The best description of this is to be found in M. Galanter: 'The Modernization of Law' in M. Weiner (ed.): *Modernization: The Dynamics of Growth* (New York, 1966) 153–65.

150. For the more specific study of protestantism, see M. Weber: *The Protestant Ethic and the Spirit of Capitalism* (New York, 1930 Parsons edn.) For a review of the literature on that period see D. Little: *Religion, Order and Law: A Study* of *Pre-Revolutionary England* (London, 1967) 226–37. Tawney: *Religion and the Rise of Capitalism* (London, 1958) considers the material afresh. Both are criticized by G. Elton, *Reformation Europe* (London, 1976 edn.) 312 ff. For a further assessment of the controversies see R.W. Green (ed.): *Protestantism and Capitalism: The Weber Thesis and its Critics* (1959). For an interesting review of relationship between 'legality' and 'political legitimacy' see R. Cotterrell: 'Legality and Political Legitimacy in the Sociology of Max Weber' in D. Sugarman (ed.): *Legality, Ideology and the State* (London, 1983) 69, esp. 75–8 and at 86. 'Western law has been dependent on the apparent fusion with *a rational system of rules* of carefully subscribed values of justice and order. The centreing of both these sets of value orientations primarily in procedural matters and their substantive implications has made this rational harmony possible.' On the view that the development of the modern legal order was essential to the growth of capitalism see D. Trubek: 'Max Weber and the Rise of Capitalism' (1972) *Wisconsin L.R.* 303; see also R. Unger: *Law and Modern Society* (New York, 1976). Weber's own views are secreted in M. Rheinstein (ed.): *Max Weber on Law in Economy and Society* (Chicago, 1968 edn.).

151. 'Modernization' has become something of an academic industry; see generally M. Wiener: *Modernization: The Dynamics of Growth* (New York, 1966); J. Gusfield: 'Tradition and Modernity: Misplaced Priorities in the Study of Social Change' (1966) *American Journal of Sociology* 351–62; C. Phillipps and M. Wainwright: *Indian Society and the Beginning of Modernization* (London, 1976); Guy Hunter: *Modernizing Peasant Societies* (Oxford, 1969); M. Singer: *When a Great Tradition Modernizes* (New York, 1972). For an interesting evaluation see R. Bendix: 'Tradition and Modernity Re-considered' (1966) 9 *C.S.S.H.* 292–348; Dean C. Tipps: 'Modernization Theory and the Comparative Study of Societies: A Critical Perspective (1973) 15 *C.S.S.H.* 199–226, 15 *C.S.S.H.* — L. Shiner: 'Tradition/Modernity: An Ideal Type Gone Astray (1975) 17 *C.S.S.H.* 245–52. Note also incisive analysis by Washbrook (*supra* n. 23) 715–17.

152. 'Discipline' is used in the same broad sense used by Foucault: *Discipline and Power* (London, 1977), at 215: 'Discipline may be identified neither with an institution nor an apparatus, it is a type of power, a modality for its exercise, comprising a whole set of instruments, techniques, procedures, levels of application, targets, it is a physics or an anatomy of power, a technology.' Galanter stresses both the institutional apparati as well as the 'discipline' developed in consequence.

153. Galanter, *infra*, Chapter 2.

154. For similar contrasts, see E.P. Walton: 'The historical school of jurisprudence and transplantation of law' (1977) 9 *J.C.L.* (IIIrd Ser.) 183. More recent research shows that the law is not really transplanted with this ease. Cf. D.Lloyd: *The Idea of Law* (1964) 220 (on India); J. Starr and J. Pool: 'The Impact of a Legal Revolution in Rural Turkey' (1974) 8 *Law and Society Review* 533; K. Lipstein: 'The reception of Western Law in Turkey' (1956) 6 *Annales de la Faculté de droit d'Istanbul* (1956) 10, 225; 'The Reception of Western Law in a Country of a Different Social and Economic Background: India' (1957–8) 8–9 *Revista del Institutio de derecho Comparado* 69, 213; K.W. Patchett: 'English Law in the West Indies' (1963) 12 *I.C.L.Q.* 922–62; Galanter, *infra*, Chapter 2.

155. Galanter, *infra*, Chapter 5. For a preliminary survey of this for the Colonial State see D.A. Washbrook (*supra* n. 23).

156. L. Rocher: 'Schools of Hindu Law' in J. Ensink and P. Gaeffke (ed.): *India Minor: Congratulatory Volume Presented to J. Gonda* (Leiden, E. Brill, 1972).

157. J. Bentham: II Works (Bowring edn. Edinburgh, 1838–43) 501 Bentham's attack on Blackstone is contained in F.C. Montague (ed.): *A Fragment of Government* (London, 1891). Bentham seems to have misunderstood Blackstone's attitude to imperative theories of law (see 1 *Commentaries on the Law of England* (London 1844) I, 44, 54) and the latter's approach generally. For reassessments of Blackstone see generally Holdsworth: 'Blackstone's Commentaries' in *Some Makers of English Law* (Cambridge, 1938); 'Gibbon, Blackstone and Bentham' (1946) 52 *L.Q.R.* 46; J.U. Lewis: 'Blackstone's Definition of Law and the Doctrine of Legal Obligation as a Link between Early, Modern and Contemporary Theories of Law' (1968) *Irish Jurist* 337; Hart: 'Blackstone's Use of the Law of Nature' (1955–7) *Butterworth's South African Law Reporter*, 169; P. Lucas: 'Ex parte Sir William Blackstone plagiarist: A note on Blackstone and Natural Law' (1963) 7 *Am. Jnl. of L. Hist.*142; Vick: 'Rebuttal of Bentham and Austin on Blackstone' (1966–7) 15 *Loyola L.R.* 71; H. Levy Ullman: *The English Legal Tradition* (1936) 156 ff; Finnis: 'Blackstone's Theoretical Intentions (1967) 12 *Nat. L.F.* 163. For a fascinating 'critical' account of the structure of Blackstone's commentaries and their significance, see Duncan Kennedy: 'The Structure of Blackstone's Commentaries' (1977) 28 *Buffalo L.R.* 205.

158. If Austin is the beginning (see his *The Province of Jurisprudence* (London, 1954 Hart edn.); *Lectures on Jurisprudence* (London, 1861)) despite what his biographer says (see W.L. Morrison: *John Austin* (London, 1980)), Bentham's *Of Laws in General* (London, 1970, Hart edn.) is a more sensitive statement of early nineteenth-century theory even though Bentham's concessions and elaborations undermine the very foundations of the theory.

159. For an excellent and highly readable account of the preoccupations of English jurisprudence see R. Pound: *Jurisprudence* (St. Paul, Minn., 1959) I, 71 ff. outlining three different stages of development. However, analytical theory was clearly political in nature. I have observed: 'So, when all is said and done, there was a clear link between the following: (a) the doctrine of separation of powers which sought to make the legislature the primary source of legal rules; (b) the Benthamite movement whereby the legislature would be the primary body responsible for innovative (change) . . . ; (c) the analytical separation of law and morals whereby the legislative monopoly for creating rules was preserved and other(s) were not to de-validate these rules by reference to other validating or de-validating criteria'. See 'The Basic Structure Doctrine — a Footnote Comment' in R. Dhavan and A. Jacob (ed.): *supra* n. 33, 160. Subsequently, theories of law developed by Hart: *The Concept of Law* (Oxford, 1961) continue to serve political ends by enticing people into believing that 'law' consists of social rules to which people internally feel they have an obligation (ibid. 79–88). This is an important shift in emphasis to develop consensual support for law in a late capitalist economy.

160. See especially Salmond: *Jurisprudence* (London, 1902). Even Dicey — otherwise quite loyal to Austin — accepts this (see his: 'Private international law as a Part of the Law of England' (1890) 6 *L.Q.R.*3). Holmes ('The Path of the Law' (1887) 10 *Harvard L.R.* 457) made the oft-quoted statement what the judges

did was, in fact, the law. Analytical theory — proceeding from a criticism of Blackstone's theory of judge made laws — seems to have diverted its attention back to judges.

161. The contribution of Hans Kelsen to developing the idea of law as an autonomous field of meaning can be found in his *General Theory of Law and State* (London, 1945); *The Pure Theory of Law* (London, 1967). For a fascinating and persuasive interpretation of his approach see J.W. Harris: *Law and Legal Science: An Inquiry into the Concepts of Legal Rule and Legal System* (Oxford, 1979).

162. Hart's work (*supta* n. 159) is a formidable re-interpretation of analytical theory. It is certainly an exercise in alluring persuasion even if it is not, as Hart claims, an 'essay in descriptive sociology' (*Preface*).

163. e.g. R. Dworkin: *Taking Rights Seriously* (Oxford, 1978). For an excellent critique of how Dworkin fails to evolve a general theory of rights from the principle of 'equal respect and concern' see Hart: 'Between Utility and Rights' in A. Ryan (ed.): *The Idea of Freedom; Essays in Honour of Isaiah Berlin* (Oxford, 1979) 77–98; see also Hart: *Essays in Jurisprudence and Philosophy* (Oxford, 1983) 121–58.

164. See K. Llewellyn: 'A Realistic Jurisprudence — the Next Step' (1930) 30 *Columbia L.R.* 431 at 432; see further W.L. Twining: 'Law and Anthropology: A Case Study of Inter-disciplinary Collaboration (1973) 7 *Law and Society Review* 561, 571–2.

165. B. Malinowski: *Crime and Custom in a Savage Society* (London, 1959) 55, 67–8.

166. E.A. Hoebel: *The Law of Primitive Man: A Study in Comparative Legal Dynamics* (New York, 1964) 28; a position which was subsequently softened in his *Anthropology of Law* (New York. 1972) 506; see further L. Pospissil: *Anthropology of Law* (New York, 1971) Chapters 3 and 4; on the distinction between 'social orders' and 'negotiated orders'; N. Tanner: 'Disputing and the Genesis of Legal Principles: Examples from the Minankabam' (1970) 26 *South West Journal of Anthropology*, 375, 395. On the need for differentiating 'law' from 'custom' note P. Bohannan: 'The Differing Realms of Law' (1965) 67 (6) *American Anthropologist*, 33–42 on law as a product of 'double institutionalization'.

167. See Gulliver: 'Negotiations as a Norm of Dispute Settlement: Towards a General Model' (1975) 7 *Law and Society Review*, 667; R. Abel: 'A Comparative Theory of Dispute Settlement Institutions in Society' (1978) 7 *Law and Society Review* 217.

168. Galanter (*supra* n. 149).

169. Galanter, *infra*, Chapter 2.

170. Galanter, *infra*, Chapter 2.

171. See Galanter: 'Justice in Many Rooms' in M. Cappalletti (ed.): *Access to Justice and the Welfare State* (Sijthoff, 1981) 147 at 169, 173.

172. Ibid. 153–61.

173. Ibid. 176.

174. Ibid. 181.

175. See Galanter: 'Reading the Landscape of Disputes: What We Think We Know and Don't Know (and Think We Know) About our Allegedly Contentious and Litigous Society' (1983) 31, *U.C.L.A.L.R.*, 4.

176. Galanter: *Competing Equalities* . . . (*supra* n. 134).

177. See M. Galanter, F.S. Palen, and J.M. Thomas: 'The Crusading Judge: Judicial Activism in Trial Courts' (1979) 52 *South Calif. L.R.* 699–741.

178. Galanter, *supra* n. 138.

179. Galanter, *supra* n. 175.

180. Galanter, *supra* n. 171.

181. For recent studies of ideology see D. Sugarman (ed.): *supra* n. 150; C. Sumner: *Reading Ideologies: An Investigation into the Marxist Theory of Ideology and Law* (London, 1979). For a different approach more conducive to Galanter's stand-point see S. Scheingold: *The Politics of Rights* (London, 1974); *The Politics of Law and Order* (New York, 1984). Galanter himself is doing some preliminary investigations into the way in which legal and 'ordinary' language captures and sustains certain ideological images of the law.

182. Galanter, *infra*, Chapter 6.

183. Galanter, *infra*, Chapter 7.

184. See Galanter 'The abolition disabilities . . .' (*supra* n. 95); *Competing Equalities . . .* (*supra* n. 134).

185. *V.V. Giri v. D. Suri Dora A.I.R. 1959 S.C.* 1318.

186. *Ibid.*

187. Galanter, *infra*, Chapter 6.

188. *Ibid.*

189. *Ibid.*

190. Galanter, *infra*, Chapter 7.

191. *Ibid.*

192. K. Marx: 'On the Jewish Question' (1843) in K. Marx and F. Engels: *Collected Works* (London, 1975) III, 146–74.

193. Note that the Constitution (Forty-Second Amendment) Act 1976 added the words 'secular' and 'socialist' to the preamble of the Constitution. For discussions of secularism, see L. Sundaram: *A Secular State for India: Thoughts on India's Political System* (Delhi 1944); D.E. Smith: *India as a Secular State* (Princeton, 1963); B.G. Tiwari: *Secularism and Materialism in Modern India* (Delhi 1964); U.K. Sinha: *Secularism in India* (Bombay, 1968); A.B. Shah: *Challenges to the States in India* (Oxford, 1964); G.S. Sharma (ed.): *Secularism: Its Implications for Law and Life in India* (Bombay, 1968); J.M. Shelat: *Secularism, Principles and Application* (Bombay, 1972); P.B. Gajendragadkar: *Secularism and the Constitution of India* (Bombay, 1971). For a more recent review of the literature see R. Dhavan: 'Religious Freedom in India' (1986), 35 *A.J.C.L.* 209–54.

194. G. Austin (*supra* n. 33).

195. Quoted Galanter: *infra*, Chapter 8; see also *Competing Equalities . . .* (*supra* n. 134) 561–2.

196. Derrett: 'The Development of Property in India *c.*AD 800–1800' (1962) 64 *Z.V.R.* 15 at 20–1.

197. See generally R. Dhavan: (*supra* n. 86) 325–43 and the literature cited there; see especially Derrett: '*Indica Pietas*: A Current Rule From Remote Antiquity' (1969) 86 *Zeitschrift der Savigny-Stiftung fuer Rechtageschicte. Rom Abt.* 37–66; and note his insistence that this is distinct from suretyship which requires an express or implied agreement (see Derrett: 'Suretyship in India: The Classical Law and its Aftermath' *Rec.Soc J. Bodin XXVIII Less Surete's Personnelles* (1972) Ch. XV, 287–319, esp. at 293, 302–4).

198. Derrett: 'The Concept of Property in Ancient Indian Theory and Practice' (Groningen, *Faculty of Law*, 1968) 15. 'The modern Indian situation where the wage owner earns less for himself than others, and where the husband and wife

(not less the latter than the former) live virtually for others than themselves can be directly traced back to these ancient propositions (about property), or rather the psychology which gave voice in and through them.'

199. Galanter, *infra*, Chapter 8.

200. For an interesting anthology on deriving 'is' from an 'ought' see W. Hudson (ed.): *The Is-Ought Question* (London, 1969); also P. Foot (ed.): *Theories of Ethics* (Oxford, 1967). On an attempt to get past the 'is/ought' problem in law, note L. Fuller's attempt in his *The Morality of Law* (New Haven, 1964) to treat 'law' as an institutional fact defined in terms of its purpose and the function it seeks to fulfill.

201. Law is obsessed with the question of the State's power over individual morality (e.g. Hart: *Law, Liberty and Morality* (Oxford, 1963) and the law enforcing consensus morality (e.g. Devlin:. *The Enforcement of Morals* (Oxford, 1885)). However, debates about 'law' and morality is concerning itself with more complex 'deontic' questions (e.g. Fried: *Right and Wrong* (Harvard, 1978)). Even so, centrality continues to be given to the relationship between the State and the individual. Cf. C. Singh: *Law, Anarchy and Utopia* (Delhi, 1986).

202. Galanter, *infra*, Chapter 10.

203. Ibid.

204. See Galanter: 'Religious Freedom in the United States: A Turning Point' (1966) *Wisc. L.R.* 216–96, attacking the views of P. Kurland: *Religion and the Law: Of Church and State in the Supreme Court* (New York, 1962) 16–18.

205. Galanter, *supra* n. 138.

206. Galanter, *supra* n. 171.

207. See D. Trubek and M. Galanter: 'Scholars in Self Estrangement: Some Reflections on the Crisis in Law and Development Studies in the United States' (1974) *Wisconsin L.R.* 1062–102; 'Scholars in the Fun House: a Reply to Professor Seidman' (1978) 1 *Research in Law and Sociology* 31–40.

208. Galanter, *supra* 177 and the literature cited there.

209. There is a considerable amount of literature on the role of the judges. The complexities of judicial decision making are nowhere more elegantly reviewed than in K. Llewellyn: *The Common Law Tradition: Deciding Appeals* (Boston, 1960). For a plea that judges must stay within neutral principles there is Weschler's celebrated lecture 'Towards Neutral Principles of Constitutional Law' (1959–60) 73 *Harvard L.R.* 1. On the principled judge see R. Dworkin: 'Hard Cases' in his *Taking Rights Seriously, supra* n. 163. For a recent plea that judges rely on an intuitive understanding of consensus see Devlin: *The Judge* (Oxford, 1979) cf. Baxi's (*supra* n. 92) more elaborate view of judges assuming what he calls a 'populist' stance. For an oversimplified political interpretation of the work of judges see J. Griffiths (*supra* n. 91). There is no shortage of recommended roles for the judiciary. Ely's *Democracy and Distrust: A Theory of Judicial Review* (Harvard, 1980) suggests that judges should concern themselves with re-inforcing the processes of participatory democracy rather than substantive issues. For the alternative approach — to which Galanter is indebted, — see A. Chayes: 'The Role of the Judge in Public Law Litigation' (1976) 89 *Harvard L.R.* 1281.

210. See Mnookin and Kornhauser: 'Bargaining in the Shadow of the Law: the Case of Divorce' (1979) 88 *Yale L.R.* 950–97; see also Galanter, *supra* n. 175.

211. Galanter, *supra* n. 177.

212. Galanter, *infra*, Chapter 10.

213. Galanter, *infra*, Chapter 11.

214. Galanter, *infra*, Chapter 10.

215. Thus, Galanter argues in *Competing Equalities* . . . (*supra* n. 134 at 529–31) that in positive discrimination matters, the Supreme Court of India appears to have followed the basic policy guidelines of the government.

216. There is a vast American literature on the subject. For a general description of the American 'export' model see R. Seidman: 'Law and Development: A General Model' (1972) 6 *Law and Society Review* 311; and more generally his *The State, Law and Development* (London, 1978). For a splendid critique of the export programmes see James Gardiner: *Legal Imperialism: American Lawyers and Foreign Aid in Latin America* (Madison, 1980). See further D. Trubek: 'Toward a Social Theory of Law and Development: an Essay on the Study of Law and Development' (1972) 82 *Yale L.J.* 1; J.D. Nyhart: 'The Role of Law in Economic Development' (1962) 1 *Sudan L.J.* 394 and Reports 394; A. Allott: 'Legal Development and Economic Growth in Africa' in J.N.D. Anderson: *Changing Laws in Developing Countries* (London, 1963) 194; H.C.L. Merrillat: 'Law and Developing Countries' (1966) 60 *A.J.C.L.* 71. Note the study of International Legal Centre: *Law and Development: The Future of Law and Development Research* (New York, ILC, 1974); Y. Ghai (ed.): *Law in the Political Economy of Public Enterprise* (New York, ILC, 1977) generally; see also the material cited *Supra* n. 207 and *infra* n. 217, 218. There is a considerable amount of more specific literature on area studies, e.g., H. Scholler and P. Breietzke: *Revolution, Law and Politics* (Stuttgart, 1976). For a review of the literature see E. Burg: 'Law and Development: A Review of the Literature and a Critique of Scholars in Self Estrangement' (1977) 25 *A.J.C.L.* 492–530.

217. For the American debate see Trubek and Galanter (*supra* n. 207); J.H. Merryman: 'Comparative Law and Social Change: On the Origins, Style, Decline and Revival of the Law and Development Movement' (1977) 25 *A.J.C.L.* 457. For a specific critique of M. Galanter see P.G. Sack: 'Paradigm Lost: Modern versus Traditional System', *Journal of the Asian Council for Law and Development* (1978–9) 9–33.

218. See especially L. Freidman: 'On Law and Development' (1969) 24 *Rutgers L.R.* 11; 'Legal Culture and Development (1969) 4 *Law and Society Review* 29; Meagher and Smith: *Law and Development Practitioner* (Final Report to the Agency for International Development, No. AID/CJD 3977, 1974). For an attack on the notion of 'culture', as not taking the analysis further see Abel (*supra* n. 56) at 205. For a more rigorous use of the term 'legal culture' see Lev: 'Judicial Institutions and Legal Culture in Indonesia' in C. Hott (ed.): *Culture and and Politics in Indonesia* (1972) 246–318.

219. Galanter, *supra* n. 138; for its application to India see *infra*, Chapter 12.

I
The Uses of Law
in Indian Studies

1

The Uses of Law in Indian Studies *

The attempt to use law as a path to the understanding of modern India may encounter scepticism from several quarters. Before proceeding to examine the possible significance of legal studies, I shall attempt to state (and label) these objections—which derive, in turn, from the nature of legal materials, from their relation to society and from the peculiar features of Indian law. First, *legal materials are normative rather than descriptive*. They are so immersed in technicalities and at the same time so given to rhetorical idealizations as to obscure rather than reveal social realities. Second, *doctrine does not necessarily reflect practice*. The actual behaviour of regulators and regulated does not necessarily bear any constant relation to doctrine; doctrine, therefore, provides no reliable index either of actual patterns of regulative activity or of the degree of conformity of the behaviour purportedly regulated. Third, *nationwide generalizations are of little value*. India is a vast and heterogeneous society. Since the law largely ignores local conditions in favour of nationwide generalizations, it is of little value in understanding local conditions and inevitably obtrudes misleading generalizations. Finally, *Indian law is foreign*. Much the greater part of modern Indian law is palpably non-Indian in origin and is notoriously incongruent with the attitudes and concerns of most Indians.

Although these objections apply with special force to Indian law, they apply to some extent to all attempts to employ law in the study of society. In this paper I shall attempt to indicate the limited force of these objections. I shall argue (1) that the law provides an instrument through which we may observe some facets of Indian society which are otherwise inaccessible or obscure; and (2) that the same foreignness and remoteness from everyday life, which arouse suspicion of its relevance, lend to Indian law a unique and compelling interest for students of India and of comparative law.

The lawyer turning his attention to India is immediately drawn to the immense reservoir of documents comprising the records made by officials involved in designing, announcing, and enforcing the law. For the lawyer, the most visible and accessible part of this body of records—the statute books and the law reports—is the law. For

* Reprinted from *Languages and Areas: Studies Presented to George V. Bobrinskoy*, University of Chicago Press, 1967.

3

his ordinary purposes of predicting the behaviour of officials and designing transactions for his clients, this equation is correct. For it is this part of the literature that is binding authority to the administrators and interpreters of the law—the statutes because of legislative power and the cases because of the doctrine of *stare decisis* which obliges courts to follow the decisions of their predecessors.

But the 'law' is more than the contents of a law library. In its broadest sense it includes—or merges indistinguishably with—all patterns of socially expected and approved rule enforcement.[1] The promulgations and decisions collected in the law library comprise what we may call the 'lawyers' law'. The law may be visualized as a continuum stretching from this official 'lawyers' law' at one end to the concrete patterns of regulation which obtain in particular localities at the other. The 'lawyers' law' must be understood in the context of the habits and attitudes of its administrators and clientele. To find 'the law' in India we must look beyond the records of the legislatures and the higher courts to the working of the lawyers and the police, to the proceedings in the local courts, to the operations of informal tribunals, and to popular notions of legality.

In order to determine what information about Indian society is contained in the contents of the law library, we must ask what is the nature of the relation between 'lawyers' law' and local practice. To what extent can we discern this relation from the materials collected in the law library? What does this relation reveal about Indian society—and about the workings of legal systems generally?

In order to utilize the body of records which comprise a law library, it is necessary for the investigator to take into account some of the peculiarities of this literature. He must equip himself to discount their characteristic biases and distortions and to interpret them in light of their generic peculiarities.

Statutes and regulations are relatively straightforward. They promulgate rules, sometimes accompanied by an (often idealized) account of the reasons which led to their enactment. Such rules reveal the shifting focus of legislative concern—the awareness of new interests deserving of protection, of a new balance between contending claims. They may also reveal a shift in rhetorical stance, in assertiveness, in claims to regulate previously immune areas. In reading a statute, we must remember that it is addressed to an audience of

---

[1] There is, of course, no exact line of demarcation between law and the many other kinds of rules that are found in a society (etiquette, technology, etc.) As Weber (1954: 20) puts it 'law, convention and usage belong to the same continuum with imperceptible transitions leading from one to the other.' We may think of law as that part of the continuum in which the sanctions applied to the neglect or infraction of rules involve the use or threat of physical force by an individual or group possessing the socially recognized privilege of so acting (Hoebel, 1954: 28).

officials and lawyers for whom its terms may have special technical meanings and who are expected to be aware of applicable canons of construction and of the context of existing regulation. But the mere text of an enactment gives us a very incomplete picture — we must see how it has been interpreted and where it has been applied or not applied.

For this we turn to the law reports which are more complex and difficult to work with. They are certainly more difficult to locate — only by threading our way through a labyrinthine system of references can we be sure we have collected all of the relevant cases. A single case is a fragment; its importance and representativeness cannot be assessed by examining it in isolation. Although sweeping language may be employed, each case decides only a few specific points as they arise in a single factual situation. It is only by comparison with cases involving other factual situations that we can estimate the thrust of the broad language. It is only by collecting a number of cases dealing with related factual situations that we can discern the 'rule' they embody and estimate the range of deviation and permissible leeway built into such rules. By collecting a set of related cases, we may form a mosaic lens through which we may view the pattern of regulation.

But what do we see in these cases? Several warnings are appropriate — some which apply to any system of recorded case law, some peculiar to the common-law system, some which apply to any 'colonial' situation, and finally some which are peculiar to India. First, the cases found in the law reports are for the most part appellate cases. They are not accounts of trials at which witnesses are presented and evidence taken, but rather accounts of proceedings in which the conduct of such a trial, having been challenged by the losing party, is reviewed by a superior court — sometimes two or three times removed from the trial court. The recorded opinion is not a narrative record of the appellate proceedings but is only the last stage of it. It tells us how the subordinate courts acted under given circumstances and records the decision of the superior court, either approving such action or ordering some modification of it. The approval or modification is rationalized in terms of generally accepted doctrine with sources in constitution, statutes or earlier case law.

In these reports of appellate-court review, we get a picture, sometimes vague but more often fairly clear, of the factual situation that obtained between the parties and often a surmise about the frequency or rarity of such situations. This factual account is an oddly distorted one — there may be great gaps in matters of little relevance to the outcome of the litigation and considerable elaboration of detail in regard to matters thought to be crucial. Since, under an adversary system of justice, the court has no independent fact-finding apparatus but is dependent on the contending parties for the presentation of evidence, the appellate court's picture of the facts is a composite made by sifting two partisan (and often incomplete and misleading) accounts. Again, the record before the court in the appeal has been strained through the sieve of the rules of evidence (which emphasize fairness rather than strict relevance and so may exclude much pertinent information) and the procedural rules which may impose their own bias into the record.

The outcome reached by the appellate court will colour the opinion writer's judgment as to which of the facts are significant. Llewellyn observes that in American courts 'the better written the opinion, the more likely the facts are to be organized in support of the judgment; and some uncomfortable facts may be omitted'.[2] But he finds that in any event most opinions 'instead of being sharply slanted, tend rather to sprawl a little in their presentation of the facts so that the contrary aspects of the situation are revealed. This may be especially the case in India due to the tendency of Indian courts, unlike their more hesitant American and British counterparts, to decide every possible issue of fact and law which arises in the case.[3] For example, where one of several alternative defences is upheld, the court will proceed to consider the others even though they are, strictly speaking, irrelevant to the decision.

No matter how exhaustive the account given by the court, the reported case is not necessarily to be taken as the entire history of the dispute. Unlike the manner in which disputes are settled by political processes or by *panchayats*, in which a variety of claims and counterclaims are balanced and accommodated simultaneously, formal judicial processes tend to isolate and settle single issues instead of whole disputes. Issues which do not exist in isolation are isolated and settled in an all-or-none fashion. Such decisions then may represent not so much settlement of a dispute as one phase of a continuing dispute. Thus records of litigation are not the full account of a dispute but parts of its history.

Appellate cases are peculiar in yet another way. The total caseload of the trial courts is itself an unrepresentative selection from the multitude of similar transactions and occurences — a selection of disputes and trouble cases in which the course of events was disturbed to an extent that inertia and informal dispute-settling mechanisms

---

[2] Llewellyn (1960: 361).

[3] The prevalence of cases with such 'alternative trains of reasoning' raises interesting questions for the whole doctrine of precedent. See, e.g. *Jaiwant Rao v. State of Rajasthan*, A.I.R. 1961 Rajasthan 250, and *Buchi Reddi v. Savitramma*, A.I.R. 1961 A.P. 305, both holding that all the lines of reasoning have equal authority, even though the previous case could have been disposed of on one point.

could not resolve. Appellate cases are an even more unrepresentative group. If we discount that possible sizable number meant merely to cause delay, expense or harassment to an opponent, the cases carried to the higher courts have several distinctive characteristics. They involve sizable interests or claimants who can afford expensive litigation. Allowing for the indeterminacy introduced by the vagaries of proof and granting that some authority can usually be marshalled to support each side, the outcome of most possible litigation is predictable. But appellate cases are those in which lawyers can forsee different possible outcomes.

Reported cases are only a small and unrepresentative fraction of all appellate cases. They are selected to be reported because they represent some modification or new application of existing law. In India about 2,500 a year out of a total of about 50,000 cases dealt with by the High Courts are reported.[4] While all the decisions of the higher courts, whether reported or not, are theoretically binding on subsequent courts, in practice only those which are reported become effectively part of the body of binding and authoritative precedent.[5]

Although many reported cases are entirely predictable, a substantial portion of the cases found in the reports reveals not only the application of pre-existing law but the reworking and transformation of the law by the judges. There are well-known indicators of modifications in legal thinking: the overruling of an accepted line of cases, the creation of legal fictions, the appearance of divergent lines of opinion. But more subtle and more characteristic is the slow reworking of old concepts to make them fit new situations, the stretching of slim precedents, the confining of uncomfortable ones — all done under the pretext of merely following authority and seldom with explicit acknowledgment that movement is taking place. Llewellyn has catalogued the multiplicity of techniques employed by common-law courts in handling precedents so as to enlarge, reduce or dispense with them.[6] He lists sixty-four techniques, many or most of which are familiar to readers of Indian law reports though they may appear in different proportions there than in American reports investigated by Llewellyn. They may even be accompanied by some peculiarly Indian techniques such as the High Court's use of Supreme Court

---

4 'A Strange Proposal: Move to Liquidate Private Law Reports', *Supplement to All-India Reporter 1956 October*, p. 14; Law Commission (1958: I, 636). The Law Commission lists a total of about 45,000 'certain categories of proceedings' in 11 (of the 18 then existing) High Courts in 1955. Id., pp. 110–11. These High Court proceedings represent only a tiny fraction of some two and one-half million proceedings of all kinds handled by the court system in 1956. Id., pp. 137 ff.

5 Law Commission (1958:I, 629–30).

6 Llewellyn (1960: 77)

dicta to ignore High Court rulings.[7] The multiplicity of doctrines of statutory construction permits comparable flexibility in the handling of statutes.[8]

Thus the judge transmutes past authority into present law by introducing change in the guise of conformity. Law is rewritten as it is reiterated; as subsequent decisions are added to the corpus of authority, the meaning and importance of old decisions change. Lawyers phrase their arguments about what the law 'is' in terms of what it has been. Lawyers, judges and law books are interested in the present meaning of a past decision — in its meaning now and for all courts and lawyers and not what it meant at the time and place of its delivery. This makes cases and secondary materials difficult for the social analyst to use. As Maitland says:

> The process by which old principles and old phrases are charged with a new content is from the lawyer's point of view an evolution of the true intent and meaning of the old law; from the historian's point of view it is almost of necessity a process of perversion and misunderstanding. Thus we are attempting to mix two different logics, the logic of authority and the logic of evidence. What the lawyer wants is authority and the newer the better; what the historian wants is evidence and the older the better.[9]

But if the law at any time is what the judges say it is, there are factors which severely limit the extent of change introduced by judicial techniques. Judges are conditioned by the ideology of *stare decisis*, by their own training and their loyalty to a limited number of known techniques for handling law and facts, by the limitations imposed by the fixed frozen record of the case and the narrowing and phrasing of issues by counsel, by the presentation of questions that are merely fragments of branches of the law, by the making of decisions in groups and the necessity of giving their reasons in a written opinion, by their insulation from public pressure and political

[7] Compare, e.g. *Kalinath v. Nagendra Nath*, A.I.R. 1959 Cal. 81, 83, where the Supreme Court dictum was followed even though the decision was on other grounds, since 'the Supreme Court Justices took a definite view on this point . . . it is binding on us,' with *Pardharadhi Rao v. Srinavasa*, A.I.R. 1959 A.P. 512, where the court carefully distinguished the Supreme Court case as 'only authority for what it actually decides and the generality of expressions in it are not intended to be expositions of the whole law but are governed and qualified by the particular facts of the case. . . .' 516. See Popkin (1962: 247 ff).

For another Indian departure, see note 3 *supra*.

[8] Llewellyn (1960: 371 ff, 521 ff).

[9] Maitland (1957: 57) In spite of their penchant for using old bottles, common lawyers are not very nostalgic or inclined to put a value on the old for its own sake. Unlike other areas of cultural life, its custodians have no pronounced antiquarian bias — nor, like art, are its different periods accorded equal status. For the lawyer, the newer supplants the old — or at least modifies it; the old has no autonomous value.

retaliation, and by the insistent pressure of professional opinion. [10]

Judicial reworking of the law is often taken as an expression of changing values and changing perceptions of society. But one must add a qualification. Judges are usually an insulated and detached group with their own peculiar perspective on society; this detachment and separation are particularly intense in India. If the judiciary is expressive of and responsive to the values of the social groups from which it is drawn, we must remember that this may be a group far removed from the law's clientele. The judge's training and experience intensify this isolation. It has been suggested that, as compared with other officials, judges as a group stood outside Indian society and its values and tended to know little of the districts. [11] This tradition of detachment continues today and is promoted by deliberate insulation of judges from public and political pressure.

Under these circumstances it is not surprising that there is a divergence between the values and perceptions of the judges and those of the law's clientele. We might expect that the law is less than fully responsive to the concerns of its clientele. To the extent that the values and perceptions of judges are representative of smaller groups within the society rather than of their clients directly, we may view the cases as a reflection of the perceptions and aspirations of these ruling groups and of the techniques they use to make actuality conform to these.[12] There may be — and in India often is — a wide divergence between these ideals and aspirations and those of the clientele. By the disposition of governmental power in their service and by the prestige of urban and official centres, these aspirations and perceptions are powerfully recommended to the clientele. Thus we can think of the courts as agents for the diffusion of these notions. Cases presenting concrete disputes arising in a local context proceed from the lowest court to the subordinate judge in the district town to the High Court in the state capital and perhaps to the Supreme Court in Delhi. Authoritative doctrine flows through this network in the opposite direction. The law reports reveal only the top end of this process — the rationalizing, urban end. Although some of the earlier stages are telescoped in the report, local concerns may be obscured by the efforts of counsel to rephrase them in terms of national and official standards. As they climb to higher courts the degree of formality in presentation and the emphasis on doctrine increase. To get a full picture of a litigation, it may be necessary

---

[10] Cf. Llewellyn (1960: 19); Maine (1895: 48).

[11] Cohn (1962: 186).

[12] Cf. the statement of a Nigerian lawyer that 'the law and the constitution of a people are an expression of the social consciousness of their leaders'. Davies (1962: 328).

to supplement the law report by the unpublished record of the trial court. As a record of the response of the urban official centre to a local situation, the law report may serve as a valuable supplement and counterpart to the field study which sees national (or regional) policy in its incidence at the local level.

In legal materials, local concerns tend to be seen through the lens of national standards. It is the fashion today for students of Indian society to eschew nationwide generalization in favour of the description and analysis of local conditions. This 'localism' serves as a valuable corrective to the ambitious generalizations which have comprised much of the literature on India; one hopes that studies of Indian society in its local detail and variation will provide a basis for more fruitful generalizations. However, legal studies can claim at least partial exemption from the imperatives of localism. For if there is no nationwide Indian system of caste, of kinship, of religion or of land tenure, there is what might be called an all-India legal culture. The student of law is concerned with doctrines and procedures which subsume diverse local conditions while being influenced by them only to the extent allowed by nationally prescribed standards. From the variety of local instances, lawgivers and courts, applying nationwide standards in accordance with nationally accepted technqiues, fashion a common doctrine which in turn has some influence on diverse local situations. To the extent that there is a nationwide legal community in India, the law reveals the 'culture' of that community—its values and attitudes and their embodiment in patterns of practice and personal relationships. The study of law can usefully supplement detailed accounts of local conditions by providing a picture of the standards and procedures injected into each local situation by the nationwide network of legal agencies. It provides us with a picture of one of the unifying networks that make India more than a congeries of localities.

Nationwide legal doctrine does not correspond with any exactness to concrete patterns of regulation and still less to the behaviour it purports to regulate. The law contains much that is mere lip service and rhetorical gloss. Legal doctrine, like religious or aesthetic, tends to be normative and didactic.[13] As in studies of religion or art, there is the problem of assessing the actual impact and effectiveness of doctrine.[14] After recognizing their value in revealing the self-image and normative principles of their compilers, there is the problem — generic to all textual studies — of assessing their value as descriptive materials.[15] However, in this respect legal texts are different from

[13] Singer (1961: 301)

[14] Id., 280.

[15] Id., 300. Cf Hoebel (1954: 42ff), who suggests that case material need be discounted to the extent the facts are subject to idealization or 'projective recasting'.

religious or aesthetic ones. First, the professionals engaged in compiling the texts are themselves parties to the situations being described — in effect we are getting their rationalization of their own action in the dispute. Second, the law differs from other textual traditions in its self-conscious concern with the authoritative character of its component elements. The law has in common with religion, institutionalized agencies for preserving and modifying its tradition. But it is distinctive in its explicit concern with the effectiveness of its doctrinal components — i.e. it is intensely self-observing on the question of its own impact and effectiveness. In a modern rationalized legal system — more so in the work of hierarchical courts than in legislatures — the law incorporates within itself a measured view of its own context. The adjudicatory mode of law lends it this special self-observing character; the report of a case inevitably involves consideration of the extent to which local practice conforms to widespread 'official' standards.

In India we must add the problem of perjury and fabricated evidence. If the things reported did not actually happen, they are apparently things that might well have happened. Cf. the observation that in India real criminals are often convicted of their crimes on false evidence.

II

The Emergence of the
Modern Legal System

2

The Displacement of Traditional Law in Modern India *

Contemporary Indian law is, for the most part, palpably foreign in origin or inspiration and it is notoriously incongruent with the attitudes and concerns of much of the population which lives under it. However, the present legal system is firmly established and the likelihood of its replacement by a revived 'indigenous' system is extremely small. The modern Indian legal system, then, presents an instance of the apparent displacement of a major intellectual and institutional complex within a highly developed civilization by one largely of foreign inspiration. This paper attempts to trace the process by which the modern system, introduced by the British, transformed and supplanted the indigenous legal systems — in particular, that system known as Hindu law.

The Foundation of the Modern Legal System

One of the outstanding achievements of British rule in India was the formation of a unified nationwide modern legal system. The word 'modern' is used here to refer to a cluster of features which characterize, to a greater or lesser extent, the legal systems of the industrial societies of the past century. Many of these features have appeared and do appear elsewhere; some of them may be absent to some degree in one or another industrial society. The salient features of a modern legal system include uniform territorial rules, based on universalistic norms, which apportion rights and obligations as incidents of specific transactions, rather than of fixed statuses. These rules are administered by a hierarchy of courts, staffed by professionals, organized bureaucratically and employing rational procedures. The system contains regular and avowed methods for explicitly revising its rules and procedures. It is differentiated in tasks and personnel from other governmental functions; yet it enjoys a governmentally-enforced monopoly over disputes within its cognizance, permitting other tribunals to operate only in its interstices and subject to its supervision. The system requires (and is supported

* Reprinted from the *Journal of Social Issues*, 24, 1968, pp. 65–91.

by) specialized professionals who serve as intermediaries between the courts and those who deal with them. [1]

In Pre-British India

In pre-British India there were innumerable, overlapping local jurisdictions and many groups enjoyed one or another degree of autonomy in administering law to themselves. [2] The existence of *dharmasastra*, a refined and respected system of written law, did not serve to unify the system in the way that national law did in the West. In Europe, local law was absorbed into, and gradually displaced by law promulgated by State authorities. Hindu law did not enjoy the political conditions for unification. But it was not only the fragmentation of jurisdictions and the extensive delegation to local authorities that obstructed development of a modern legal system. The relative absence of written records, of professional pleaders and of appeals made even local unification difficult. Furthermore, the respective authoritativeness of governmental, *sastric*, and local components was not visualized in a way to provide either the techniques or the ideology for the ruthless supersedure of local law. The system allowed for change, but did not impose it; it allowed the old to remain alongside the new. The relation of the 'highest' and most authoritative parts of the legal system to the 'lower' end of the system was not that of superior to subordinate in a bureaucratic hierarchy. It was perhaps closer to the relations that obtain between Paris designers and American departmental store fashions or between our most prestigious universities and our smaller colleges than to anything in our own legal experience. Instead of systematic imposition, of 'higher' law on lesser tribunals, there was a general diffusion by the filtering down (and occasionally up) of ideas and techniques, by conscious imitation and by movement of personnel.

The Moghuls and other Muslim rulers had, in cities and administrative centres, royal courts which exercised a general criminal

---

[1] Galanter (1966b). 'Modernization' here refers only to the development of the features mentioned above or the sustained movement towards these features. Although, in many cases, the importation of 'Western' law seems to serve as the stimulus for such development, it does not imply 'Westernization'. Nor do I mean to imply that the processes of modernization proceed relentlessly until they produce a legal system which corresponds to the model in every detail. As society becomes modernized in other spheres, new kinds of diversity and complexity generate pressures for differentiation and flexibility in the law. Modern societies develop new methods of making law flexible and responsive — e.g. administrative agencies, arbitration, juvenile courts. Modern law as depicted here is not a destination, but a focus or vector towards which societies move. But the very forces which support this movement and which are released by it deflect it from its apparent destination.

[2] On this period see Altekar (1958); Cohn (1961); Derrett (1965, 1961a, 1964a); Gune (1953); Kane (1930–41. 1950); Mookerji (1958); Sarkar (1958).

(and sometimes commercial) jurisdiction and also decided civil and family matters among the Muslim population.[3] These courts operated according to Muslim law — at least in theory, for the application of *Shari'a* was qualified by custom and royal decree, by corruption and lack of professionalism, and by arrangements allowing considerable discretion to the courts of first instance. While a hierarchy of courts and a right of appeal existed, it seems that the activity of these higher courts fell short of any sustained and systematic supervision of the lower courts. Hindus were generally allowed their own tribunals in civil matters. Where these matters came before royal courts, the Hindu law was applied. The government's courts did not extend very deep into the countryside; there was no attempt to control the administration of law in the villages. Presumably, the Hindu tribunals proceeded as before Muslim rule, except that whatever ties had bound these tribunals to governmental authority were weakened; there was no appeal to the royal courts.

The 'Expropriation' of Law

In undertaking to administer the law in the government's courts, staffed with government servants (rather than to exercise a merely supervisory control over administration of law by non-governmental bodies), the British took the decisive step toward a modern legal system,[4] initiating a process that might be called the 'expropriation' of law.[5] This expropriation, which made the power to find, declare and apply law a monopoly of government, came about in slightly different ways and at different times in different places. But the general movement was the same.

Three distinctive, if overlapping, stages can be discerned in the development of the modern Indian legal system. The first, the period of initial expropriation, can conveniently be dated from Warren Hastings' organization in 1772 of a system of courts for the hinterland of Bengal.[6] This period was marked by the general expansion of government's judicial functions and the attrition of other tribunals, while the authoritative sources of law to be used in governmental courts were isolated and legislation initiated. The second period, which began about 1860, was a period of extensive codification of the law and of rationalization of the system of courts, while the sources of law became more fixed and legislation became the dominant mode of

---

[3] On this period, see Ahmad (1941); Sarkar (1958).

[4] The alternative is exemplified not only by earlier Indian law, but such arrangements as the *millet* system of the Ottoman empire under which each religious community was required to administer its own law to itself — a system which continues in some degree in parts of the Middle East today.

[5] Weber (1958: 83).

[6] Misra (1961, 1959: chap. 5, 6); Patra (1961).

modifying the law. This period lasted until independence, after which there was a further consolidation and rationalization of the law and the development of a unified judicial system over the whole of India.

The Law Before 1860

The law applied in the courts before 1860 was extremely varied. Parliamentary charters and acts, Indian legislation (after 1833), Company Regulations, English common law and ecclesiastical and admiralty law, Hindu law, Muslim law, and many bodies of customary law were combined in a bewildering array.[7]

It was a fundamental and persisting British policy that, in matters of family law, inheritance, caste and religion, Indians were not to be subjected to a single general territorial law. Hindus and Muslims were to be governed by their personal law, i.e. the law of their religious group. In other cases, the judges were instructed to decide 'according to justice, equity, and good conscience.' This puzzling formula, whatever its original meaning[8] was the medium for the uneven application of some indigenous law and for the importation, sometimes uncritical, of a great deal of English law.

There were, prior to 1860 numerous attempts to reorganize and reform the courts and to systematize and reform the law[9] including some reforms which changed Hindu law. However, there was no major progress towards simplifying and systematizing the law until the Crown took over the governing of India from the East India Company in 1858. The quarter of a century following the takeover by the Crown was the major period of codification of law and consolidation of the court system. During this period a series of Codes, based more or less on English law and applicable, with minor exceptions, throughout British India, were enacted. By 1882, there was virtually complete codification of all fields of commercial, criminal and procedural law.[10] Only the personal laws of Hindus and Muslims were exempted. While Hindu and Muslim laws previously had been applied to a variety of topics besides the listed ones, they were now confined (with minor exceptions) to the personal law matters (family law, inheritance, succession, caste, religious endowments). The codes themselves do not represent any fusion with indigenous law;[11] there is no borrowing from Hindu, Muslim or customary law, although there is occasional accommodation of

---

[7] Morley (1850: Vol. I, lxii, xcvii); Rankin (1946: ch. 10, 11); Patra (1961: chap. 8); Ilbert (1907: ch. 3).

[8] Derrett (1963a).

[9] Morley (1850: Vol.I, intro.); E. Stokes (1959: ch. 3); Desika Char (1963: 278–92).

[10] W. Stokes (1887–8); Acharya (1914).

[11] Bryce (1901: 107, 117).

local rules and there are adjustments and elaborations of the common law to deal with kinds of persons and situations and conditions found in India.[12]

The Transformation of Indigenous Law

What happened to indigenous law as a result of the formation of the modern legal system? First, its administration moved from 'informal' tribunals into the government's courts; second, the applicability of indigenous law was curtailed; third, the indigenous law was transformed in the course of being administered by the government's courts.

The most striking impact of the provision of governmental courts was the shift of dispute-settlement from local tribunals (and local notables) to the government's courts. Nineteenth century (and later) observers speak of a flood of litigation, sometimes with the implication that these disputes would have been peaceably settled or indeed would never have arisen without the availability of official courts. In the absence of information about the quantity of disputes and litigation in traditional India, it seems reasonable to regard most of this litigation as the mere transfer of old disputes to new tribunals.[13] These new tribunals and their strange methods had a powerful allure. Maine speaks of the 'revolution of legal ideas' inadvertently produced in the very course of attempting to enforce the usages of the country. This evolution, he found, proceeded from a single innovation — 'the mere establishment of local courts of lowest jurisdiction' in every administrative district.[14] The availability of these courts, with their power to compel the attendance of parties and witnesses, and, above all, with their compulsory execution of decrees, opened the way for 'the contagion . . . of the English system of law'.[15]

The common law courts undertook to deal with the merits of a single transaction or offence, isolated from the related disputes among the parties and their supporters. The 'fireside equities' and qualifying circumstances known to the indigenous tribunal were excluded from the court's consideration. In accordance with the precept of 'equality before the law', the status and ties of the parties, matters of moment to an indigenous tribunal, were deliberately ignored.

12 Lipstein (1957: 92ff).
13 This supposition is compatible with observations of more recent instances of disputes moving from traditional to governmental tribunals. See Cohn (1955), Beals (1955). Cf. the observations of Frederick John Shore, that in fact British rule *decreased* the number of tribunals available (1837: Vol. 11, 189).
14 Maine (1895: 70–1).
15 Maine (1895: 74), cf. Derrett (1961a: 18).

And, unlike the indigenous tribunals which sought compromise or face-saving solutions acceptable to all parties, the government's courts dispensed clear-cut 'all or none' decisions. Decrees were enforced by extra-local force and were not subject to the delays and protracted negotiations which abounded when decisions were enforced by informal pressures. Thus 'larger prizes' were available to successful litigants and these winnings might be grasped independently of the assent of local opinion. The new courts not only created new opportunities for intimidation and harassment and new means for carrying on old disputes, but they also gave rise to a sense of individual right not dependent on opinion or usage and capable of being actively enforced by government, even in opposition to community opinion.[16]

Traditional Tribunals

Traditional tribunals still functioned, though certain subjects (e.g. criminal law) were withdrawn from their purview. On the whole, these tribunals lost whatever governmental enforcement their decisions had previously enjoyed. The caste group was now treated as a private association. While it thus enjoyed an area of autonomy, it no longer could invoke governmental enforcement of its decrees. At the same time, the sanctions available to the indigenous tribunals declined in force. The new opportunities for mobility, spatial and social, provided by British rule not only increased transactions between parties beyond the reach of traditional sanctions, but also made outcasting and boycott less fearsome. With their own sanctions diminished, their ability to invoke governmental support limited, and the social relations necessary for their effectiveness disrupted, the indigenous tribunals declined as the government courts flourished.

The movement of disputes into the government courts in India has not been definitely charted. We might visualize it, borrowing Bailey's term, as a 'moving legal frontier'.[17] At first the village lay beyond the reach of the modern legal system, ruled by its traditional tribunals according to its customary rules.[18] With the impingement of new regulations and the arrival of new forms of wealth and power from the outside, sooner or later some party in the village found it both feasible and advantageous to resort to the governments' courts for what it could not obtain from village justice.[19] Other parties were obliged to defend themselves in court. As more learned to use the official courts, the authority of village

[16] Cohn (1959); Rudolph and Rudolph (1965).
[17] Bailey (1957: 4–5)
[18] Marriott (1955: 186ff).
[19] Bailey (1957: 262ff); Cohn (1955: 66); Beals (1955: 90); Woodruff (1953: 298).

tribunals was displaced. Over time the modern system encroached on the traditional system: court law replaced village law on more topics of law for more groups over more territory.[20] With this 'expropriation' of independent legal 'estates', the government's monopoly on making, finding and applying the law was extended.

The Search for Indigenous Law

It was early acknowledged that Indians should be ruled by their own laws. Hastings's plan, which provided the model for the other *mofussil* systems set out to apply indigenous law. The British assumed that there was some body of law somehow comparable to their own, based on authoritative textual materials to be applied by officials according to specified procedures to reach unambiguous results. However, there was no single system in use, but a multiplicity of systems; and within these there was often no fixed authoritative body of law, no set of binding precedents, no single legitimate way of applying or changing the law. Yet these British assumptions and expectations about Hindu law had a powerful effect upon it and ultimately proved to be self-fulfilling prophecies.

The 'Sastra', Custom and British Law

In their search for authoritative bodies of law, the British made collections and translations of ancient texts and recent commentaries. However, Indian law proved strangely elusive.[21] Maine speaks of the 'vast gaps and interspaces in the Substantive Law of India'.[22] India was 'a country singularly empty of law'.[23] In the written *sastra* 'large departments of law were scarcely represented'.[24]

It was soon recognized that *sastra* was only a part of the law and that in many matters Indians were regulated by less formal bodies of customary law. But even the customary law was not sufficient. For when custom was recorded and the quasi-legislative innovative role of the tribunals that administered it was restricted, it did not supply 'express rules in nearly sufficient number to settle the disputes occasioned by the increased activity of life and the multiplied wants which result[ed] from . . . peace and plenty. . . .'[25] The need to fill the felt gaps was ultimately to lead to statutory codification

20 This does not imply that traditional norms and concerns are displaced by official ones. On the contrary, it appears that these traditional attitudes outlast traditional legal practice and are responsible for the inspired manipulation of 'modern' law for purposes foreign to the law. Cf. the observation that most Indian court cases are 'fabrications to cover the real disputes' (Cohn 1959: 90).

21 Hunter (1897: 371).

22 Maine (1890: 209, 'Minute of 1st October, 1868').

23 Maine (1890: 225, 'Note of 17th July, 1879').

24 Maine (1895: 51); Derrett (1959a: 48ff); Derrett (1964a: 109–10).

25 Maine (1895: 75).

on the basis of English law. But, in the meantime, courts, empowered to decide cases in accordance with 'justice, equity and good conscience', filled the interstices of *sastra* and custom with 'unamalgamated masses of foreign law'.[26] Although there was some attempt to draw the most suitable rule from other sources[27], in most cases the judges were inclined to assume that English law was most suitable.[28]

Even where Indian rules were available, their application by the British transformed them.[29] Mere restatement in English legal terminology distorted the Hindu and customary rules.[30]. English procedure curtailed some substantive rights and amplified others.[31] The British insisted upon clarity, certainty and definiteness of a kind foreign to Hindu tradition.[32] Neither the written nor the customary law was 'of a nature to bear the strict criteria applied by English lawyers'.[33] Rules seemed vague and requiring of definition, and this was accomplished by English methods. The mere process of definition had the effect of creating rights of a kind that did not previously exist.[34] Comparing the effect of English legal method in the Supreme Courts and the Sudder Courts, Maine observed that:

> At the touch of the Judge of the Supreme Court, who had been trained in the English school of special pleading, and had probably come to the East in the maturity of life, the rule of native law dissolved and, with or without his intention, was to a great extent replaced by rules having their origin in English law-books. Under the hand of the judges of the Sudder Courts, who had lived since their boyhood among the people of the country, the native rules hardened and contracted a rigidity which they never had in real native practice.[35]

One of the remarkable and unanticipated results of the British administration of Hindu law was the elevation of the textual law over lesser bodies of customary law. The sources of personal law were assumed to be not the customary law that prevailed among most Hindus (and Muslims) on most matters, but the highest and most authoritative bodies of textual law. It was assumed that the Hindu law could be ascertained from sacred books. Hastings' plan 'took Orthodox Brahmanic learning as the standard of Hindu law'.[36]

[26] Maine (1895: 76).
[27] Derrett (1959b).
[28] *Waghala v. Sekn Marlvodin*, 14. S.A. 89 (1887); Twining (1964).
[29] Derrett (1961a: 20, 21–2).
[30] Derrett (1961a: 41).
[31] Derrett (1961a: 40).
[32] Derrett (1961b: 112).
[33] Maine (1895: 37).
[34] Maine (1895: 167).
[35] Maine (1895: 45).
[36] Derrett (1961b: 80).

It was later acknowledged that according to the Hindu law, where there was a conflict between custom and *sastra*, the custom overrode the written text[37]; nevertheless, the texts were elevated to a new supremacy over custom. While some more widespread and long-standing customs gained recognition, 'the most distinct effect of continued judicial construction . . . has been . . . greatly to extend the operation of semi-sacred collections of written rules . . . at the expense of local customs which had been practiced over small territorial areas'.[38]

While the British courts may have strengthened some customs by impeding the traditional methods by which orthodox standards spread to new groups,[39] the rules of evidence provided the mechanism for the disappearance of legal effectiveness of much customary law. Custom was unwritten and therefore difficult to prove in court. Yet the British courts, with their heritage of common-law hostility to local customs, applied requirements for proving the existence of a custom that were onerous to Indian litigants. To prevail over the written law a custom must be 'proved to be immemorial or ancient, uniform, unvariable, continuous, certain, notorious, reasonable (or not unreasonable), peaceful, obligatory and it must not be immoral nor opposed to an express enactment . . . or to public policy'.[40] The difficulties in meeting these requirements combined with the general assumption that Hindus were ruled by *dharmasastra* to extend the *sastric* rules to many groups which had previously been ruled by their own customs.[41]

Even where an explicit attempt was made to preserve customary law, the *sastric* law advanced over it.[42] Custom recorded for the purpose of applying it in the courts, was changed in the process of recording it.[43] From a body of orally transmitted percepts and pre-

---

37 *Collector of Madura v. Mootoo Ramalinga* (1868) 12 *M.I.A.* 397.

38 Maine (1895: 208); Derrett (1961b: 101); Gledhill (1954: 578).

39 Derrett (1956: 237).

40 Kane (1950: 44); For the courts' treatment of custom see Kane (1950: 22–6); Roy (1911); Jain (1963).

41 The most striking elaboration of this view is found in the works of J.H. Nelson (1877; 1880; 1886). See also Derrett (1961b). Similar developments in the elevation of Roman over customary law in Europe are described in Smith (1927: 35); on the similar role played there by rules of evidence, see Bryce. (1901: 106). While the number of topics ruled by Hindu law has been restricted, the portion of the population ruled by it continues to increase even today under the rubric of 'Hinduization of tribals' (see *Chunku Manjhi v. Bhabani Majhan*, A.I.R. 1946 Patna 218.

42 Derrett (1961a: 29).

43 In the Punjab, custom was regarded as the primary rule of decision on certain specific matters (Rattigan, 1953). But even here, custom was recorded and its administration becomes almost indistinguishable from statute and case law (See *Lohare v. Civil Judge*, A.I.R. 1964 Raj. 196). On the method and impact of recording, see Gledhill (1960).

cedents, subject to variable interpretation and quasi-legislative innovation at the discretion of village notables or elders, it became
a body of fixed law to be construed by a professional court. Variable
sanctions, imposed with an eye to the total situation of the parties,
were replaced by the compulsory and drastic execution of the court's
decrees. Judicial enforcement of custom rigidified it and stripped
it of its quasi-legislative character.[44] Official courts were and are
reluctant to permit the creation of new binding custom.[45]

Sastric Law was Rigidified

Customary law then, was rigidified, restricted in scope and replaced
by *dharmasastra*.[46] What was the effect of the courts on *sastric* law?
To ascertain the Hindu and Muslim law, the courts appointed law
officers — Muslim *moulavis* and Brahmin *pundits* — to select and interpret the relevant portions of the Hindu and Muslim law for the
English judges. At the same time, the British set about collecting
and translating authoritative books in the hope of making the Hindu
law more accessible and certain. Dissatisfaction with the work of
the law officers, the growth of a body of translated texts, digests and
manuals prepared by the British, as well as a growing body of precedent from the courts themselves, led eventually to the elimination
of the law officers as intermediaries between the courts and the
Hindu law. With the general reorganization of the legal system in
the 1860s, the posts of the law officers were abolished and the common-law judges undertook to administer the law directly from the
existing corpus of materials. Derrett observes that 'the *dharmasastra*, as a living and responsible science died when the courts
assumed full judicial knowledge of the Hindu law in 1865 . . .'.[47]

Derrett tells us that the 'death-sickness' began when, in their quest
for 'clarity, certainty and finality in terms foreign to Hindu tradition',[48] the British attempted to treat Hindu law as if it could be
made to assume a fixed form. They insisted on a certainty and
consistency alien to Hindu jurisprudence, which depended on expansive judicial discretion. '. . . [T]he *sastra* . . . offered the judge
discretion, not only in choice of a rule of law from permissible
alternatives, but also in manipulating the judicial procedure, e.g.

[44] Lawson (1953: 19).

[45] See e.g. *Daivanai Achi v. Chidambaram'* A.I.R. 1954 Mad. 657; *Sankaran Nam-
boodiri v. Madhavan*, A.I.R. 1955 Mad. 579.

[46] For an account of parallel development within Muslim personal law, Rankin,
(1940); Ali, (1938). The elevation of textual law over custom culminated in the
Muslim Personal law (Shariat) Application Act, 1937 (Act XXVI of 1937), which
abrogated custom with specified exceptions (Fyzee, 1955).

[47] Derrett (1961b: 94).

[48] Id., 112.

in the admission of witnesses, etc.'[49] But while such discretion and flexibility were necessary to the working of the traditional system, they constituted an intolerable deficiency to the British, who 'had no means of inserting themselves into the tradition which would have enabled sound discretion to be exercised'.[50]

In their effort to make Hindu law more uniform, certain and accessible to British judges and to check the discretion of the *pundits*, the courts relied increasingly on translations of texts, on digests and manuals, and on their own precedents. Regard for precedent as such was foreign to the Hindu system.[51] Introduction of the rule of *stare decisis* diminished the flexibility of Hindu law by ruling out innovations to meet changes in community sentiment.[52]

Earlier, *sastra* had changed and developed by successive commentaries and had maintained its flexibility by its complex and discretionary techniques of interpretation. British administration not only dissipated these techniques but also narrowed the selection of authoritative texts. Courts were not to consult the whole of *sastric* science, but only those commentators accepted in the locality, a view which led to the elaboration of partially artificial 'schools' of Hindu law. Any further development by commentary and reaction was impeded.[53]

> [A]s the influence of the pandit gradually wanes in the courts we see the latter coming to rely more and more on the older, more narrowly defined *dharmasastra* works and less and less on the miscellaneous and more recent works which the good pandit would frequently rely upon. The pandit as a professor of a living science was rejected for the more or less dead treatises which would head the pandits' list of references.[54]

With its innovative technique stripped away, *sastric* law, like customary law, became more rigid and archaic[55] as well as more uniform and certain. Yet judicial precedent and legislation provided new means of growth and development.[56]

---

[49] Id., 76.
[50] Derrett (1961b: 76); Derrett (1961a: 33); Morley (1850: clxxvii ff).
[51] Derrett (1961b:83).
[52] Derrett (1961a: 48).
[53] Maine (1895: 46–7).
[54] Derrett (1961b: 99).

[55] A striking instance of this is to be found in the increased emphasis on the *varnas*, or four great classes into which Hindu society is theoretically divided by the *sastric* texts. *Varna* distinctions received scant attention from the courts during the early years of British rule, but became a major factor in the administration of Hindu law after the courts undertook to administer it without intermediaries and directly from the texts.

[56] Sarkar (1958: 366–90); Gledhill (1954).

The Impact of the Modern Legal System

Let us consider some of the effects on Indian society of the modern legal system with its regular hierarchies of courts applying codified English law and rationalized indigenous law. We noted before the spread of a sense of individual right independent of local usage or opinion and enforceable by reference to standards and agencies beyond the locality or the group. The new system provided new avenues of mobility and advancement within Indian society.[57] There were new methods for conflict, acquisition and pursuit of status.[58] Powerful persons and groups on the local scene possessed new weapons for intimidating and harassing their opponents. But the local underdogs could now carry the fight outside the local arena by enlisting powerful allies elsewhere. Persons and property were freed from hereditary prescriptions, making possible a wider range of 'market' transactions.

The legal system also provides new channels for the dissemination of norms and values from governmental centres to towns and out to villages. The legal system is a hierarchical network, which radiates out from the cities and through which authoritative doctrine flows outward from governmental centres. By the prestige of urban and official centres, and by the disposition of governmental power in their enforcement, elements of this doctrine might be powerfully recommended. New methods of group activity and new images of social formation are presented.[59]

Modern Law: A Unifying Element

The modern legal system may be viewed as an important unifying element. While previously there were wider networks of marriage, ritual activity, pilgrimage and economic and military activity, until the advent of the modern system, law and justice were in good part purely local concerns.[60] Today, while India has no single nation-wide system of caste, kinship, religion or land-tenure, there is an all-India legal system which handles local disputes in accordance with uniform national standards. This legal system provides not only a common textual tradition but also a machinery for insuring that this tradition is applied in all localities in accordance with nationally prescribed rules and procedures rather than dissolved into local interpretations.

[57] Kumar (1965); Cohn (1960); On speculation in lawsuits see Report of the Civil Justice Committee (1925).
[58] *Subrao Hambirrao Patil v. Radha Hambirrao Patil* I.L.R. 52 Bom. 497 (1928).
[59] McCormack (1963, 1966); Conlon (1963); Maine (1895: 9, 38).
[60] Cohn (1959: 88).

With this system goes what we might call an all-India legal culture. Its carriers are all persons who are connected with the courts, but primarily the numerous lawyers.[61] With their skills in manipulating the legal system, they serve as links or middlemen between official centres and rural places, disseminating official norms, rephrasing local concerns in acceptable legal garb, playing important roles in devising new organizational forms for forwarding local interests (e.g. caste associations, political parties, economic interest groups).[62] In spite of differences of region, language, caste and religion, they share a common legal culture and they are able to put this culture at the service of a wide variety of local interests. In a situation where local concerns and interests get expression by

[61] *Report of the All-India Bar Committee* (1953); Cohn, (1961: 625ff). Lawyers are not the only intermediaries who carry official law to the wider society; there are also social workers, administrators, police, etc. And, of course, the petition-writers, clerks and touts, who often act as intermediaries between villagers and urban lawyers [Mack (1955); Srinivas (1964: 94); Chattopadhyay (1964: 81ff); Law Commission of India (1958: Vol. I, 577 ff); Cohn (1965: 103)].

In absolute numbers, India has the second largest legal profession in the world (after the United States). In proportion to its population, there are fewer lawyers in India than in Western common-law countries, but many more than in other new states. The Indian figures are in the same range as many continental civil-law countries. (In these comparisons the Indian figures are somewhat understated, since the proportion of children is higher in the population of India than in those of the wealthier countries.) But a rough idea may be gathered from the following figures, which represent the number of persons per lawyer in selected countries:

United States (1960)	728
Canada (1961)	1,366
Italy (1957)	1,601
Great Britain (1959)	2,105
West Germany (1958)	3,012
India (1952)	4,920
Egypt (1964)	5,768
France (1958)	5,769
Japan (1960)	14354
Nigeria (1964)	22,765
Indonesia (c. 1960)	c.100,000

The figures for the United States, Canada, Italy, Great Britain, West Germany, France and Japan are taken from T. Hattori (1963). The Indian figures are based upon the Report of the All-India Bar Committee (1953). Figures for Egypt and Nigeria were supplied by the Commonwealth Library of the American Bar Foundation. The Indonesian figure is a calculation based upon Lev (1965: 183, 189).

[62] Sir Ivor Jennings, observing that the Constituent Assembly was dominated by lawyers, contends that 'the lawyer-politician has . . . played a more important part in Indian politics than in the politics of any country in the world' (Jennings, 1955: 24). In 1953, lawyers comprised 26 per cent of the Lok Sabha (Lower House) and 29 per cent of the Rajya Sabha (Upper House). Cf. approximately 60 per cent of the U.S. Congress (85th Congress); 20 per cent of the British House of Commons (1955); 14 per cent of the French National Assembly (1951); 11 per cent of the West German Bundestag (1957) [McCloy (1958: 5–6)].

representation at centres of power, rather than in the traditional way of enjoying a sphere of autonomy, the lawyers are crucial agents for the expression of local and parochial interests at the same time that they rephrase these interests in terms of official norms. Thus the modern legal system provides both the personnel and the techniques for carrying on public business in a way that is nationally intelligible and free of dependence on particular religious or local authority. It thus provides one requisite for organizing Indian society into a modern nation-state.

Constitutionalism

The formation of an independent Indian nation provided a basis for further integration and consolidation of the modern legal system. With the coming of independence, enclaves previously outside the legal system were integrated into it. A layer of constitutionalism was superimposed on the existing legal system and structure of government. The Constitution (1950) established India as a secular federal republic with a parliamentary system in the British style and a strong central government. The framers of the Constitution rejected the various proposals to construct a government along 'indigenous' lines.[63] The Constitution established powerful legislatures at the centre and in the states. It also established a unified judiciary covering the whole of India under a Supreme Court as a court of final appeal in all cases.

The Constitution includes a bill of Fundamental Rights, which are enforceable by the judiciary[64] and to which all governmental regulation and all laws in every part of India must conform. Government is enjoined by these provisions to be indifferent to particularistic and ascriptive characteristics (e.g. race, religion, caste, place of birth, and sex) in its dealings with citizens, whether as electors, employees or subjects.[65] A wide range of private conduct, involving the asser-

[63] The proponents of 'indigenous' systems of government (of both orthodox and Gandhian persuasions) were severely disappointed with the Constitution, which did little to dismantle complex bureaucratic government, to re-assert the virtues of village autonomy or to express dedication to a life of purity in Hindu terms. The whole effort managed to deposit only three provisions in the Constitution, all in the Chapter Directive Principes: prohibition, an item of uplift with religious overtones that had long absorbed social reformers (Art. 47); a commitment to laws abolishing cow-slaughter (Art 48); and, most important, a promise to organize self-governing village *panchayats* (Art. 40). On the constituent assembly's choice, generally, see Austin, (1966). On schemes and pleas for 'indigenous' alternatives, see AVARD 1962; Agarwal (1946); Sharma (1951).

[64] Constitution of India, Part III; Cf. Art. 32 and 226 on the wide judicial powers in this area.

[65] Part III of the Constitution of India. Cf. Art. 325 *State of Rajasthan v. Pratap Singh*, A.I.R. 1960 S.C. 1208; *Sanghar Umar v. State*, A.I.R. 1952 Saur. 124; *Nain Sukh Das v. State of U.P.*, A.I.R. 1953 S.C. 384;

tion of precedence or the imposition of disabilities — including venerable usages which had previously enjoyed religious and some-times legal sanction are outlawed.[66] Governmental enforcement of rights based on caste position, heredity, vicinage and the like is forbidden.[67]

To serve as a guide to the legislatures, the Constitution contains a set of non-justiciable 'Directive Principles of State Policy'[68] which call for the reconstruction of Indian society and government along the lines of a modern welfare state. Accordingly, the central legislature and the legislatures of the several states have released a flood of legislation aimed at economic development and social reform, extending governmental regulation to many areas of life previously immune from official control.[69] Extensive regulation of landholding, religious endowments, caste practices and family law by central and state governments has supplanted governmental re-cognition of local rules of unofficial or parochial provenance.

The Hindu Code

The extension and consolidation of the modern features of the legal system can be observed in the treatment of two basic institutions of Hindu society — the family and the caste. Among the Directive Principles is a commitment that the State 'secure to the citizens a uniform civil code throughout the territory of India'[70] which contemplates the complete abandonment of the personal-law system. Although no such unification of the laws of Hindus and Muslims has yet been undertaken, the Parliament in 1955–6 passed a series of Acts known collectively as the Hindu Code, which effect a wholesale and drastic reform of Hindu law.[71]

Where earlier legislation introduced specific modifications into the framework of *sastric* law, the Code entirely supplants the *sastra* as the source of Hindu law. Hindu social arrangements are for the first time moved entirely within the ambit of legislative regulation; appeal to the *sastric* tradition is almost entirely dispensed with. The Code turns away from the *sastra* by abandoning *varna* distinctions

[66] Constitution, Arts. 17, 15[2], 23[1].

[67] See *Gazula Dasratha Rama Rao v. State of Andra Pradesh* A.I.R. 1961 S.C. 564; *Bhau Ram v. Baij Nath*, A.I.R. 1962 S.C. 1476; *Aramugha Konar v. Narayana Asari*, A.I.R. 1958 Mad. 282; see further Articles 25 and 26 of the Constitution of India and Subra-manian (1961).

[68] Constitution of India, Part IV.

[69] In the Constitution's first eight years, some 600 Acts were passed by the Central Parliament (in addition to 89 Ordinances, 21 Regulations and 62 Presidential Acts). During the four years 1953–6, the State Legislatures passed 2,527 Acts, of which 275 dealt with land reform (Law Commission of India, 1958; note 114 at Vol. 1,30).

[70] Constitution, Art. 44.

[71] Derrett (1963b, 1957, 1958); Levy (1961).

and the indissolubility of marriage, the preference for the extended joint family and for inheritance by males only and by those who can confer spiritual benefit. It favours instead greater individualism, emphasis on the nuclear family, divorce and equality of *varnas* and sexes. Very few rules remain with a specifically religious foundation.

The Code marks the acceptance of Parliament as a kind of central legislative body for Hindus in matters of family and social life. The earlier notion that government had no mandate or competence to redesign Hindu society has been discarded. For the first time, the bulk of the world's Hindus live under a single central authority that has both the desire and the power to enforce changes in their social arrangements. It has been pointed out that, throughout the history of Hinduism, no general and sweeping reforms were possible, just because of the absence of centralized governmental or ecclesiastical institutions.[72] Reformers might persuade others and they might gain acceptance as a sect; but there was no way for them to win the power to enforce changes on others. They could supplement existing practice but they could not supplant it, because there were no levers which could be grasped to accomplish across-the-board changes. The modern legal system has made possible enforcement of changes among all Hindus by a powerful central authority.

The Code subjects Hindus to a degree of uniformity unprecedented in Hindu legal history. Regional differences; the schools of commentators; differences according to *varna*; customs of locality, caste, and family; many special statuses and estates, and (largely) distinctions of sex have all fallen by the wayside. Some narrow scope is allowed for custom, but for the first time a single set of rules is applicable to Hindus of every caste, sect and region.

Reform and Unification

Much the same might be said of constitutional provisions and legislative enactments regarding caste.[73] Here, too, there is the assertion of broad regulative power by the government and curtailment of the autonomy of the component groups within Hinduism. This power is exercised to eliminate disparities of law and custom and to impose uniformity of rights and equality of opportunities. *Sastric* notions, and legal categories (*varna*, pollution) are discarded and Western or modern categories substituted. In both instances emphasis falls on eradicating the barriers within Hinduism and promoting an integrated Hindu community nity — and eventually a non-communal society. Finally, in both

---

[72] Pannikar (1961: 72, 79ff).
[73] Galanter (1961a, 1963b, 1966a).

instances, we have the Western-educated elite using the law to impose its notions. As in many areas of Indian life, the law in regard to the family and caste does not represent a response to the felt needs of its clientele or an accommodation of conflicting interests and pressures. Rather the law is the expression of the aspirations of the most articulate and 'advanced' groups, which hope to use its educational as well as its coercive powers to improve the unenlightened. Deliberate social change was not unknown before the coming of the British. On the contrary, Hindu law contained its own techniques for deliberate and obligatory innovation and these continued to be used into the early part of the British period. The revolutionary principle fostered by British rule was not the notion of deliberate social change, but rather the notion of the unit which might legitimately introduce and be the subject of such changes. The recent legislation visualizes a single national community — or at least a community embracing all Hindus, transcending divisions of region, caste and sect.

Thus the present legal system provides a unifying element in India in a way that neither Hindu nor Muslim law ever did. Muslim law never went deep enough; it was never applied to disputes among Hindus. *Dharmasastra* tolerated diversity, preferring unification by example, instruction and slow absorption rather than by imperative imposition. Change was piecemeal rather than comprehensive. In contrast, the new legal system provides machinery (and the ideology) for legislation to be enforced throughout the society. Such a system, along with mass communications, makes possible unprecedented consolidation and standardization of Hinduism, as well as of Indian society generally.

Traditional Law in the Modern System

What, then, is the role of Hindu law in the Indian legal system today? The *dharmasastra* component is almost completely obliterated. While it is the original source of various rules on matters of personal law, the *sastra* itself is no longer a living source of law; these rules are intermixed with rules from other sources and are administered in the common-law style, isolated from *sastric* techniques of interpretation and procedure. In other fields of law, *sastra* is not used as a source of precedent, analogy or inspiration. As a procedural-technical system of law — a corpus of norms, techniques and institutions — it is no longer functioning. There seems to be little nostalgia to revive particular *sastric* rules,[74] (which would, in any event, be

---

[74] Very considerable portions of *sastra*, with their emphasis on graded inequality, would fail to meet present constitutional requirements — and would hardly be likely to appeal to India's present rulers.

administered in the common-law style); the pleas for an 'indigenous system' are for the directness, cheapness and simplicity of local law, not for the complexities of *dharmasastra*.

The local customary component of Hindu law is also a source of rules at a few isolated points, but it, too, has been abandoned as a living source of law. There is but one significant attempt to promote such indigenous law, by devolving certain judicial responsibilities to the local elective village *panchayats*.[75] But these elective *panchayats* are quite a different sort of body than the traditional *panchayat*.[76] It is suggested that rather than inspiring a resurgence of local law, they may instead effect a further displacement of local law by official law within the village.[77]

The traditional method of relating the authoritative 'official' law to local customary law has definitely been supplanted. The Indian legal system is now equipped with machinery for bringing local law into line with national standards.[78] Once such a mechanism is present, local law can survive only by taking on the character of modern law—it must become certain, definite, consistent, obligatory rather than discretionary or circumstantial.

The Gap Between 'Higher Law' and Local Practice

Every legal system faces the problem of bridging the gap between its most authoritative and technically elaborate literary products at the 'upper' end of the system and the varying patterns of local practice at the 'lower' end. It must decide on allowable leeways—how much localism to accommodate, how to deflect local to general standards. Hindu law solved these problems by willingly accommodating almost unlimited localism; it was willing to rely on acceptance and absorption through persuasion and example. These methods are too slow and irregular to appeal to a ruling group which aspires to transform the society radically and to build a powerful and unified nation. Even where specifically Hindu norms are made the

---

[75] Law Commission (1958: Vol. II, 874–925); Malaviya (1956); Report of the Study Team on Nyaya Panchayats (1962).

[76] Retzlaff (1962: 23ff); Luchinsky (1963a: 73); Robins (1962).

[77] See *Jai Kaur v. Sher Singh*, A.I.R. 1960 S.C. 1118; *Nain Sukh Das v. State of U.P.* A.I.R. 1953 S.C. 384; *Sanghar Umar v. State*, A.I.R. 1952 Saur. 124.

[78] It should be noted that this machinery is more insistent in India than in, say, the United States, where juries and locally elected prosecutors and judges introduce a check on uniformity and provide enclaves for localism. Again, the unified judiciary, the competence of the Indian Supreme Court in matters of state law, the high estimation put on its *dicta* as well as its holdings, litigants' direct access to higher courts, the frequency of appeals, and the practice of higher courts entering their own orders instead of remanding — all of these incline the system to a high degree of centralization.

basis of legislation — e.g. in prohibition and anti-cow slaughter laws — these norms are not implemented by the old techniques. Enforcing these matters by legislation, courts and the police, stands in striking contrast to allowing them simply to be adopted gradually by various groups in the society. Such change still takes place, but it operates outside the legal system. While the harsher British methods have displaced the methods of persuasion and example from the legal system itself, they persist alongside it in the form of propaganda, education and the widespread tendency to imitate urban and official ways.[79]

But the demise of traditional law does not mean the demise of traditional society. Traditional notions of legality and methods of change still persist at a sub-legal level — e.g. in the area of activities protected by the doctrine of 'caste autonomy', in the form of accepted deviance, and in arrangements to evade or ignore the law. The modern legal system may provide new possibilities for operating within traditional society. Official law can be used not only to evade traditional restrictions, but to enforce them.[80] Traditional society is not passively regulated by the modern system; it uses the system for its own ends. Traditional interests and groupings now find expression in litigation, in pressure-group activity and through voluntary organization.

Two Political Idioms

Morris-Jones (1963) speaks of two contrasting political idioms or styles in contemporary India: the modern idiom of national politics with its plans and policies and the traditional idiom of social status, customary respect and communal ties, ambitions and obligations. He notes that 'Indian political life becomes explicit and self-conscious [modern] idiom. . . . But this does not prevent actual behaviour from following a different path'.[81] Similarly, all contact with the legal system involves the translation of traditional interests and concerns into modern terms in order to get legal effectiveness. For example, at the touch of the official law, a caste's prerogatives become the constitutionally protected rights of a religious denomination[82]; a lower caste's ambitions become its constitutional right to equality; property can be made to devolve along traditional lines, and land-reforms can be frustrated by transactions in good legal form.[83]

79 Marriott (1955: 72).
80 Srinivas (1964: 90); Siegal and Beals (1970: 408); Cohn (1965: 98–9, 101).
81 Morris Jones (1963: 142).
82 *Sri Venkataramana Devaru v. State of Mysore*, A.I.R. 1958 S.C. 255.
83 Derrett (1964b); Luchinsky (1963b); Ishwaran (1964).

Traditional interests and expectations are thus translated into suitable legal garb, into nationally intelligible terms.[84] But the process of translation opens new possibilities for affiliation and alignment; new modes of action. If we regard tradition not as a stationary point, a way of remaining unchanged, but as a method of introducing and legitimizing change, we can say that the modern legal system has displaced traditional methods within the legal system itself while it has supplemented them outside it.

A Dualistic Legal System

India has what we might call a dualistic or colonial-style legal system — one in which the official law embodies norms and procedures congenial to the governing classes and remote from the attitudes and concerns of its clientele. Such systems are typical of areas in which a colonizing power superimposes uniform law over a population governed by a diversity of local traditions. However, legal colonization may occur from within as well as from without, as in Turkey,[85] Japan[86] and in India since the departure of the British. The colonial legal situation prevails wherever there is unresolved tension between national and local, formal and popular law.[87] In a relatively homogeneous society, the law can be visualized as the expression of widely shared social norms. In a heterogeneous society (differentiated horizontally by culture, or vertically by caste or class), the law expresses not primarily the aspirations and concerns of the society, but those of the groups that formulate, promulgate and apply the law. A gap between the official law and popular or local law is probably typical of most large political entities with intensive social differentiation. To some extent this colonial legal situation obtains in most modern societies.[88] But it is present with special force in the so-called new states. In the nineteenth and early twentieth centuries, the poorer parts of the earth were the scene of a reception of foreign law unprecedented in scope

---

[84] 'The use of the courts for settlement of local disputes seems in most villages almost a minor use of the courts. In Senapur, courts were and are used as an arena in the competition for social status, political and economic dominance in the village. Cases are brought to harass one's opponents, as a punishment, as a form of land speculation and profit making, to satisfy insulted pride and to maintain local political dominance over one's followers. The litigants do not expect a settlement which will end the dispute to eventuate from recourse to the State Courts' (Cohn, 1965: 105).

[85] 'The Reception of Foreign Law in Turkey', (1957).

[86] Takayanagi (1963).

[87] The colonial legal situation then stands midway between those systems where official law is reflective of, and well integrated with, popular law because it has been precipitated out of that law (or because it has completely absorbed and digested local law); and those where it is reflective of a well integrated folkways because no remote official law has ever differentiated itself institutionally from folk or popular law.

[88] Priestly (1962: 196–7); Dewey (1946: 116–17); Maine (1895: 59–60).

(even by the reception of Roman law in medieval Europe). In India, the incorporation of large blocs of common law and civil law in the nineteenth century was followed by the reception of new constitutional models in the twentieth century and by a post-independence wave of reform and rationalization. This process of borrowing, consolidating and modernizing national legal systems seems to involve certain common trends: application of laws over wider spatial, ethnic and class areas; replacement of personal by territorial law; the breakdown of corporate responsibility and the growth of individual rights; increasing generality and abstraction; greater specialization and professionalism, secularization, bureaucratization and replacement of moral intuition by technical expertise. In almost all of the newer countries, the legal system is comprised of these modern elements in uneven mixtures with traditional ones and the discrepancy between the different components of the legal system strongly felt. This multi-layered legal situation involves common processes of the displacement of local by official law and seems to be accompanied by common discomforts.[89]

Failure of Revivalism

A certain irreversibility in this process of forming a modern legal system, even where it is based upon foreign sources (at least as long as a unified political power retains control of the law), seems indicated by other instances of the reception of complex law based upon foreign sources, as in the reception of Roman law in Western Europe or the massive borrowing of civil law in nineteenth-century Japan. This irreversibility is confirmed by the very limited success of revivalist movements. Attempts to purify and reconstruct Irish law[90] fared no better than present attempts in Pakistan[91] and Israel[92] which have so far not succeeded in bringing about any fundamental changes in their respective legal systems. In Ireland, Israel and Pakistan, there is, if anything , more common law in the broad sense, i.e. law of the modern type, than before independence. In India, where the proponents of indigenous law are less attached to *dharmasastra* than nostalgic for the 'simplicity' of local-customary law — and where they tend to be persons who find detailed consideration of the law uncongenial — any change in this direction is even more unlikely.[93]

[89] Smith (1927: 35–6).
[90] Moran (1960: 31–5, esp. 33); Takayanagi (1963: 31).
[91] Maududi (1960: esp. Part I); Coulson (1963).
[92] Kahana (1960); H. Cohn (1958); Yadin (1962).
[93] It should be recalled that the similar distaste for the law of former colonial rulers found in the early history of the United States is not to be observed in more recent American evaluations of our common law heritage. As India feels safely distant from her colonial past, a similar embrace of her legal heritage is at least a possibility.

One may compare the fate of British law with that of the English language as a medium of public business and civil life. In general, colonial languages appear to recede from their former preeminence, while the tide of law continues to advance. Strangely, the law seems more separable from its origin, relatively easy to borrow and hard to discard. Bryce found in the spread of Roman and British law 'a remarkable instance of the tendency of strong types to supplant and extinguish weak types in the domain of social development'.[94] But what made British law a 'strong type' was not the superiority of the norms it embodied or the elegance with which the system was elaborated. It should be noted that, unlike the civil law which spread widely by voluntary adoption, common law spread only by settlement or political dominion.

One observer noted that '. . . the spirit of English law which settled down on our legislative centres [in India] was that of a period when the law itself was the most technical, the least systematic and the least founded on general, equitable and coherent principles, that the world has ever seen'.[95] The 'strength' of British law lay in its techniques for the relentless replacement of local law by official law, techniques by which it accomplished its own imposition half inadvertently. And this imposition seems to be enduring in a way that language is not. An official language does not become a household language; each generation must recapitulate the painful process of estrangement. The official language does not necessarily gain at the expense of the household languages; on the contrary, we find in India an enrichment and development of indigenous languages during British rule. However, official law of the modern type does not promote the enrichment and development of indigenous legal systems; it tolerates no rivals; it dissolves away that which cannot be transformed into modern law and absorbs the remainder; it creates a numerous class of professionals who form the connecting links of the nation-state and a vast array of vested rights and defined expectations which disincline those holding them to support or even conceive drastic changes.

94 Bryce (1901: 122).
95 Baden-Powell (1886: 372).

3

The Aborted Restoration of 'Indigenous' Law in India *

Traditional law — Hindu, Muslim and customary — has been almost entirely displaced from the modern Indian legal system. Today, the classical *dharmaśastra* component of Hindu law is almost completely obliterated. It remains the original source of various rules of family law. But these rules are intermixed with rules from other sources and are administered in the common-law style, isolated from *śāstric* techniques of interpretation and procedure.[1] In other fields of law, *dharmaśastra* is not employed as a source of precedent, analogy or inspiration. As a procedural-technical system of laws, a corpus of doctrines, techniques and institutions, *dharmaśastra* is no longer functioning.[2] This is equally true of Muslim law.[3] The local customary component of traditional law is also a source of official rules at a few isolated points, but it too has been abandoned as a living source of law.[4]

'Legal system' and 'law' are employed here in a narrow (but familiar) sense to refer to that governmental complex of institutions, roles and rules which itself provides the authoritative and official definition of what is 'law'. In contemporary India, as in other complex societies, there are myriad agencies for making rules and settling disputes which lie outside of the legal system as narrowly

* Delivered at the annual meeting of the Association for Asian Studies, Boston, Mass., March 30, 1969.

[1] On the fate of *dharmasastra*, see Derrett (1961a): Galanter (1968).

[2] An exception must be made for a small but unknown quantity of private recourse to exponents of *dharmasastra*. (I am indebted to Professor V. V. Deshpande for allowing me to see an unpublished memorandum on the functioning of *dharmasabhas* in contemporary India.) Similarly, some Muslims consult *muftis* for advice on matters of *Shari'a*.

[3] While more textual Muslim law has been preserved in the family law area, it is equally detached from its former procedural and institutional setting. Muslim law is not discussed in the remainder of this paper because there has been no serious thought of reviving it. The Law Commission felt that it was unnecessary to consider Muslim law in the context of claims for a revived indigenous system; see Government of India, Ministry of Law (1958: 1, 26). More recently, attention has focused on recurrent proposals to abolish the existing separate Muslim law in the personal law fields in favour of a Uniform Civil Code, as directed in the Constitution.

[4] On the role of custom in the courts, see Jain (1963); Kane (1950); Roy (1911).

and authoritatively defined. Many matters are regulated by traditional legal norms; tribunals of the traditional type continue to
function in many areas and among many groups, but without governmental force.[5] The point here is that they have been displaced from
the official system, powerfully influenced by it and in many cases
entirely supplanted by it.[6] The official legal system comprises laws,
techniques, institutions and roles which are, with few exceptions,
modifications of British or other western models.

The first part of this paper examines briefly the failure — or
perhaps containment is more accurate — of post-independence
attempts to replace the present legal system with revived indigenous
law. The second part attempts to explain this failure and to suggest
its implications for the comparative study of legal systems.

The dichotomy between the official law and popular legality[7]
has been the theme of continuing stream of criticism from administrators, nationalists and students of Indian society, who have emphasized the unsuitability of British-style law in India.[8] As a scholarly
British District Officer plaintively concluded in 1945: ' . . . we proceeded, with the best of intentions, to clamp down upon India a
vast system of law and administration which was for the most part
quite unsuited to the people. . . . In Indian conditions the whole
elaborate machinery of English Law, which Englishmen tended to
think so perfect, simply didn't work and has been completely per-
verted.'[9] Administrators and observers have blamed the legal system
for promoting a flood of interminable and wasteful litigation,[10] for

[5] For a useful summary and analysis of the ethnographic data, see Cohn (1965).
For detailed treatment, see Srinivas (*c.* 1964); Nicholas and Mukhopadhyay (1962);
Ishwaran (1964); Berreman (1963). Extensive data on the functioning of Orissa
tribunals can be found in Bose (1960). For an account of tribunals in an urban setting,
see Lynch (1967).

[6] For some instances of total eclipse of traditional tribunals, see Fox (1967); Gough
(1960).

[7] For analysis of the dissonance between British and Indian notions of legality, see
Cohn (1959); Rudolph and Rudolph (1965).

[8] There was, of course, an even broader stream of eulogy and appreciation of
British legality and of its Indian personnel. Many recent examples may be found below.

[9] Moon (1945: 22).

[10] Shore (1837:II, 187–215) attributes excessive litigation to the inadequacies of
the British legal system and to British ignorance of Indian customs. Henry St. George
Tucker, a Director of the East India Company, complained in 1832 that 'the natives
of these provinces, to whom the duel is little known, repair to our courts as to the
listed field, where they may give vent to all their malignant passions' (Tucker, 1853:
21). For a contrary view, see Hunter (1897: 342–3) who finds that the supposed
proneness to litigation of Indian peasants is due to the fact that rights in land are so
widely spread (compared to England) and that British courts offer a vent for 'the pent
up litigation of several centuries'. It is 'only a healthy and most encouraging result
of three-quarters of a century of conscientious government'. For the first time, Indians
'are learning to enforce their rights'.

encouraging perjury and corruption, and generally exacerbating disputes by eroding traditional consensual methods of dispute-resolution. The indictment was familiar by the mid-nineteenth century:

. . . in lieu of this simple and rational mode of dispensing justice, we have given the natives an obscure, complicated, pedantic system of English law, full of 'artificial technicalities', which . . . force them to have recourse to a swarm of attorneys . . . that is . . . *professional rogues* . . . by means of which we have taught an ingenious people to refine upon the quibbles and fictions of English lawyers. . . . The course of justice, civil as well as criminal, is utterly confounded in a maze of artifice and fraud, and the natives, both high and low, are becoming more and more demoralized. . . .[11]

In the nationalist movement, there were similar complaints, issuing in proposals for the restoration of indigenous justice. There was hostility to the courts as an agency of British control, and the civil disobedience movements of 1920–2 and 1931 included attempts to boycott the official courts and to organize truly Indian tribunals which would work by conciliation, relying on moral persuasion rather than coercive sanctions.[12] The misgivings of some nationalists about the legal system were succinctly expressed by a Gandhian publicist in 1946, who accused the British system of working havoc in India by replacing quick, cheap and efficient *panchayat* justice with expensive and slow courts which promote endless dishonesty and degrade public morality.[13] Existing law, he said, is too foreign and too complex; this complexity promotes 'criminal mentality and crime'. In their place he would have *panchayats* dispense justice at the village level, thereby eliminating the need for lawyers and complex laws.[14]

The Constituent Assembly (1947–9) contained no spokesmen for a restoration of *dharmaśāstra*, nor for a revival of local customary

11 Dickinson (1853: 46). Cf. Shore (1837: II. 236 ff); Connell (1880: chap. 3): P. N. Bose (1917); Moon (1945: especially chap. 2). See also the passionate but diffuse denunciation of Das (1967), which includes charges of foreignness, though the critique seems based more heavily on natural law than *dharmasastra* notions.

12 Gandhi's attacks on lawyers for enslaving Indians by co-operation with the British legal system, his plans for indigenous arbitration courts, etc. are collected in Gandhi (1962: sections 4, 5). The 1920–2 and 1931 proposals for arbitration courts may be found in Malaviya (1956: 227 ff., 774). Earlier proposals for boycott of courts were made by B.G. Tilak and Sri Aurobindo in 1907. See de Bary (1958: 772, 727).

13 Agarwal (1946: 97, 100, 131). (In his Foreword to this volume M. K. Gandhi said 'There is nothing in it which has jarred me as inconsistent with what I would like to stand for'.)

14 Agarwal conceded that there should be higher courts of a professional type, but does not indicate what effect their law would have on the *panchayats* (1946: 98-9).

law as such. An attempt by Gandhians and 'traditionalists' to form a polity based on village autonomy and self-sufficiency was rejected by the Assembly, which opted for a federal and parliamentary republic with centralized bureaucratic administration.[15] The only concession to the Gandhians was a Directive Principle in favour of village *panchayats* as units of local self-government.[16] The existing legal system was retained intact, new powers granted to the judiciary and its independence enhanced by elaborate protections. All in all, the Constitution amounted to an endorsement of the existing legal system.

In the early years of independence there was much open discussion of the need for large-scale reform of the legal system. There was some outspoken criticism that the system was entirely unsuited to Indian conditions and should be radically altered or abandoned.[17] In 1952 a member of the All-India Bar Committee discerned 'a large volume of opinion in the country that this legal system is entirely foreign to the genius and the traditions of the people of this country who need a simpler, quicker and cheaper . . . system than the present dilatory and costly system'.[18]

The most prominent and politically potent of these critics were Gandhians and socialists within the ruling Congress party, who supported a revival of *panchayat* justice. 'Among the depredations caused by the British, the destruction of the village system of decision over disputes, and the consequent imposition of British legal forms and courts, would have a pride of place'.[19] Restoration of *panchayats* was proposed as one phase of the reconstruction of India's villages, in which faction and conflict, bred by colonial oppression, would be replaced by harmony and conciliation. Critical discussion focused almost exclusively on adjectival law—on court administra-

[15] On the rejection of the village alternative, see Austin (1966: chap. 2); AVARD (1962).

[16] Art. 40 provides: The State shall take steps to organize village *panchayats* and endow them with such powers and authority as may be necessary to enable them to function as units of self-government.

[17] During this period the legal system was also under attack from another direction by those who regarded the courts as obstacles to rapid reform and development and saw lawyers as agents of delay and obfuscation in the service of narrow private interests. Such sentiments were present in the highest places. E.g. in the debate over the First Amendment, Prime Minister Nehru complained: 'Somehow . . . this magnificent Constitution that we have framed was later kidnapped and purloined by the lawyers'. Parliamentary Debates XII-XIII (Part II) Col. 8832 (May 16, 1951). Proposals that judicial processes be replaced by administrative tribunals from which lawyers were to be excluded were common and frequently acted upon. See Law Commission 1958: 671 ff. This 'left' criticism tended to share with the Gandhian its unflattering estimate of lawyers.

[18] C. C. Shah at the All-India Bar Committee (1952: 45).

[19] Malaviya (1956: 773).

tion (delay, expense, corruption), complexity of procedure, unsuitability of rules of evidence, the adversarial rather than conciliatory character of the proceedings, and (occasionally) the nature of penalties. Although different personnel and procedures might be expected to entail different bases of decision, there was virtually no discussion of 'substantive law'.

This movement had little support among legal professionals.[20] Lawyers and judges agreed that the system displayed serious defects — perjury, delay, proliferation of appeals, expense. But they did not attribute its shortcomings to its foreignness.[21] Except for a measure of decentralization for petty cases, they rejected the notion that the remedy lay in a return to indigenous forms. Almost without exception, the profession viewed the Anglo-Indian law as a most beneficent result of the British connection: '. . . . the British period gave us a rule of Law beneficial to our interests. If at all we are beholden to anything British, it is their system of Justice and Jurisprudence, that have taken an abiding and glorious place in the life of our country'.[22] While many lawyers called for re-examination and even radical reformation of the legal system, almost none could perceive any advantage in reversion to pre-British models. 'The reform of our Legal Profession and our Legal system does not lie in that way of 'Village Panchayat Revival'. It is a suicidal policy that will lead only to factions and anarchy'.[23]

Proposals for an indigenous system were among the many matters taken up by the Law Commission in its full-dress survey of the administration of justice in 1958. The Commission found that even a brief depiction of the ancient system

[20] The most prominent exception was the eminent advocate (later Home Minister and Governor) Dr K. N. Katju. See Katju (1948); Malaviya (1956: 781). Dr Katju was also a member of the Congress Village Panchayat Committee which proposed a cautious programme of giving *panchayats* small-cause powers and emphasized their conciliatory aspects of settling disputes by persuasion and advice. All-India Congress Committee (1954: 40).

[21] Some, indeed, saw fault in the departures from the foreign model. Thus an experienced Punjab judge observed:

I have no doubt that the judicial machinery imported from England and set up here with slight modification can work efficiently if some of the dirt and grit can be eliminated. . . . Indeed the few departures from the British way and the local modifications are, to a great extent, responsible for the lack of public confidence in the Courts. . . . The defects in our judicial system are not the defects of a foreign institution planted in conditions wholly unsuited to its healthy growth, but arise from changes and modifications introduced for an ulterior purpose (Khosla, 1949: 70–1, 87).

[22] Ramachandran (1950: 53). (It should be noted that this essay was awarded the Gold Medal at the Sixth Session of the Madras State Lawyers Conference in Coimbatore, May 1950.) Cf. Govinda Menon (1951: 91); Jagannadh Das (1955).

[23] Ramachandran (1950: 53); Rajamannar (1949); Shah at the All-India Bar Committee (1952: 45); Misra (1954); Ayyar (1958: 330–1).

shows how unsound is the oft-repeated assertion that the present system is alien to our genius. It is true that in a literal sense the present system may be regarded as alien. It is undoubtedly a version of the English system modified in some ways to suit our conditions. . . . But it is easy to see that in its essentials even the ancient Hindu system comprised those features which every reasonably minded person would acknowledge as essential features of any system of judicial administration, whether British or other. . . . We can even hazard the view that had the ancient system been allowed to develop normally, it would have assumed a form not very much different from the one that we follow today.[24]

The commission notes that the attraction of the indigenous system lay not in the intricacies of classical textual law but in the simplicity and dispatch of popular tribunals that applied customary law. But it finds it unthinkable that such courts could be expected to cope with the complexities of the law in a modern welfare state:

No one can assert that in the conditions which govern us today the replacement of professional courts by courts of the kind that existed in the remote past can be thought of. . . . We cannot see how the noble aims enshrined in the Preamble to our Constitution can ever be realized unless we have a hierarchy of courts, a competent judiciary and well-defined rules of procedure.[25]

While rejecting any fundamental change in system, the Commission indicates that to a limited extent it might be possible to utilize some of the simple features of judicial administraation that obtained in the past in the form of judicial *panchayats*. Reviving or establishing popular village tribunals had been recommended by many bodies throughout the British period and such tribunals had been established in some places. After surveying their accomplishments the Law Commission recommended the establishment of *panchayats* with simplified procedure and exclusive jurisdiction over petty matters.[26]

In the late 1950s the Government adopted the policy of community development, whereby elective village *panchayats* were established as instruments of village self-government in the hope that they would increase initiative and participation in economic development. The eager promotion of these administrative *panchayats* secured the acceptance of judicial *panchayats* in almost all states.[27] Either the administrative *panchayats* themselves or allied bodies, elected directly

---

[24] Government of India, Ministry of Law (1958: I, 29–30).

[25] Id., 30–1.

[26] Id., chap. 43.

[27] The basic policy study is Ministry of Law (Government of India, 1962). For a concise survey of developments, see Ministry of Food, etc. (Government of India, 1966).

or indirectly, were given judicial responsibilities in specific categories of petty cases. Almost uniformly lawyers were barred from appearing before these tribunals.

The establishment of judicial *panchayats* was officially urged in the hope of resolving the alienation of the villager from the legal system: '. . . by reviving panchayats and moulding them on the right lines we will be taking a much needed step in the direction of making law and administration of justice reflect the spirit of the people and become rooted once again in the people'.[28] Although the establishment of these judicial *panchayats* derived sentimental and symbolic support from the appeal to the virtues of the indigenous system, it should be clear that these new tribunals are quite a different sort of body than traditional *panchayats*. The new judicial panchayats are selected by popular election from clear territorial constituencies,[29] they are fixed in membership, they decide by majority vote rather than a rule of unanimity; they are required to conform to and apply statutory law; they are supported by the government in the compulsory execution of their decrees; these decrees may be tested in the regular courts.[30]

As might be expected, judicial *panchayats* enjoy little favour with bar and bench. They are largely ignored and disdained by lawyers and have been strongly criticized by judges.[31] In reviewing their work on appeal, courts have tolerated some departure from ordinary judicial procedures,[32] but they have also restricted the powers and discretion of *panchayats*.[33] In particular, the exclusion of lawyers in cases where a party has been arrested for a crime has been held unconstitutional by the Supreme Court.[34]

The reception of judicial *panchayats* by villagers awaits systematic

[28] Ministry of Law (Government of India, 1962: 13).

[29] The very selection of the *village panchayat* as the unit to have governmental support implies a great change from the traditional system in many places, where there were functioning caste tribunals in addition to, or instead of, village ones. The emphasis on the *village* panchayat represents an attempt to recreate an idealized version of the traditional society — an ideal not only based upon a picture of the older society that emphasizes harmony and unity, but also infused with the designers' animus toward communal units.

[30] On the contrast of the new statutory *panchayats* with traditional village *panchayats*, see Retzlaff (1962: 23 ff.); Luschinsky (1963b: 73).

[31] E.g., *Marwa Maghani v. Sanghram Sampat*, A.I.R. 1960 Punj. 35; *Venkatachala Naicken v. Panchayat Board*, 1952 M.W.N. 912.

[32] E.g. *Khachu Jagganath v. State of Madhya Pradesh*, A.I.R. 1964 M.P 239; *Shrikishan Kaskavam v. Dattu Shwaam*, A.I.R. 1953 Nag. 14.

[33] E.g., *Ram Prakash v. Nyaya Panchayat*, A.I.R. 1967 H.P. 4 (*Panchayat* cannot try party for insult to bench); *Lohare v. Civil Judge*, A.I.R. 1964 Raj. 196 (*Panchayat* must adhere to rules regarding size of bench); In re *S. Rengaswamy*, A.I.R. 1964 Mad. 435 (self-help by *panchas* to affect decree not permissible).

[34] *State of Madhya Pradesh v. Shobharam*, A.I.R. 1966 S.C. 1910.

study. Little is known, e.g. about the kinds of cases *panchayats* are hearing,[35] the classes that are using them, the law being applied, the impact of revisions by regular courts,[36] the development of local expertise.[37] It is often claimed that *panchayats* have reduced litigation in the countryside and it is clear that they have to some extent relocated it and probably made it less expensive.[38] But it is not clear that there has been any strong movement away from the courts in favour of *panchayats*.[39]

Like the traditional *panchayats*, the statutory ones seem to face severe problems of establishing their independence of personal ties with the parties,[40] of enforcing their decrees,[41] and acting as expeditiously as it was hoped they would.[42] A recent survey found it 'remarkable that rural respondents favoured a greater degree of

[35] Some useful works on the early years include Samant (1957); Purwar (1960); Government of Rajasthan (1964).

[36] The sheer quantity of intervention by the higher courts is not entirely clear. Purwar (1960: 228, 236) reports that over a six-year period in U.P. there was recourse to higher courts in 6.7 per cent of the cases decided by *panchayats*. But since many cases were compromised, transferred, or disposed of *ex parte*, it appears from his figures that 11 per cent of cases in which judgment was rendered were taken to higher courts. This may be higher, but it is certainly not lower than the general prevalence of appeals in the court system. See Government of India, Central Statistical Organization (1968: 526).

[37] E.g. it was reported that in Travancore-Cochin 'a class of professional agents attached to Village Courts has emerged. . . . ' All-India Congress Committee (1954: xxxiii). The Study Team on Nyaya Panchayats later identified this class as composed of lawyers' clerks (Government of India, Ministry of Law, 1962: 105).

[38] Such claims must be assessed in the context of a long-run decline in civil litigation in India. A preliminary examination of official statistics suggests that civil litigation in India has been declining since the 1930s. In 1961 the Indian courts decided 47 original civil cases per 10,000 population; in 1931 the courts of undivided India had decided 147 original civil cases per 10,000 population. Although criminal cases seem to have more than kept pace with population growth, the total per capita litigation in India was apparently lower in 1961 than at any time in this century.

[39] E.g. in a survey of government officers in the field conducted by the Rajasthan Study Team, almost two-thirds did not believe *nyaya panchayats* had led to a reduction litigation (Government of Rajasthan, 1964; 336)

[40] Robins (1962: 245). Malaviya (1956: 432). But cf. Ministry of Law (Government of India, 1962: 37).

[41] Purwar (1960: 225); Ministry of Law (Government of India, 1962: 105).

[42] Purwar (1960: 220); Government of Rajasthan (1964: 107). Cf. the urging of the Congress Village Panchayat Committee that cases be decided 'at one sitting' (All-India Congress Committee 1954: 38). It should be noted that the expeditiousness (and cheapness) of indigenous justice are at least partly legendary. For example, under the Mahrattas, cases concerning land tenure might take from two to twelve years and perhaps half of the cases were never decided. See 'Lunsden's report on the Judicial Administration of the Peshwas' (1819), reprinted in Gune (1953: 373 ff.). As to expense, see Gune (1953: 86, 131).

supervision and control by the Government ... over their functioning'.[43]

There is little reason to think of *panchayats* as a reassertion of local legal norms or institutions. It has been pointed out that administrative *panchayats* have tended to act as downward channels for the dissemination of official policies rather than as forums for the assertion of local interests as locally conceived. It is submitted that this is the case with judicial *panchayats* too. Rather than inspiring a resurgence of indigenous local law, they may serve as agencies for disseminating official norms and procedures and further displacing traditional local law by official law within the village.[44]

Judicial *panchayats* invite comparison with Paul Brass's findings about Indian medicine, which he sees undergoing a dual modernization, in which the growth of 'modern medicine' in the western style is accompanied by a 'revival' of the indigenous schools of medicine. He finds that this revival is really another stream of modernization which he calls 'traditionalization', i.e. a movement that uses traditional symbols and pursues traditional values, but engages in technological and organizational 'modernization'.[45] Village *Panchayats* fit Brass's model of 'traditionalization': the traditional *panchayat* symbolism and values of harmonious reconciliation and local control and participation are combined with many organizational and technical features borrowed from the modern legal system— statutory rules, specified jurisdiction, fixed personnel, salaries, elections, written records, etc. The movement to *panchayats* then is not a restoration of traditional law, but its containment and absorption; not an abandonment of the modern legal system, but its extension in the guise of tradition.

Why was the movement for indigenous justice so readily contained? Why did the proponents of indigenous law settle for what is hardly more than the marginal popularization of existing law?

[43] G. S. Sharma (1967: 19). On the need for such external controls, see Nicholas and Mukhopadhyay (1962: 17, 24).

... despite the form of justice in these proceedings [in a traditional village *panchayat*] the odds appear to be strongly against the defendant. Everyone, including the 'judge' is trying for a conviction. ... [T]he interest of the villagers is generally in raising money for the village fund, for trials 'represent an important source of income for the village collectivity'.

[44] Indeed, one may hazard the speculation that eventually the deficiencies of the *panchayats* will be remedied by training them in the law. Cf. Malaviya (1956: 432). The Study Team proposed paid and trained secretaries 'familiar with the law to be administered by the Nyaya Panchayats. ...' Ministry of Law (Government of India, 1962: 109–10). Cf. their proposals for training *panchas* (Id.: 65 ff).

[45] Brass (1969).

Let us look at the actors, goals and issues in the 'dispute'.

First, the present legal system was supported by a numerous and influential class of lawyers (not to mention the many ancillary personnel who live off the legal system) who were fully committed to the system, had a heavy personal stake in its continuance, and were genuinely convinced of its general virtue[46] and that revivalism was a threat to them as well as a mistake for the country.[47] On the other hand, there was no organized body of carriers of the proposed alternative, no educational institutions to produce them and no existing group whose occupational prospects might be advanced. (In all of these, traditional law stands in contrast to *ayurvedic* medicine.)[48] The proponents of indigenous law tended to lack qualification as experts while the lawyers were generally recognized as competent authorities on legal reform.

Second, the proponents of an indigenous system presented no vivid alternative. Contrast, for example, movements for replacing one language with another, where there is an alternative that is palpable to all and clearly promises advantage, symbolic if not tangible, to many. Here, the proponents themselves were not moved by a lively sense of what the alternative might be. In part this reflected the absence of any plausible candidate — this was a restorationist movement without a believable pretender!

Dharmaśāstra, of course, was one alternative, an elaborate and sophisticated body of legal learning. But any proposal in this direction would run foul of some of independent India's most central commitments. It would violate her commitment to a secular state, insuring equal participation to religious minorities. Furthermore, *dharmaśāstra's* emphasis on graded inequality would run counter to the principle of equality and would encounter widespread opposition to the privileged position of the higher castes. Indeed, the one area where *dharmaśāstra* retained some legal force, Hindu family law, was in the early 1950s being subjected to thoroughgoing reform

[46] An eminent advocate (now Vice-President) who had been a member of the 1958 Law Commission recalled five years later that '. . . after a comparative study of the various systems prevailing in other countries [w]e reached the conclusion that the British system which we had adopted was the best. This system secured greater and more enduring justice than any other system'. He went on to warn that 'Ideas from the foreign countries may be borrowed and adopted in our system. But it will be dangerous to introduce innovations which will result in radical changes. They may not fit in with our system which had been a part of our national life for a long time' (G. S. Pathak in Sen *et al.*, 1964: 80–1).

[47] On the size, eminence and influence of the legal profession, see Galanter (1968–9) and the various contributions to *Law & Society Review* (1968–9).

[48] In spite of the parallels on the consumer side (cf. footnote 59), the professional organization of 'traditional' law remains strikingly in contrast with that of traditional medicine, with its parallel and imitative professionalization. On the contrast, see Galanter (1973).

which largely abandoned the *śāstra* in favour of a Hindu law built on modern notions.[49] Thus, it is hardly surprising that none of the leading documents supporting *nyaya panchayats* even mentions *dharma-śāstra*. While few would condemn it (as had an earlier generation of reformers)[50] claims on its behalf were limited to the symbolic and intellectual levels.[51] It was not an available alternative for practical application.

Nor was the local customary component of traditional law a likely candidate. Customary law with its innovative, quasi-legislative element restored would be a formidable counterweight to national unity, mutual intelligibility, free movement and interchange. It had no evident capacity to contend with nationwide problems and projects. Although there were vestiges of such traditional law extant, there was no pronounced widespread admiration for its contemporary representatives. Indeed they were in at least as bad a popular repute as the courts.[52]

Yet another alternative, a creative synthesis of Indian and Western, blending the best of both into a new system adapted to India's needs and aspirations, was easy to call for, but hard not only to produce, but to portray.[53] Such a call was not without appeal to

[49] On these reforms, see Derrett (1958); on the crucial role of lawyers in producing them, see Levy (1968–9).

[50] The prominent exceptions are in neo-Buddhist, Scheduled Caste and 'non-Brahmin' (DMK) circles where Manu and *dharmaśāstra* are negative symbols. See e.g. Borale (1968). For a famous example of an earlier view that *dharmaśāstra* was suited only to a stagnant and slave society, see Sankaran Nair (1911: 216).

[51] Thus a High Court judge calls for a renaissance of *dharmaśāstra* study to check the excessive rights-consciousness of India's educated classes and because its emphasis on duty is more consonant with socialism and India's urgent needs (Dhavan 1966: 102–3, 136–7). More common is the assimilation of *dharmaśāstra* to the modern system by stressing its adaptiveness, change over time and the extent to which it anticipated features of the modern system.

[52] For example, Berreman (1963: 271–2, 281–2) reports that villagers have little faith in the objectivity of *panchayats* and avoid using them, especially in property disputes, on grounds that they would decide wholly in terms of self-interest. Contemporary *panchayats* seem to experience the same difficulties with tutored witnesses as do the official courts. Srinivas (*c.* 1964: 42, 95); Lynch (1967: 154). And most notably these tribunals may be oppressive to the poor and powerless and unable to enforce decisions opposed to the interests of powerful personages or factions. Srinivas (*c.* 1964: 66); Nicholas and Mukhopadhyay (1962); Hitchcock (1960: 262).

[53] Inaugurating a seminar on jurisprudence, the Governor of Rajasthan projected such a synthesis into the future:

... the Indian law of today ... is not a spontaneous growth at all, but an exotic growth, planted on Indian soil forcibly, by foreign rulers, not because this suited us but because it suited them. ... There are symptoms of a revolt against such foreign impositions. The spirit of the country seeks to go back to the fountainhead of its life. ... Fifty years hence ... India may have switched on to an entirely new system of laws, based on legal principles which have intimate contacts with the spiritual and cultural life-currents of the country. ... We should not for a moment

lawyers, but they were disinclined to abandon the existing system pending its arrival.[54]

Third, the revivalist cause was not attached to any concrete grievance that could mobilize popular support. Court delay and expense were not adequate issues for this purpose. It was hard to attach organized personal or political ambition to them. Apparently the symbolic gratifications to be had from restoration of indigenous law were not sufficiently appealing to any significant sections of the population. On the other hand, the interests that were threatened were concrete and tangible and defended by organized and articulate groups.

We may ask then why actors, goals and issues were in such short supply, compared, for example, to linguistic changeover movements or even the *ayurvedic* movement described by Brass. Ironically it appears that the answer is that the legal system was so thoroughly domesticated. That is, indigenization on the ideological/programmatic level failed because the law had become 'indigenous' on the operational/adaptive level.

The law and the society had adapted to each other in several ways. The law itself underwent considerable adaptation. British institutions and rules were combined with structural features (e.g. a system of separate personal laws) and rules (e.g. *dharmaśāstra*, local custom) which accorded with indigenous understanding. The borrowed elements underwent more than a century and a half of pruning in which British localisms and anomalies were discarded and rules elaborated to deal with new kinds of persons, property and transactions.[55] By omission, substitution, simplification, elaboration, the law was modified to make it 'suitable to Indian conditions'.[56]

ignore Dharmasastra. Let us rather seek to derive sustenance from it. I am not asking you to boycott the light that comes from the West: I want you to blend it with what has come down to us from our own past (Sampuranand, 1963: 4).

[54] 'It would be folly to throw away what we have acquired and start a search for something which may prove elusive and which may result in atavistic retrogression. . . . We must retain the present system as long as we are unable to replace it . . . with a superior one . . . that is at the same time more acceptable to . . . our sense of justice . . . our common man' (Misra, 1954: 49). Cf. Ramachandran (1950: 53).

[55] Special adaptations of common law to suit Indian conditions include, e.g. in the criminal law the elaborate protection of religious places and feelings, the different treatment of bigamy, adultery, false evidence and defamation; in contract law, treatment of duress and agreements in restraint of trade. See Lipstein (1957: 74–5); Acharyya (1914: Lecture III); Setalvad (1960: chaps. 2–3). Examples of distinctive 'Indian common law' may be found in Setalvad (1960: 59–60); Acharyya (1914: 38, 136).

[56] Derrett's (1969) provocative assessment of the carryover of traditional elements in contemporary Indian law includes a series of interesting examples of ways in which modern legislation (on, e.g. safety, welfare and employment) gives expression to traditional normative concerns.

The numerous body of legal professionals were, almost without exception, so thoroughly committed to the existing system that it was difficult for them to visualize a very different kind of legal system.[57] Its shortcomings were seen as remediable defects and blemishes, not as basic flaws which required fundamental change. To lawyers the system seemed fully Indian. This sense of being at home is expressed by an eminent Attorney General:

> For over a hundred years distinguished jurists and judges in India have, basing themselves upon the theories of English common law and statutes, evolved doctrines of their own suited to the peculiar need and environment of India. So has been built up on the basis of the principles of English law the fabric of modern Indian law which notwithstanding its foreign roots and origin is unmistakably Indian in its outlook and operation.[58]

The lawyers not only disseminated the official norms, but served as links or middlemen, putting the law in the service of a wide variety of groups in the society and providing new organizational forms for forwarding a variety of interests. The lawyers were the carriers of what we might call an all-India legal culture which provided personnel, techniques and standards for carrying on public business in a way that was nationally intelligible. Thus the legal system and the lawyer supplied much of the idiom of public life. To the lawyers and the nationally oriented educated class of whom they formed a significant part, the legal system was the embodiment and instrument of the working principles of the new India — equality, freedom, secularism, national uniformity, modernity.

In a very different way, villagers are also at home in this legal system. At least, they are neither as radically isolated from the system nor as passive as they appear to some critics of the present system. Villagers are, as Srinivas has observed 'bi-legal'; they utilize both 'indigenous' and official law in accordance with their own calculations of propriety and advantage.[59] Summing up the ethnographic evidence, Cohn finds that:

· [57] Rowe (1968–9). In the course of several dozen interviews I conducted with Indian lawyers in 1965–6 and in 1968, I could uncover no sentiment for fundamental changes. The defects perceived in the present system were expense, delay and corruption. These were seen as the result of human failing, not as a result of the nature of the legal system. Defects were never attributed to the foreignness of the system, even by those lawyers who were devout Hindus or were connected to Hindu communal political organizations.

[58] Setalvad (1960: 225).

[59] Srinivas (*c.* 1964); Ishwaran (1964). Cf. Berreman (1963: 271) on villagers picking and choosing among tribunals. Similar 'forum shopping' among alternative treatment systems and popular syncretism in combining them has been observed in Indian consumption of medical services (Leslie, 1970). A recent study in Ghaziabad by T. N. Madan (1969) found that over 80 per cent of his respondents preferred allopathic medicine, but over two-thirds combined treatments from various schools.

Even though there are inadequacies of procedure and scope for chicanery and cheating, the lack of fit with indigenous jural postulates notwithstanding, the present court system is not an alien or imposed institution but part of the life of the village. Looked at from the perspective of the lawyer's law and that of the judges and the higher civil servants, the ability of some peasants to use the court for their own ends would appear a perversion of the system. However, looked at from the ground up it would appear that many in the rural areas have learned to use the courts for their own ends often with astuteness and effectiveness.[60]

The displacement of indigenous law from the official legal system does not mean the demise of traditional norms or concerns. The official system provides new opportunities for pursuing these, at the same time that it helps to transform them.[61] There is no automatic correspondence between the forum, the motive for using it and the effect of such use; both official courts and indigenous tribunals may be used for a variety of purposes.[62] Official law can be used not only to evade traditional restrictions, but to enforce them.[63] Resort to official courts can be had in order to disrupt a traditional *panchayat*,[64] or to stimulate it into action. Official law can be used to vindicate traditional interests;[65] caste tribunals may be used to promote change.[66]

Not only are villagers capable of using the official courts for their own ends, but they have assimilated many elements of official law into the workings of indigenous tribunals. Nicholas and Mukhopadhyay, in their study of two Bengal villages, report that 'almost all persons have had some experience in . . . the government court, and the form of village legal proceedings is modelled after this experience. Stress is laid upon evidence such as eye-witness accounts, written documents, markings of injury, correct description of a stolen article'.[67] Srinivas's studies indicate that even in villages where there

---

[60] Cohn (1965: 108–9).

[61] On the role of law (and lawyers) in providing new modes of caste organization and activity, see McCormack (1966). More generally, Derrett (1968) argues that in spite of superficial discontinuities, modern legal concepts and institutions provide a vehicle for the authentic continuation of Hindu tradition.

[62] 'The use of the courts for settlement of local disputes seems in most villages almost a minor use of the courts. In Senapur, courts were, and are, used as an arena in the competition for social status, political and economic dominance in the village. Cases are brought to harass one's opponents, as a punishment, as a form of land speculation and profit making, to satisfy insulted pride and to maintain local political dominance over one's followers. The litigants do not expect a settlement which will end the dispute to eventuate from recourse of the State courts' (Cohn, 1965: 105).

[63] E.g. Srinivas (*c.* 1964: 90).

[64] E.g. Lynch (1967: 153).

[65] Derrett (1964b); Ishwaran (1964: 243); Luschinsky (1963a).

[66] Cohn (1965: 98–9, 101).

[67] Nicholas and Mukhopadhyay (1962: 21).

is little recourse to government courts, the form of the dispute within a *panchayat* seems affected by official models in drafting of documents, keeping of records, terminology and procedure.[68]

Traditional law, either absorbed into the official system or displaced from it, has been transformed along the lines of the official model. As I have argued elsewhere, the attrition of traditional law resulted not from the normative superiority of British law, but from its technical, organizational and ideological characteristics, which accomplished the replacement and transformation of traditional law half inadvertently.[69]

Though spoken in accents grating to some, the present system is India's legal vernacular. Like many Indian languages it is characterized by functional diglossia. Overlapping formal and colloquial varieties form a multiplex medium through which interests are pursued and issues are perceived by various groups and strata. Its replacement would require not only a concrete alternative and specialists to implement the change, but powerful political support. Not only are alternatives and specialists in short supply, but the political support that could accomplish such a change is unlikely to assemble around this issue. Therefore, a qualitative change on the scale of the shift from pre-British to British legal institutions (or of a linguistic changeover) is highly unlikely.

The legal system then is 'indigenous' in quite different ways for the lawyer, the nationally-minded educated classes, and the villager. It is most palpably foreign to those sophisticated urbanites who attempt to view it through the eyes of the villager. (And, perhaps to the sophisticated foreigner who attempts to see it through Indian eyes.)

In the sense of unease and dismay that Indian law provokes in otherwise disparate observers[70] we may discern several components: its lack of an indigenous or autochthonous quality; its lack of congruence with under lying social norms; its internal disparities (i.e. the law in operation is remote from the law on the books). Modern India is measured against societies in which law is supposedly an accurate and coherent expression of social values — Britain for

---

[68] Srinivas (*c.* 1964: 46); Ishwaran (1964); Lynch (1967: 153–4).

[69] Galanter (1966a, chap.2 above). The introduction of new opportunities into India's compartmented society generated numerous disputes that were not resolvable by the earlier decentralized dispute-settling mechanism, which relied on local power for enforcement and enjoyed only intermittent and remote external support. British law and courts fostered and filled a demand for near-at-hand authority that could draw upon power external to the immediate setting of the dispute.

[70] See footnotes 11–13. For a recent estimate of the Indian situation see von Mehren (1963a, 1963b). Cf. Derrett (1969: 11). For a broad comparative assessment, see Pye (1966: chap. 6).

British administrators,[71] traditional India for the Gandhians, the 'West' for comparative scholars. The Indian situation is perceived as deficient or even pathological; prognoses range from stress and demoralization to rigidity and obstruction of development.

The Indian experience provides an occasion for questioning the familiar notions that underlie these judgments, notions of what is 'normal' in legal systems; that law is historically rooted in a society, that it is congruent with its social and cultural setting, and that it has an integrated purposive character. These notions express expectations of continuity and correspondence, of present with past, of law with social values, of practice with precept; expectations which are in part projections of the working myths of modern legal systems.[72]

The Indian experience suggests a set of counter-propositions. It suggests that neither an abrupt historical break nor the lack of historical roots prevents a borrowed system from becoming so securely established that its replacement by a revived indigenous system is very unlikely. It suggests that a legal system of the modern type may be sufficiently independent of other social and cultural systems that it may flourish for long periods while maintaining a high degree of dissonance with central cultural values. It suggests that a legal system may be disparate internally, embodying inconsistent norms and practices in different levels and agencies.

These counter-propositions point to the need for some refinement of familiar notions of what legal systems are normally like. Specifically they point to the desirability of disaggregation: we need to find ways of asking how various parts of the legal system are related to different sectors of the society.[73]

If a legal system need not be historically emergent from its society, what are the mechanisms by which it becomes 'rooted'? How does it secure acceptance and support from crucial sections of the popula-

[71] Cf. Bernard Cohn's observation that British administrators in India had often left England after their schooldays and preserved a rather idealized picture of English life against which they measured the Indian reality.

[72] They may also derive support from theories generated by the study of small and relatively homogeneous societies. Hoebel (1965: note 54) suggests that the theory which holds that law is the expression of jural postulates, which are in turn expressions of fundamental cultural postulates, 'is limited to autochthonously developed legal systems. It obviously cannot apply to a tyranically or conquest-imposed system that has no roots in the local culture'. This seems to point to the rather startling conclusion that law in such places as India has a fundamentally different character than in primitive societies or in the West.

[73] See, e.g. Morrison's (1972) description of the articulation of professional activity with kinship and Khare's (1972) description of the selective insulation and 'restricted mutual validation of lawyer's law and traditional institutions.

tion? If a legal system can persist without pervasive support from other social institutions (or global agreement with cultural norms), what are the specific links that connect it with other institutions and norms and what are the mechanisms that maintain its segregation from them? If a legal system is not itself a normative monolith, what are the mechanisms that permit a variety of norms and standards to flourish? How are widely disparate practices accommodated? It is submitted that the discontinuities observed in the Indian case should not be dismissed as exceptional or pathological, but should be taken as the basis for hypotheses for probing some of the general characteristics of legal systems that are often obscured in our view of societies closer to home.

4

Panchayat Justice: An Indian Experiment in Legal Access*

(with Upendra Baxi)

Introduction

India's massive attempt to provide judicial access to hundreds of millions of villagers through the promotion of Nyaya Panchayats[1] (village courts) is theoretically provocative as well as practically important. Yet it remains virtually unexamined in India as well as unknown abroad.

Traditional Panchayats

Panchayats, which have existed in India for thousands of years, are a characteristic and distinctive institution of Indian civilization. Literally the term means the 'coming together of five persons', hence, a council, meeting or court consisting of five or more members of a village or caste assembled to judge disputes or determine group policy. Although Indian civilization contained refined and respected bodies of legal learning — the *Dharmasastra* (Hindu law) and later Muslim law as well — and although there were royal courts in administrative centres, these did not produce a unified national legal system of the kind that developed in the West. The textual law influenced but did not displace the local law. The government's

* Reprinted from Mauro Cappelletti and Bryant Garth, eds., *Access to Justice*, Vol. III *Emerging Issues and Perspectives* (Milan: Giuffre; Alphen aan den Rijin: Sijthoff and Noordhoff, 1978). This paper derives in part from an earlier contribution of Professor Baxi, published in *18 Journal of the Indian Law Institute* 375–430 (1976). The authors would like to express appreciation to Catherine Meschievitz of the University of Wisconsin for her capable assistance in the preparation of this paper.

[1] A note on terminology: Panchayati Raj (regime of *panchayats*), abbreviated in this paper as PR, refers to the Indian government's policy of promoting elective village *panchayats* as units of local self-government. The judicial branch of these are *nyaya* (Sanskrit for justice) or (in some states) *adalati* (Persian for court) *panchayats*.. We use the term Nyaya Panchayats throughout to refer to such judicial tribunals established by government policy and abbreviate it as NP.

law did not penetrate deeply into the countryside. Throughout most of Indian history there was no direct or systematic State control of the administration of law in the villages where most Indians lived.[2]

In pre-British India there were innumerable, overlapping local jurisdictions, and many groups enjoyed some degree of autonomy in administering law to themselves. Disputes in villages and even in cities would not be settled by royal courts, but by tribunals of the locality, of the caste within which the dispute arose, or of guilds and associations of traders or artisans. Or, disputes might be taken for settlement to the *panchayat* of the locally dominant caste or to land-owners, government officials or religious dignitaries.

Some *panchayats* purported to administer a fixed body of law or custom; some might extemporize. In some places and some kinds of disputes, the process was formal and court-like. Some *panchayats* were standing bodies with regular procedures, but many of these tribunals were not formal bodies but more in the nature of extended discussions among interested persons in which informal pressure could be generated to support a solution arrived at by negotiation or arbitration.[3] These tribunals would decide disputes in accordance with the custom or usage of the locality, caste, trade or family. Custom was not necessarily ancient or unchangeable; it could be minted for the occasion. The power of groups to change customs and to create new obligatory usages was generally recognized.

Rulers traditionally enjoyed and occasionally exercised a general power of supervision over all these lesser tribunals. In theory, only the royal courts could execute severe punishments. These lesser tribunals could pronounce decrees and invoke royal power to enforce them. But while some adjudications might be enforced by governmental power, most depended on boycott and excommunication as the ultimate sanctions. Community enforcement of these sanctions therefore, had to reflect a high degree of consensus.[4]

The Constitutional Setting

India is a highly stratified and heterogeneous society. It is a rural society and an overwhelmingly poor one. In 1971, 438 million out of a population of 547 million lived in about 570,000 villages. Agriculture accounts for over 50 per cent of the national product.

---

[2] On the pre-modern Indian legal system see Derrett (1968); Sen-Gupta (1953); Ahmad (1941); Gune (1953); Kane (1930–62). On the transition to the present system see Galanter (chap. 2; Derrett (1968); Rudolph and Rudolph (1965).

[3] Thus Cohn (1965: 90) prefers to regard the *panchayat* as a 'process' rather than a body.

[4] Further discussion of traditional dispute handling in *panchayats* in rural India is in Cohn (1965); id. (1961); Rudoph and Rudolph (1965); Hitchcock (1960); Srinivas (1962: 112–19); Nicholas and Mukhopadhyaya (1962).

Most people in rural areas live in dire poverty.[5] The efforts to provide access to justice that we shall discuss must be seen in the context of the need for development to eliminate or at least reduce this poverty, as well as for integration of a diverse population. Such efforts at development have taken place under a constitution which establishes the Western type of competitive liberal democracy through adult suffrage, guaranteed fundamental rights and judicial review.[6] Political democracy, as many observers have noted, provides a setting (and indeed a vehicle) for economic development that did not exist when Western societies embarked on industrialization and which differs from that obtaining in most nations that have gained independence since 1947.

The Indian Constitution, drafted during the turbulent period 1947–9, explicitly gives the Indian State a vast mandate for economic and social planning. It is the overarching obligation of the state to pursue the 'welfare of the people' through 'the promotion of a social order' in which 'justice, economic, political, and social, shall inform all aspects of social life' (Article 38). The guarantees of Fundamental Rights (Part III) and the enunciation of the Directive Principles of State Policy (Part IV) as 'fundamental to the governance of the country', imposing a duty on the state to apply them in the 'making of laws', together represent the values and aspirations of the constitutionally desired social order.[7] Contrary to Gandhian pleas for a decentralized, village-based, democratic format, it is the State which is entrusted with the massive task of social and economic planning and development.[8]

---

[5] Estimates of rural poverty vary, but even the most conservative suggest a very high incidence of poverty. One generally accepted estimate is that about 45 per cent of rural people are poor at a poverty line of Rs. 15 monthly per person, which (at 1960–1 price levels) was just adequate to provide a nutritional minimum of 2,500 calories per day.

[6] The probability that courts may be activated to delay or defeat certain developmental measures has led the Indian Parliament to assert its supremacy in the sphere of constitutional amendment, including a readiness to use the unique device of the Ninth Schedule, which immunizes the statutes inserted therein from any challenge on the ground that they violate certain Fundamental Rights. *See also* note 23 *infra*. This has, of course, led to a revision of initial constitutional assumptions. Liberation from the constraints of judicial review may be justified by an appeal to the need for rapid social change and even as a *necessary* condition for such change. Even if this is so, such liberation does not constitute a sufficient condition for planned social change. The extent to which such liberation has in fact generated legislative activity and accompanying executive action remains an urgent matter for empirical study.

[7] Austin (1966); Baxi (1967).

[8] Austin (1966: 26–46); Baxi (1967: 339–44); Galanter (chap.3). The constitutional preference for a liberal competitive polity as a framework for planned social transformation has resulted in what Gunnar Myrdal has aptly described as the 'Third World of Planning'. Indian planning does not follow the historic Western model in which state intervention arose as a consequence rather than a precondition

In a sense, India's planning ideology and practice has been one long story of seeking *access* to the village people, and their participation in the formulation and attainment of development goals. But all this had to be achieved in conditions different from Western or socialist societies. By and large, India had to create an institutional infrastructure in her quest for public participation.[9] The Indian quest for public participation (or the State's access to its people for the tasks of development) has led to remarkable experimentation in the form of co-operative movements in village areas, programmes of community development, and attempts at 'democratic decentralization' through the system of Panchayati Raj (PR). All this has no doubt led to a growing politicization of village life, a transformation of caste functions, a proliferation of bureaucracy, and an emergence of new bases of power and influence in competition with the existing structures of domination. Although informed opinion remains sharply divided on the scope, rate and adequacy of change and the suitability of the forms of polity within which the transformation is taking place, contemporary India has certainly been transformed by these attempts. Our account of *panchayats* as a means of access to justice can only be understood in the context of these attempts to transform village life and link the villages to the polity. But these in turn are built on a history of pre-independence efforts to promote self-rule in the villages.

Pre-independence Attempts to Revive Panchayats

Continuing efforts to reorganize rural self-government through *panchayats* include the Mayo Resolution of 1870 on Decentralization,

---

for development. It also differs from the socialist prototype where 'state enterprise and collectivism' is the rule by its democratic emphasis. Myrdal (1968: 738–40). Another key difference between India and 'Western countries as they stood on the threshold of' industrialization is her 'commitment to egalitarianism', which is an 'integral part of the ideology of planning'. This ideology is a 'radical variant of the modern Western concept of the advanced welfare state. Economic development is defined as a rise in the levels of living of the masses'. Social equality is a cherished goal. Id. at 741.

9 The dilemma is an acute one:

The South Asian countries are in a hurry and need the modern infrastructure in order to mobilize popular support for planning and development. They cannot wait for an infrastructure to emerge from below. . . . There is no choice but to create the institutional infrastructure by government policy, and to spur its growth by government intervention. . . . The question is: Can this be done? And if it can be done — without a monolithic state, a disciplined ruling party and a network of cadres — will not the resulting infrastructure nevertheless tend, at least initially, merely to channel a stream of influence from above and thus contradict the entire ideology of democratic planning? . . . [I]s not the ideal of democratic planning rather an illusion that will weaken the entire effort? Id., 869.

Lord Ripon's famous resolution of 1882, the Report of the Royal Commission on Decentralization, the Government of India Resolution of 1915, and the Montague-Chelmsford Report of 1918. All these efforts were 'nowhere intended to reproduce the characteristics of the old-time panchayats'.[10] Indeed, they were based on the view that revival of ancient *panchayats* was neither necessary nor possible.

The 1911 census had directed particular attention to the question of the persistence of the 'ancient' *panchayat* system. The Bombay census found no evidence of 'such an organization as village panchayat', adding that the 'myth' of their existence has 'probably arisen from the fact that a village is generally if not invariably formed by families of one caste'. The census for the United Provinces similarly concluded that the *panchayat* was 'entirely an organ of caste government'. The government accepted this evidence[11] and proceeded to form *panchayats*, as units of rural local government primarily for the purpose(s) of providing 'a rudimentary municipal framework for large villages and small towns' and/or to form a 'simple judicial tribunal'.[12]

The development of village government from 1920 to 1947 consisted primarily of the creation of *panchayat* bodies blending 'municipal/administrative' and 'judicial' functions. Except in the United Provinces (U.P.), village councils were elected bodies; in some the franchise was limited by literacy tests. There were no doubt certain *ex officio* members including the village headman, nominees of government, certain classes of landlords (in Bombay), and — after 1939 — women, communal groups and untouchables. The development of *panchayats* varied from one province to another according to such factors as taxing powers, 'enthusiasm or indifference' shown by officials or people, and ecological conditions. Professor Tinker, in an authoritative survey, concluded that:

> The expansion of village councils did not fulfil expectations in most provinces. A complete network of village authorities was built up in Bengal, and was later established throughout wider areas of Madras; one quarter of the rural population of the United Provinces was brought within the panchayat's orbit; in Punjab, Bombay and C.P. [Central Provinces] they covered only about one-tenth or fifteenth of the countryside, and in other provinces village councils affected only an insignificant fraction of people.[13]

The administrative tasks entrusted to *panchayats* were really minor (upkeep of country roads, village streets, minor sanitation, lighting

10 Tinker (1954: 99).
11 Id.,
12 Id., 197.
13 Id., 199.

by oil lamps, etc.), though in the area of education, especially in Madras and Bengal, some major responsibilities did lie with rural bodies.[14] Some other functions like water supplies and medical (usual *ayurvedic*) services were also significant and aroused some enthusiasm.

Judicial functions seem pre-eminent in the available accounts of the *panchayat* institutions. These functions were sometimes performed by special village courts (as in Madras) or by 'ordinary territorial panchayats. . . . The ordinary panchayats of C.P., Punjab and U.P. were mainly occupied with judicial work.' In Bengal 'there were special union benches to try criminal offenses and union courts to which civil suits might be taken'; by 1937 'there were 1,521 union benches and 1,338 union courts'. The village courts were usually elected by villagers though in some cases there was indirect election and nomination.[15]

Professor Tinker, in particular, has drawn our attention to the pre-eminently judicial character of some of the *panchayats*. Under the U.P. Panchayat Act, 1920, the 'principal function of the panchayat was to act as a petty court'; and the Bombay Village Panchayat Act of 1920 was 'broadly similar in purpose' though 'quite different in detail'.[16] Judicial functions were pre-eminent in the ordinary *panchayats* of Central Provinces, Punjab and the 'union' courts in Bengal and 'village courts' in Madras. The dimensions of judicial activity of *panchayats* are indicated in the following fragmentary account. In U.P., the *panchayats* disposed of 122,760 cases in 1925 (two-thirds being civil cases); the subsequent pattern is one of steady decline: 1931, 91,476 cases; 1936, 85,399 cases; 1937, 67,233 cases. In Bengal, by contrast, the union benches and courts handled an increasing volume of disputes: 120,000 in 1929 and 174,000 in 1937. There was a considerable amount of judicial work in other provinces, but not on 'such a huge scale'.[17] Income from fines was among the important sources of revenue for village *panchayats*, which had during this period no taxing powers.

Constitution-making and Village Panchayats

The draft constitution of India did not contain any reference to villages and was subjected to the criticism that 'no part of it repre-

---

[14] During the Dyarchy years (1919–35), the ideal variously defined as 'rural development' or 'uplift' or 'rural reconstruction' was espoused in every Indian province. The 'movement' aimed at rejuvenation of village life in all its aspects — through *ad hoc* committees; the district boards and *panchayats* had very little share in it' and when the co-operation of village authorities was called for, they were utilized as agents of officially sponsored schemes, rather than approached as autonomous local bodies.'

[15] Id., 206–7.

[16] Id., 117.

[17] Id., 207.

sents the ancient polity of India'. Dr Ambedkar, the chief drafts-
man, vigorously defended the omission of villages, and ruffled the
feelings of many in the Constituent Assembly by stating bluntly:
'I hold that those village republics have been a ruination of India. . . .
What is a village but a sink of localism, a den of ignorance, narrow-
mindedness and communalism?' In response, Mr H. V. Kamath
dismissed Ambedkar's attitude as that of an 'urban high-brow' and
insisted that 'sympathy, love and affection' toward 'our village and
rural folk' was essential for the 'uplift' of India. Mr. T. Prakasam
pleaded for a modernized system of *panchayats* which will give 'real
power to rule and to get money and expend it, in the hands of the
villagers'. Professor N. G. Ranga asked, 'Without this foundation-
stone village panchayats, how would it be possible for our masses
to play their rightful part in our democracy?'[18]

Gandhi himself had urged a different form of polity for India.
Gandhi stated his ideal of village *swaraj* (self-rule) in pragmatic as
well as poetic terms. The ideal village will be self-sufficient, he said,
in food and cotton; it will have its reserve for cattle and playgrounds
for children, its own waterworks, and sanitation arrangements (to
which he attached very great importance). 'The government of the
village will be conducted by the Panchayat of five persons' annually
elected by adult villagers. The 'panchayats will be the legislature,
judiciary and executive combined'; there will be 'no system of
punishments in the accepted sense' as 'non-violence with its techniques of satyagraha and non-co-operation will be the sanction of the
village community'. Gandhi concluded his 'outline' of village
government thus:

> Here then is perfect democracy based upon individual freedom. The
> individual is the architect of his own government. The law of non-violence
> rules him and his government. He and his village are able to defy the
> might of a world.[19]

Neither Gandhi nor his followers were unaware that they were
positing an ideal picture of village self-sufficiency and democracy
which had no prospect of acceptance. They knew, as well as the
modernists, that even if such village republics had existed in the
distant past, or indeed even up to Moghul rule, they were vastly
affected by the 'anarchy' following the dissolution of the Moghul
empire, the growth of transport and communication, the spread of
commerce and the organization of markets, patterns of revenue
settlement, the penetration of the bureaucracy, the introduction of
British justice, irrigation, roads, education, etc.[20]

[18] Venkatranigah and Patabiraman, eds. (1969: 248–53); Austin (1966).
[19] Venkatranigah and Patabiraman, eds. (1969: 247–8).
[20] Id., 5–9; Tinker (1954: 1–26).

Gandhi compared *gram-rajya* to *Ram-rajya* (i.e. self-rule by villagers, to the righteous polity of Lord Rama). His advocacy of this kind of polity on the eve of constitution-making served certain clear purposes. He was 'making it possible for the traditional elite to assert its values both in the Congress Party and in the Constituent Assembly, thus hoping to create an atmosphere in which over-Westernization of the projected political system can be consciously corrected'.[21] Gandhi was in this view 'aiming to create a healthy division within the party, one wing of which will at least be dedicated to bringing about the social, moral and individual transformation from below while the organizational and political wings engaged themselves in a frontal assault on the traditional society'.[22] The Constitution as it emerged did include certain village-oriented Directive Principles of State Policy.[23]

Article 40 obligates the state to 'take steps to reorganize village *panchayats* and endow them with such powers and functions as may be necessary to enable them to function as units of self-government'. Subsequent developments in the area of democratic decentralization, though not realizing the Gandhian utopia, owe much to the overall ideology.

Democratic Decentralization

Since the adoption of the Constitution there have been continuing efforts at 'democratic decentralization' — insistence on public participation by villagers in formulating and implementing, at various levels, planned social change. Those efforts have generated the formation of rural credit and service co-operatives, co-operative farming experiments, community development programmes, agricultural extension services and, beginning with 1959, the institution of PR.

Article 40 has been among the most vigorously implemented provisions of the Indian Constitution. The directive to organize village *panchayats* and to empower them to function as units of self-

[21] Baxi (1967: 343).

[22] Id.

[23] Article 48 urges the State to 'endeavour to organize agriculture and animal husbandry on modern and scientific lines'. To promote the well being of agricultural workers, Article 42 exhorts the state 'to endeavour to promote cottage industry on an individual or cooperative basis in rural areas'. The state is directed, by Article 46, to 'promote with special care' the 'educational and economic interests of the weaker sections of the people, and in particular of the scheduled castes and tribes', who shall be protected against injustice and 'all forms of exploitation' (On the status of these Directive Principles, *see* note 41 *infra*). Article 31A, in a thoroughgoing derogation of the fundamental right to property guaranteed by Articles 19 and 31, enables the state to immunize certain laws against challenges of violation of fundamental rights by placing them in a special Ninth Schedule to the Constitution.

government saw a steady fulfillment in the period 1949–59, which was followed by the introduction of the PR system in the period 1959–62. Shortly after the passing of the Constitution, a number of states which did not have statutory village *panchayats* enacted legislation providing for them; and states which had had *panchayat* legislation proceeded to strengthen it. *Panchayats* were given a wide variety of developmental and regulatory (local governmental) functions. The number of *panchayats* increased from 14.8 thousand to 164.3 thousand during the first ten years of independence.[24] The Five Year Plans emphasized the role of *panchayats* in community development — the *panchayats* were represented on bloc advisory committees; modest funds were made available to them for specific projects, and they were considered to be the principal agencies for creating public awareness and participation in developmental work.

The adoption of the *Balwantray Mehta Committee Report* (1958)[25] on democratic decentralization led to the creation of a three-tier system of PR. The three levels of reorganized local government system are: *gram panchayat* (village), *panchayat samiti* (bloc) and *zilla parishad* (district). Each of the two higher levels is indirectly elected from the tier below and also draws membership from legislators, co-operative officials and others. There were in India in 1965, 219,694 *gram panchayats*, covering 99 per cent of villages and a population of more than 406 million. In addition, there were at that time 3485 *samitis* and 246 *parishads*.[26]

The village *panchayats*, the primary units of the three-tier PR systems, cover populations varying from 250 in Uttar Pradesh to an average of more than 15,000 in Kerala. The membership on these *panchayats* ranges from 5 to 31. The *panchas* (members) are elected by all adult residents on the legislative electoral rolls in the *panchayat* area. In most states this election is by secret ballot by *gram sabhas* (village assemblies);[27] special representation is provided for women

24 Jain (1968: 129–39).

25 Balwantray Mehta Report (1957).

26 Maddick (1970).

27 Although not originally envisioned, the adult residents of *panchayat* areas have now been constituted in all states (except Madras and Kerala) into statutory *gram sabhas*, whose functions are to discuss (in some cases approve) the annual budget of the *gram panchayats*, its annual administrative reports, proposals for taxation and major developmental programmes. A minimum number of *gram sabha* meetings are also statutorily prescribed. The size of a *gram sabha* varies from 250 to 5000, though the average number of people comprising it may be around 300. Available literature suggests that this institution has not been successful in evoking substantial public participation; indeed, it is clear that it has failed in fulfilling even a modicum of its prescribed functions. The size of the *gram sabhas*, lack of proper scheduling of its meetings, the dominance of the *sarpanch* (chairman of *panchayat*), *samiti* representatives,

and members of Scheduled Castes and Tribes.[28] The *sarpanch* (chief)
of the *panchayat* is elected directly by the *gram sabha* in some states
and indirectly by members in others. The *panchayat samiti*, the middle
tier of the PR system, generally consists of *sarpanchas* of the *panchayats* within a bloc of villages or of members elected indirectly by
sarpanchas.[29] The key office of the *samiti*‘ its chairman or *pradhan*,
carries considerable power and prestige. The apex body, the *zilla
parishad*, generally consists of the representatives of the *samitis*. Except
in Maharashtra the *parishad* is not directly elected. It draws its
membership from diverse sources: *panchayats*, *samitis*, members of
state legislatures and parliament, co-operatives, women, Scheduled
Castes and Tribes, urban local bodies and people with 'special experience in administration.' The functions of these bodies are varied
and wide, but it should be noted that with the establishment of Nyaya
(judicial) Panchayats (NP), the village *panchayats* lost their adjudicatory powers. This loss marks a historic break with the *panchayat*
system as it existed prior to independence, an aspect we examine
later in this paper.

Panchayats are subject to the powers of the state government which,
on paper, are extensive.[30] How far PR has attained the objectives

---

development officials and local notables, and overall indifference by the villagers are
some of the notable reasons for this result. Maddick (1970); Mathur, Narain and
A. Sinha (1966: 140–75); Jain (1968: 167–9).

28 The Committee on the Status of Women has found representation inadequate
and ineffective and has suggested the creation of statutory women's *panchayats* to deal
with special problems of women with a view to enlarging their participation in rural
development and to promoting equality. This recommendation is still under consideration. 'Towards Equality: Report of the Committee on the Status of Women'
(1974: 385).

29 The average physical area covered by the bloc is about 566 square miles, although there are wide regional variations in size (from 105 square miles to 2,837
square miles): the average number of *panchayats* per bloc is about 40. Jain (1968: 202);
Maddick (1970: 5). In almost all states, members of legislative assemblies are members
of the *samiti* in each region, in some cases local members of Parliament also serve on
the *samiti*. In addition to the provision for co-option of the Scheduled Castes and
Tribes, some states also provide for the co-option of persons 'experienced in social
work, development and administration'. This latter provision has often been criticized
as interfering with the elective character of the *samiti* and with its efficient functioning.
Jain (1968: 200–8); Maddick (1970: 90–8).

30 It can determine the jurisdictional area of *panchayats*: it frames rules and by-laws
concerning all important matters relative to the *panchayats*' functions (conduct of
business, personnel, administration, assets and liabilities, records, budget, accounting,
audit, etc.). It may call for records and reports. The government also has the power
of supervision and suspension of *panchayats*, presumably to be exercised in consultation
with *samiti* and *parishads*. This latter power does not seem to have been exercised
frequently. *Panchayats* also depend upon the government for grant-in-aid. Similarly,
panchayats depend on major decisions of the *samiti* — for allocation of resources or
approval of budgets. The fact that *sarpanchas* are members of the *samiti* creates structural linkages between *samitis* and *panchayats*, but it also creates further structural

of democratic decentralization remains a matter of serious and sustained debate in a vast and growing literature, official and scholarly. Different images of PR give rise to differing assessments. The contemporary political scientist sees in the PR movement a massive endeavour to mobilize people's participation in a *political* rather than (as in the earlier community development programme) in a *bureaucratic* mode. The contemporary Gandhian judges PR in terms of movement towards the ideal of consensual communitarian democracy, without parties and professional politicians. The growing politicization and pervasive factionalism seem to him a perversion of the PR ideals, whereas the political scientist regards this as not merely inevitable but indeed commendable.

Political images provide only one basis for evaluation of PR. *Panchayats* may also be envisioned as units of local rural government and as developmental units.[31] Since PR institutions respond to the central elements of most of those images, analytical disaggregation (necessary for impact analysis) is very difficult. Nor, of course, do available empirical studies support any major India-wide generalizations. But studies with specific focus do suggest some basic problems in regard to the functioning of the PR institutions. For example, Henry Maddick, viewing PR specifically as rural government, finds that while the 'basic structure' of PR is sound, there are widespread problems. Relations with government departments, 'variable as they are, are nothing like as good as they might be. Staffing systems vary from the ill-considered, as in Madhya Pradesh, to the careful approaches of Rajasthan or Maharashtra'. Financial allocation is 'generally far from satisfactory', given the wide range of functions entrusted to the PR institutions.[32] The problem of resources for developmental work thus continues to affect the fulfillment of the objective of PR.

The impact of competitive politics on PR institutions is harder to assess, but it increasingly provides the principal context for the future development of PR—whether in terms of administration or development programmes or financial assistance. The enormous increase in the number of *gram panchayats* functioning on the principle of direct elections has meant a greater role for party and factional politics. In villages where elections have been contested, the proportion of people actually voting is reported to range from about

distance between *sarpanch* and *panchas* (other members) within *panchayats*. There is some evidence of (and considerable anxiety over) *samiti* hegemony over *panchayats* impeding the democratic decentralization principle.

[31] Narain (1969); Haldipur and Paramhansa, eds. (1970: 26).

[32] Maddick (1970: 307–8); see also Jain (1968: 610–28); Mathur and Narain, eds. (1969: 24 and 28); Mathur, Narain and Sinha (1966: 165–7); Haldipur and Paramhansa, eds. (1970: 25–30).

70 per cent to 90 per cent. Surveys indicate that elections are fair.[33] Election studies also indicate that voting is frequently influenced by factional factors cutting across caste and kin: upper caste-lower caste alliances to offset dominance of other castes are not uncommon,[34] though. caste loyalties do play a significant role.[35]

Some studies show the emerging pattern of rural leadership. Significant numbers of *panchas* are young people with comparatively better education and greater political awareness. The monopoly of leadership by certain groups is being disturbed, although poorer and weaker sections of the community still do not occupy effective leadership roles. Also, while emerging leaders are mostly agriculturist, there is a tendency to shift over to business. Membership of political parties, mostly the ruling party, is a feature of the emerging rural leadership. Conceptions or images of leadership also seem to be changing; there seems to be greater emphasis on constructive and developmental tasks.[36]

Electoral politics at the *panchayat* level also has led to patronage, electoral bargains, and a certain degree of corruption. Since the PR institutions are conduit pipes for funds and grants, the struggle for PR positions is naturally keen; it has been often suggested that the dominant sections in villages capture power for their own ends.[37]

A recent empirical study disclosed that 97 per cent of the villagers in the areas under study were of the view that the working of PR institutions (including, in this instance, NP) 'have encouraged such evils as crimes, theft, personal jealousies, favouritism, litigations, feuds and insecurity of life and property at the village level.' All this has contributed to a 'loss of faith' in PR.[38]

---

[33] Indian Institute of Public Opinion (1961: 8–9); Grangrade and Sanon (1969).

[34] Grangrade and Sanon (1969: 33); Retzlaff (1962); Patnaik (1960).

[35] Research on the socio-economic background of the elected leaders shows that a large proportion of them belong to upper castes and upper economic strata. It has been suggested that

> the only logical explanation for this . . . is that the upper classes have been supported by the lower classes . . . [since] the general pattern of this support was based on group alliances. This is probably inevitable when the traditional pattern of leadership is giving way to a more representative leadership. Haldipur and Paramhansa, eds. (1970: 89).

[36] Jain (1968: 283–311); Bhat (1970: 217–26). See also Singh (1969).

[37] A senior official of the Government of India wrote in 1970:

> There was also marked corruption in the techniques of elections; and with stakes becoming higher, competing groups went to extreme lengths (not excluding kidnapping and murder) to ensure victory at election. All this in turn had its impact on the morale of officials working in the system; the good ones becoming confused and dejected, and the bad ones jumping on the winning bandwagon, without much compunction. Vepa (1970: 251).

[38] Singh (1972: 112). See also Bhat (1970: 216–25); Weiner (1967: 306–13).

Nevertheless, the tendencies towards coercive politics highlighted in the preceding paragraph do not appear to be the dominant features of electoral politics for PR as a whole, if the available literature is any guide. On the other hand, factionalism may be rampant; but the growth of factionalism may open up possibilities of political mobilization through alliances and bargains across established lines.[39]

Micro-level studies indicate significant and continuing changes arising out of the political dimension of PR. Although its scope and direction may vary, it has been 'universally acknowledged that political consciousness' of the people 'has increased under *Panchayati Raj*. An average villager is more conscious of his rights today than before. He has also developed a self-confidence.' PR institutions have also 'provided training ground in democracy to rural people who have so far been denied access to the avenues of power'. Indeed, in this sense, the PR is a 'revolutionary step', even though the emerging rural leadership may still, in some cases, be more a 'silent spectator than an active partner in the deliberations and working of the panchayati raj institutions'. Politicization has also meant that at the bloc level administration is getting 'somewhat demo-craticised':

> [T]he average villager is nearer the administration today than in the pre-panchayat period. He now moves about with greater self-confidence to office of the bloc development officer and not with the drooping spirit of a person who is in quest of a bakshish [tip, handout]. The . . . administrator also does not treat the villager with indifference or contempt.[40]

These findings in a study on PR in Rajasthan in 1966 point to trends that are confirmed by many other studies.

This very fragmentary account of an ongoing massive social transformation, favouring values of access, participation and democratic decentralization of power in the development process should leave no doubt that PR has introduced into the political and administrative systems greater responsiveness to people's needs and demands. It has also engendered strong expectations of change and has introduced, however variably, incremental changes in structures and patterns of domination. At the same time, the dilemmas in structuring access are severe. The levels of funding available for *panchayats* limit their capacity to achieve results, and this latter affects the institutionalization (and legitimization) of PR. Development bureaucracy is assisted, as well as frustrated, by power politics. The greater

[39] Weiner (1967); Nicholas (1965: 21); Bailey (1965: 15). Even the more extreme type of pervasive factionalism (see Siegel and Beals (1960: 394–417) may gradually yield to co-operation in search of a new basis for power. Beals (1970: 57).

[40] Mathur, Narain and Sinha (1966: 290).

the discrepancy between the rhetoric and the reality of change, the greater are the prospects of either alienation or organized restiveness —both, to some extent, dysfunctional to planned development. The tension between centralized planning and the drive to popular participation in planning change persists. Against this background we turn, finally, to our primary subject: the judicial organ of the village *panchayats*.

Nyaya Panchayats: Experimentation in Legal Access for The Village Population

The Formation of Nyaya Panchayats

While article 40 of the Constitution enjoins the State to organize village *panchayats*, another Directive Principle (Article 50) directs it to take steps to separate the judiciary from the executive.[41] Apart from the states which already had a system of village courts at the time of the adoption of the Constitution (Madras, Mysore, Kerala), only a few states (Madhya Pradesh, Uttar Pradesh) implemented Article 50 upon the adoption of the Constitution by creating separate NP. In the period following the adoption of the *Balwantray Mehta Committee Report* (1959) and the reorganization of the village institutions both as local government and developmental agencies, many more states established NP as separate judicial bodies, thus fulfilling the Directive Principle of separation of the judiciary from the executive.

The ideology of separation of judicial from the executive power, embodied in Article 50, was clearly one impulse that led to the creation of NP in states which did not have such separate bodies. This ideology also influences, as we shall shortly note, the structuring of NP. But the creation of judicial *panchayats* was not entirely a function of this ideology. As *panchayat* institutions were reorganized

[41] The Directive Principles of State Policy in Part IV of the Constitution are declared to be 'fundamental in the governance of the country', and the State is under a duty ' to apply these principles in making laws'. The Directives are not enforceable, like the Fundamental Rights, in a court of law; in other words, they furnish no cause of action, nor confer any legislative power. On the other hand, the Directives are a part of the Constitution of India and can be amended only in accordance with the provisions of the Constitution. The courts have, with varying degrees of success, used the Directives to aid in constitutional and statutory interpretation. For example, the ambit of Fundamental Rights has often been formulated with the help of the Directive Principles. See Austin (1966: 75–98); Baxi (1967: 345–74); Baxi (1967).

The Directive Principles, together with the Fundamental Rights, state the essential features of the constitutionally desired social order. The problem of the juridical 'inferiority' of the Directive Principles in relation to the fundamental rights has been sought to be settled by the 42nd. Amendment to the Constitution, which now provides that the State may, in implementing the Directive Principles, override the guarantees of Fundamental Rights contained in Articles 14, 19 and 31.

and oriented to a wider range of functions, it was felt that considerations of efficiency in performance of the assigned developmental and governmental tasks required relief from the judicial workload. We have already noticed that in the period 1920–47 the village *panchayats* — for example, in U.P., Bombay and Bengal — were already engaged in a substantial amount of adjudication work. The Law Commission's *Fourteenth Report* testified to the volume of this work in the years following independence. In U.P., for example, judicial *panchayats* heard, for the period 15 August 1949 to 31 March 1956, 1,914,098 cases, of which 1,894,440 cases were disposed.[42] Clearly, then, as a measure of planning efficient allocation of workload, the establishment of NP must have been felt essential. However, this separation was accompanied by some apprehension that 'without judicial authority the Panchayats would become ineffective bodies and would fail in seeking people's participation.'[43]

Apart from the ideology of separation and considerations of efficient division of labour, the creation of NP can be seen as illustrating two other concerns. First, their creation testifies to concern for providing easy legal access to the village population. Second, at the same time, it also represents a massive attempt by the state to displace (as effectively as it could) the existing dispute processing institutions in village areas — be they *jati* (caste) institutions, territorially based secular institutions or special dispute processing institutions established under the auspices of social reformers (such as the Rangpur People's Court).[44] The NP seek to do this by retaining procedural flexibility and lay adjudicators, thus co-opting the very features of the institutions they seek to displace. On the other hand, the NP, as integral parts of the administration of justice, are characterized by principles of formal organization and of judicial oversight and control which do not 'mesh in' with the organization of justice by village communities. In the very structuring of NP, therefore, inheres the drive to extend State power and constitutionalist ideology to the countryside, and this creates incompatibilities and dilemmas to which there is no easy solution.

Although the establishment of NP derived symbolic support from an appeal to the virtues of traditional *panchayats*, it should be emphasized that these new tribunals are in many ways very different from their traditional counterparts. Their membership is fixed rather than flexible and is based, indirectly, on popular election rather than social standing; their constituencies are territorial units rather than functional or ascriptive groups; they decide by majority vote

[42] *Law Commission Report* (1958: 875, 896–902).
[43] Jain (1968: 194).
[44] Baxi (1976: 52).

rather than by rule of unanimity; they are required to conform to and to apply statutory law; they are supported by the government in the compulsory execution of their decrees; these decrees may be tested in the regular courts.[45]

Constitution and Composition of NP

Legislative details concerning the constitution and composition of NP vary so that any general account of NP may be somewhat inaccurate as regards a particular region. For present purposes, however, it may be convenient to state the main features of NP organization which have now become more or less generally established. NP are established for a group of villages, usually an area covering 7 to 10 villages. NP usually cover a population of 14,000 to 15,000 villagers. A member of an NP must be able to read and write the state language, must not suffer from any disqualifications described in the statute, and must not hold an office of *sarpanch* or be a member in the *samiti, parishad,* or state or union legislature. The rules regarding appeals in disputed elections are the same as those which apply to *gram panchayats.* The NP has a chairman and secretary elected by its members; one-third of its members retire every second year.

Almost all states have adopted election as a method of constituting NP. Each *gram panchayat* (itself an elected body) elects members for NP. Some states combine the method of elections with nomination; thus, in U.P. members of *panchayats* nominate a person from among themselves to membership of the NP; such nomination may also be by consensus. The nominations are then screened by a subdivisional officer and forwarded to the District Magistrate who, as the chairman of the Advisory Committee established for the purpose, shall ultimately appoint members of NP. It has been pointed out that in practice the District Magistrate (and the Committee) follow the advice of the sub-divisional officer, 'who in turn relies on the advice of minor government officials such as the village accountant and the panchayat secretary', a procedure which 'may result in favoritism.'[46]

Another combination of nominative and elective principles is furnished by the Bihar legislation in which the *panchayat* courts, comprising the *sarpanch* and eight other *panchas,* are chosen in three different ways. The *sarpanch* is directly elected by the *gram sabha;* four *nyaya panchas* are also directly elected, but from four wards into which *panchayats* are divided for election purposes; the remaining four are chosen from the *gram panchayat* by the NP *sarpanch* and the

---

[45] Cf. the contrast of the statutory *gram panchayats* with traditional village *panchayats* in Retzlaff (1962: 23ff); Luchinsky (1963).

[46] Robins (1962).

four elected members in a joint meeting. The chairman of the *panchayat* has no direct role in this process of constituting NP. Provision also exists for the representation of Scheduled Castes and Tribes in NP.[47] Finally (without being exhaustive), one might mention the states of Kerala and the Union Territory of Delhi as furnishing extreme ends of the spectrum: in the former all *nyaya panchas* are nominated; in the latter all of them are directly elected.

The method of the constitution of NP has been a subject of some controversy. The Law Commission, in its *Fourteenth Report*, expressed itself against the principle of government nomination of *nyaya panchas*. It felt that nominated *panchas* may not 'command the complete confidence of the villagers'; nominated *panchas* may be impartial, but the nominating officer may lack 'first-hand knowledge of local conditions'; in that event, 'the freely expressed will of the villagers, in substance, [would] be replaced by untrustworthy recommendations of sub-ordinate officials.' The nominees would 'tend to act in a manner which will command the approval of the appointing authority rather than discharge their functions in a true spirit of service to the village community'. Although the Commission did not, in principle, support an elected judiciary, it did not regard NP as judiciary in the proper sense of the term, but rather as 'tribunals' who have to 'inspire the confidence of villagers'.[48] The Study Team on Nyaya Panchayats in 1962, endorsing those views, concluded:

> [T]he system of nomination in any form has to be ruled out. Villagers must be given a free hand and the choice lies between the system of direct elections and indirect elections. The method of indirect elections seems to afford for the time being the best solution and of the various possible methods of indirect elections, the best seems to be the type in which each of gram panchayats in the Nyaya Panchayat circle elects a specified number of persons to serve on the Nyaya Panchayat.[49]

There are very few studies of NP in action, with the result that it is not possible to assess the correctness of the foregoing assertions concerning the worthwhileness or otherwise of the election of NP. To be sure, nomination of *nyaya panchas* may degenerate into mechanical endorsement of 'untrustworthy recommendations of subordinate officials', as the experience of U.P., for example, seems to indicate.[50] But, as the Commission itself acknowledged, outright nomination of village court members in Kerala 'seems to have worked satisfactorily' for the reason of 'smallness of the area and the consequent

[47] Pillai (1974); Bastedo (1969: 194–5).
[48] *Law Commission Report* (1958: 912–13).
[49] *Study Team Report* (1962: 125).
[50] Robins (1962: 239).

ease with which the higher officials are in a position to choose persons respected in villages as the panchas.'[51]

However, one of the few empirical studies of NP discloses the dominant role of the *sarpanch* of the *gram panchayat* in the process of indirect elections of NP. Since each *panchayat* generally elects one *nyaya pancha*, politics and patronage do intrude in elections. Very often the person elected by the *panchayat* is a man 'with ridiculously low level of education . . . quite incapable of grasping the niceties of the law' he has to administer. Such persons were 'reported to have political aspirations which they sought to fulfil by getting into Nyaya Panchayat. . .'. Often the *sarpanch* accommodated the defeated *panchayat* members in NP. So dominant is the influence of the *sarpanch* that 'a look at the antecedents of the members of NP would convince anyone that they were strong supporters (and in one case, a relative) of the Sarpanch'.[52]

The dominance of *panchayats* by *sarpanches* has emerged as a general structural feature of PR in action. Hence these Rajasthan findings may well be replicated in other regions. When elections to NP are overtly political or perceived to be such, it is unlikely that NP will inspire the confidence of the villagers let alone symbolize 'the freely expressed will of the villagers'. Indeed, the recent report of the High Powered Committee on Panchayati Raj in Rajasthan recommends the abolition of the NP altogether, partly on the ground that they have 'not been able to inspire public confidence'.[53] Similarly, the Maharashtra Evaluation Committee on Panchayati Raj finds entrustment of judicial functions to NP 'on the basis of democratic elections or otherwise' both 'out of place and unworkable' and also recommends the abolition of NP.[54]

If indirect election of *nyaya panchas* involves politics and patronage and reduces the level of public confidence in, and use of, NP, it does not seem likely (though there are no studies on this point) that direct election by the *gram sabha* will necessarily lead to better results. On the other hand, nomination of *nyaya panchas* by state officials may avoid such results (despite the experience in Kerala), but at the cost of diluting the ideology and symbolism of NP.

Jurisdiction

NP have civil and criminal jurisdiction, but the former is more limited than the latter. Civil jurisdiction is normally confined to

[51] *Law Commission Report* (1958: 914).

[52] Mathur, Narain & Sinha (1966: 177, 182–5); cf. Hitchcock (1960), where an extremely close relationship between *gram panchayat, sarpanch* and *NP sarpanch* was dominated by the latter.

[53] *Rajasthan Report* (1973: 43).

[54] *Maharashtra Report* (1971: 203).

pecuniary claims of the value of Rs 100 (about U.S. $12) (which may by agreement among parties be raised to Rs 200 involving money due on contracts not affecting interests in immovable property, compensation for wrongfully taking or damaging property and recovery of movable property. In some states the civil jurisdiction extends to the recovery of minimum wages or arrears for maintenance (e.g., Kerala). Also, in certain states (e.g. Bihar) civil jurisdiction of NP may be enlarged by an agreement between the parties to do so 'irrespective of the nature and value of the suit'.[55] Nevertheless, it appears that in most cases the civil jurisdiction is confined to pecuniary claims related to property.

The criminal jurisdiction is comparatively extensive and covers a substantial range of offences under the Indian Penal Code as well as the special statutes (e.g. Cattle Trespass Act, Gambling Acts, Prevention of Juvenile Delinquency Act). The range of offences triable under the Code varies from state to state, but NP have jurisdiction to try offences such as criminal negligence or trespass, nuisance (including water pollution), possession or use of false weights and measures, theft, misappropriation (subject to a pecuniary limit — in some cases as low as Rs 25 to 50), intimidation, prejury, attempt to evade a summons, etc. The NP are authorized to levy fines (ranging from Rs 25 to 100), but they have no power to sentence offenders to imprisonment, substantively or in default of fine. NP are usually empowered to pay compensation to the injured out of fines thus imposed. They have also the power to admonish in certain cases. The state government retains power to enhance the jurisdiction of NP as well as to diminish it if there is 'admission of miscarriage of justice',[56]

Emphasis on the amicable settlement of disputes is an important aspect of the NP ideology. Accordingly, conciliation is emphasized over adjudication in some state legislation. In Bihar and Kerala it is obligatory on NP to first resort to conciliation in all matters, including criminal cases; in Rajasthan conciliation is permissible though not obligatory. In the latter state, however, field studies have shown a 'surprising feature' — 'conciliation is resorted to not so much at the stage of judgment but at the review stage'. The NP reopen the suit for consideration in cases of non-compliance; in one instance 33 out of 34 suits decided in a year were thus reviewed and settlements reached. While such settlements may be desirable in principle, the processes of review after decision and non-compliance are 'likely to affect the reputation' of NP; the practice is also against 'the accepted canons of judicial administration'.[57]

55 Pillai (1974: 34–62); Bastedo (1969: 197).
56 Pillai (1974: 56–7).
57 Mathur, Narain and Sinha (1966: 186–7).

Organization of Nyaya Panchayat Work

Two essential features of lay adjudication, simplicity of procedures and flexibility of functioning, are realized in the design of NP. The NP are not encumbered by the need to follow the elaborate rules of civil or ciminal procedure or the laws of limitation of evidence. Complaints may be made orally or in writing; no legal representation is allowed, although in some civil matters parties may be represented by an 'agent'. Parties are heard by NP in a fairly informal manner. At the stage of reaching a decision, parties are asked to absent themselves; *panchas* confer among themselves and arrive at a decision which is pronounced in open court. A judgment is written which, after being read out in open court, is signed (or thumb-impressed) by the parties, signifying the communication of judgment to them. Witnesses, if any, are examined on oath or solemn affirmation.

The NP share these features, more or less, with other community dispute institutions. But of necessity they depart from the model of non-state dispute institutions in certain major respects. First although the statutes setting court fees do not apply to NP, minor court fees are levied (e.g. 50 paise or a rupee for a criminal complaint or execution order; 5 to 10 per cent of the value of the claim in a civil matter). Second, NP have power to issue summons (though not warrants) and to proceed *ex parte* in case of a recalcitrant defendant/respondent. Third, NP have the power to levy execution through attachment orders pursuant to unfulfilled decrees, although this power is confined to certain kinds of movable property (for example, bullocks, cows, seed grains, tools, etc. are excluded from attachment orders). Fourth, the judgments of NP are written and maintained as a part of official records, as is the gist of the depositions by parties and witnesses. Fifth, the record-keeping functions also include maintenance of registers of civil and criminal matters, court fees and fines, summons and notices, and expenses of witnesses (paid for by parties calling for them). Sixth, the higher judiciary has powers of control and oversight. The subdivisional or district magistrate can transfer a case from one NP to another. They may intervene in a particular case on the ground that miscarriage of justice is imminent or has occurred. Seventh, despite the recommendations of the government's Study Team on NP, the magistracy has the power to entertain appeals from NP decisions in most states.[58] In addition to the right of appeal, parties also have a privilege to apply for revision of a NP decision.[59] The Study Team favoured availability of the latter only.

[58] *Study Team Report* (1962: 128); Pillai (1979: 75–81).

[59] Revision is the right of a high court or authority to call for the record of any case decided by a court subordinate to the court, and in which no appeal lies herein. If

All these features, cumulatively, distinguish NP rather sharply from the community (non-state) dispute-processing institutions. Additional distinctions are imposed by the need to constitute benches of NP. Traditional *panchayats*, it should be recalled, might consist of standing bodies of village or caste leaders who heard disputes arising within their respective groups. Or they might be open bodies, incorporating all interested parties into the discussion and reaching consensus by negotiation and mediation. In contrast, the chairman of NP constitute benches which may be 'as numerous as the number of villages'. The eminently practical idea of appointing benches has created certain problems in the functioning of NP. The Rajasthan study highlighted the considerable scope for misuse of this power by the chairman. He 'may put himself in all benches excepting the one functioning in his village'; he may also 'pack' them with his 'favorite *panchas*', excluding some *panchas* altogether from the NP work. The discretion of the chairman is only limited by the rule that a *nyaya pancha* is not to serve on a bench which functions in his own village. In the abstract, this rule serves the value of impartiality; on the other hand, it makes the NP approximate a court of law more closely. It has been rightly observed that when a villager

> has only outsiders as Nyaya Panchas, the atmosphere is very similar to that of a court of law where a man could indulge in all sorts of falsehood, sometimes even under oath with a sense of impunity and without fear of social conscience operating against him.[60]

The bureaucratic component of NP, howsoever slight it may appear in comparison to the court, poses another set of problems. The Rajasthan study highlights three distinct problems. *First,* the NP require a minimum staff, which may comprise a peon and a clerk, the latter working as a secretary. It was found that the recruitment was affected by favouritism: no proper norms were followed. Also, the salary scales were quite low, and this was aggravated by the fact that even the legally due salaries were not paid regularly (in one case a secretary of NP left in 'disgust' after he could not realize his salary for eight months!). Second, the NP derive their revenue primarily from contributions of the village *panchayats*, but the contributions become inadequate and irregular as the *panchayats* themselves experience financial difficulties. Indeed, the *panchayats*

---

the high court feels that the subordinate court exercised a power not held, failed to exercise a power properly held, or acted improperly in the exercise of the power, the reviewer may uphold, revise, reverse or remand the case. Procedural consistency is deemed imperative, but the court may decide not to intervene if substantial justice has been done in spite of legitimate error. *A.I.R. Commentary, The Code of Civil Procedure* Sec. 115, 111–15 (8th. ed. 1968).

[60] Mather, Narain and Sinha (1966: 186–7).

accepted the NP scheme with the expectation that 'their share of the fines imposed by nyaya panchayat would total more than they would be asked to contribute.' when it became clear to village *panchayats* that 'their calculations were far from correct, they became lukewarm to the whole scheme.' In any event, the income from fines accruing to NP does not appear to be substantial. In the Rajasthan study, it was discovered that against the estimated income of Rs 500 the income actually realized by a NP was only Rs 2![61] A share of court fees is made available to the NP but only after cumbersome procedures have been compiled with. Third, this dismal financial position affects not just the conditions of the administrative staff of the NP but also the work of the *nyaya panchas* themselves. The *panchas* have to do some (and even considerable) travelling to attend to their judicial work. The Rajasthan study found that no travelling allowance or related expenses were paid to the *panchas*. The low rate of participation by the *panchas*, often causing a lack of quorum in the benches, was partly related to this lack of minimum finances. Of course, conflicting social and economic commitments, as well as the absence of the 'glamor of executive authority' (enjoyed by village *panchayat* members) may have also contributed to this result. The lack of a minimum financial base for NP is nonetheless an important factor affecting the performance of its bureaucratic as well as judicial tasks.

Workload and Disposition

No aggregate information is available on the number of NPs, their personnel, workload, or disposition patterns. The Law Commission of India found, in 1958, that the workload of NP (and village courts) was substantial. We emphasize the need for more up-to-date figures on the incidence of recourse to NP. Whether recourse is affected by such factors as: nominative versus elective NP, enforcement patterns, financial position, caste composition and so forth must await richer data. We have available to us information on Uttar Pradesh for the period 1950–6 and 1960–70 and on Rajasthan for the period 1961–2 — an obviously highly unsatisfactory state of affairs. We proceed *faute de mieux* with the available data.

Detlef Kantowsky in a study of NP in Uttar Pradesh noted that the total number of cases going before NP in U.P. is in relative decline. Taking the year 1957–8 as the norm (when 202,116 cases were filed, of which 82 per cent were heard; of the latter, 83 per cent were disposed of, 3 per cent were appealed, and 41 per cent cent of those disposed of were settled by compromise), he finds that the percentage index of cases filed before NP in the years 1963–6

[61] Id. at 189–93.

has declined (1963–4, 194,625 (96 per cent); 1964–5, 159,840 (79 per cent); 1965–6, 122, 923 (61 per cent). The number of cases filed in the Benaras District NP showed a similar marked decline: from 2,671 cases in 1963–4, NP in the district handled only 1,339 cases in 1965–66 (registering nearly a 50 per cent decline). He also finds that the rate of appeals from NP decisions increased in the period 1963–6.[62]

This decline in the number of cases filed is corroborated by state reports on the administration of justice in Uttar Pradesh. Table 1 shows the figures for the number of NP and the number of cases filed and pending for 1950–6, 1961–6 and 1969–70. There is some minor discrepancy with Kantowsky's figures but the drastic decline in cases, from 633,502 in 1950–1, with 8,543 operating NP, to 91,107 in 1960–1 (8,662 NP in operation), to only 35,865 in 1969–70 (8,727 NP in operation), reflects the same trend.

The rate of filings per NP has fallen to one-fourth of its peak in the 1950s.[63] Figure 1 displays the average filings per *panchayat* for Uttar Pradesh as tabulated in Table 1.[64]

FIGURE

Cases Filed per Nyaya Panchayat in Uttar Pradesh 1951–70

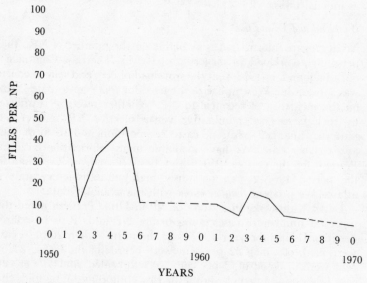

[62] Kantowsky (1968: 149).

[63] *Report on the Administration of Justice* (1961–6, 1969–70); Purwar (1960: 220). See Table 1 supra.

[64] *Report on the Administration of Justice* (1961–6, 1969–70).

TABLE 1

Workload of Nyaya Panchayats in Uttar Pradesh 1950–70

Year*	No. of NP	Civil		Criminal		Total Filings	Filings per NP	Source
		Filed	Pending	Filed	Pending			(a)
1950–51	8,543					633,502	59.3	202
1951–52	8,543					123,363	14.4	202
1952–53	8,543					302,548	35.4	202
1953–54	8,543					347,790	40.7	202
1954–55	8,543					371,519	43.4	202
1955–56	8,543					105,376	12.3	202
								(b)
1960	8,583	82,321				—	—	1961: 6,45
1961	8,662	53,523	22,774	37,584	20,852	91,107	10.51	1961: 6,45
1962	8,662	49,164	22,275	32,579	18,384	81,743	9.43	1962: 6,51
1963	8,700	93,994	25,350	71,792	17,059	165,786	19.05	1963: 7,57
1964	8,626	83,627	19,684	55,873	15,424	139,500	16.17	1964: 6,54
1965	8,687	55,000	11,898	36,904	11,451	91,904	10.57	1965: 7,46
1966	8,680	35,362	19,498	31,854	8,059	67,216	7.74	1966: 8,47
1967	—	—	—	—	—	—	—	
1968	8,580	—	—	—	—	—	—	1969: 7
1969	8,833	22,600	4,025	15,671	2,780	38,271	4.33	1699: 7,53
1970	8,727	22,912	4,766	12,953	4,293	35,865	4.10	1970: 8,65

* Year runs 1-1-50 to 31-3-51, thereafter each year begins at 31-3 of the year. The filing rate is prorated to 12 months.

(a) V. PURWAR, *Panchayats in Uttar Pradesh* (Lucknow, Tej Kumar, 1960).

(b) REPORT ON THE ADMINISTRATION OF JUSTICE, UTTAR PRADESH (Allahabad).

(The higher rates in 1950–1 may be due at least in part to an accumulation of cases between enactment and commencement of NP, and those in 1952–4 may reflect a surge of cases related to land reform legislation.)[65]

The decline in NP activity does not appear matched in the experience of other dispute processing institutions. The business of the regular state courts increased steadily throughout this period. Table 2 shows available figures for the period 1960–70.

TABLE 2

Cases in Civil and Criminal State Courts in Uttar Pradesh 1960–70

Year	Civil Cases in Subordinate Courts (a)	Criminal Total No. of Persons on Trial	Source (b)
1960	74,958	—	1961: 41
1961	75,174	760,704	1961: 8,41
1962	71,912	789,670	1962: 8,47
1963	71,068	845,982	1963: 10,53
1964	71,371	879,687	1964: 9,51
1965	81,908	895,859	1965: 9,43
1966	85,114	873,023	1966: 10,43
1967	—	—	
1968	88,762	—	1969: 49
1969	91,859	1,001,249	1969: 11,49
1970	86,749	1,037,896	1970: 12-13,60

(a) Includes paid sub-divisional tribunals (courts of *munsiffs*), small cause courts, district courts and chief courts of the districts.

(b) *Report on the Administration of Justice, Uttar Pradesh* (Allahabad).

No direct relationship between the regular state courts' increased activity and the decline in NP activity can be established with the available data.

The apparent low workload of NP in Uttar Pradesh is matched by low figures provided by Bastedo's fieldwork in Bihar. Fifteen NP in a single bloc heard only 202 cases over a sixteen year period, or an average of 1.25 cases per year for each *panchayat*. This was one of three blocs where NP had been instituted. Bastedo noted an increase in regular court activity throughout the same period.[66]

[65] Purwar (1960: 220).
[66] Bastedo (1969: 205–6, 210).

A final note to the discussion of low and declining workloads in NP is provided by the meager but interesting statistics for West Bengal. West Bengal instituted NP in 1958, holding general elections for *gram, anchal* and NP positions. Each *anchal panchayat*, or middle level administrative *panchayat* in the PR system, was to constitute its NP and thus there was a potential for 2,926 NP to match the 2,926 *anchal panchayats* in that state. However, NP did not progress that rapidly. No further elections have been held for NP. In 1965–6 there were 49 NP operating. Of these 49, 46 heard 937 cases in that year.[67] By 1973 only 13 more NP were in operation.[68]

A profile of disposition rather than of workload is disclosed in the Rajasthan study. In 1961–2 the total number of suits in the NP studied was only 121. The following table provides a profile of the time to dispose of a portion of the cases (measured from the time of filing to the delivery of judgment):[69]

TABLE 3

Time from Filing to Judgment in Rajasthan NP cases

Time Taken (Days)	Cases
301	1
244	1
150-182	6
110-150	7
50-100	14
50	1
25-50	8
15-25	2
14	1

Source: R. Mathur, I. Narain, & A. Sinha, Panchayati Raj in Rajasthan (New Delhi, Impex India, 1966).

Significantly, of the 94 criminal cases in the period 5 June 1961 to 31 March 1962, 51 were tried, 43 remained pending; in the period 1 April 1962 to 31 March 1962, 13 of the 55 criminal cases remained pending. On the other hand, for the first period 5 out of 6 civil suits were completed; in the second period 14 out of 45 suits were completed.[70] Similar disposal and pending rates are evident in the U.P. data in Table 1 and 2.

[67] Basu (1967: 45).
[68] Mukherjee (1974: 42).
[69] Mathur, Narain, and Sinha (1966: 201–20).
[70] Id. These figures are compatible to some extent with work in some *panchayat* courts in the years 1954 and 1955. The *Law Commission Report* yields the following information.

Even if the data base is very limited, we see here the problem of 'delay' and arrears, so familiar (and sometimes overwhelming) in urban courts. This 'delay' in settlement of cases may be due to a whole variety of factors: lack of minimum training of the *panchas* in the law they are to administer, lack of quorum in NP, inefficiency of secretarial staff, different attitudes towards conflict resolution on the part of *panchas* as well as parties, etc.

Obviously, only the most tentative generalizations are possible from this data. Despite the informality and flexibility of procedure, considerable time lag does occur between the institution of suits and their disposal, and the arrears of workload in some cases are substantial. The delays and arrears may or may not compare favourably with the disposal rate in magistrate's courts, but that they occur in any magnitude suggests that the administration of lay justice deviates from the ideals that inspired it. Recourse to NP and their overall legitimacy may, in turn, be diminished by perception of this deviation.

Official Evaluations of NP System

While the Law Commission (1958) and the Study Team on Nyaya Panchayats (1962) saw a bright future for NP, two recent evaluations of PR recommend the abolition of the NP altogether. The Maharashtra Committee's *Report on Panchayati Raj* (1971) finds that out of 3,446 NP in the state only 723 are reported to be actively functioning (i.e., 86 per cent are not functioning); the rest are 'moribund and ineffective'.[71] Of the 14 per cent NP which are functioning, most are in the Vidarbha Division. Aside from this piece of information, the Committee does not have much by way of empirical

State	Total No. of Cases for Disposal		Pending at the Close of Year	
	1954	1955	1954	1955
Andhra	2,468	2,454	622	579
Bihar	31,842	39,908	5,753	7,785
Madhya Pradesh	47,506	46,698	20,022	18,456
Madras (5 districts)	2,331	598	766	135
Orissa	2,033	3,468	303	539
Rajasthan	66,044	53,909	11,664	10,311
Uttar Pradesh	132,977	179,918	11,481	16,461

Law Commission Report (1958:899).

[71] *Maharashtra Report* (1971: 202).

analysis to support its recommendations. And it is not clear just
how the Committee assesses the 'effectiveness' of NP. True, the
Collectors[72] were asked to advise the Committee as to how many
NP were effective in their districts and to state reasons why they
were ineffective. But if any reasons were recorded, the Report does
not mention them, except very generally (e.g. improvement of
transport and communications has meant that effective administration has reached 'much lower territorial levels'; NP are open to
influence and pressures which may distort the course of justice).
No information is contained in the Report to suggest that the government has exercised its statutory power to revoke any or all powers
of NP for reasons of incompetence or miscarriage of justice.

Be that as it may, the Maharashtra Committee is emphatic that
'entrustment of judicial functions, even of petty character, to such
institutions set up at the village level on the basis of democratic
elections or otherwise, seems out of place and unworkable'. The
Committee is also concerned that community dispute institutions
functioning in villages 'on the basis of common consent rather than
on the strength of any law' may be adversely affected by NP, indeed
to such an extent that the NP might 'destroy all possibilities' of
continuing or reviving (in the future) community dispute processing
institutions.[73] The Committee would favour either a full fledged
extension of the state legal system through trained judicial functionaries or leave the voluntary community effort, 'which must be
considered commendable', unimpeded by intrusion of the NP
system. Indeed, the Committee stressed:

> We consider it a fortunate circumstance that these bodies (NP) have not
> yet come into vigorous existence and not much damage is done. It would
> be better to withdraw these steps when there is enough time and in our
> opinion the whole concept of Nyaya Panchayats needs to be now dropped
> and the bodies abolished [*sic*].[74]

The Rajasthan Committee also concludes that the NPs are 'neither
functioning properly nor have they been able to inspire confidence
in the people'. No purpose would be served, it reports, by 'continuous flogging of a dead horse'. The reasons contributing to this
'tragic result' are pithily stated; 'the Nyaya Panchayats in Rajasthan
are today languishing for want of funds, secretarial assistance,
adequate powers as also the people's faith in them.' Indeed, in its
visits and responses to questionnaires (excluding responses received

---

[72] The Collector is the chief administrative officer of a district (a subdivision of the
state).

[73] *Maharashtra Report* (1971: 42–3).

[74] Id. 203.

from the chairmen of NP) the Committee found 'almost a unanimous demand' for the abolition of NP.

However, the Rajasthan Committee seems to feel that the failure of NP is due entirely to the separation of the judiciary from the executive at the 'grassroots level'. It, accordingly, recommends that the functions of NP be 'entrusted to a sub-committee of Gram Panchayat', having five members including the *sarpanch* as the *ex officio* member acting as Chairman. One of the four members must be a woman, and one member should be from the Scheduled Castes or Tribes. The rule debarring a *nyaya pancha* from hearing cases from his area/ward should be applied to this subcommittee.[75]

Neither of the two high powered committees refers to the recommendations of the 1962 *Report of the Study Team on Nyaya Panchayats*, which had drawn attention, albeit inconspicuously, to the need to provide the *nyaya panchas* with 'the basic amenities such as, a convenient place to hold their meetings or adequate stationary funds for recording proceedings and for other purposes or requisite funds for meeting various contingencies incidental to their work',[76] including travelling and out-of-pocket expenses for *nyaya panchas*. It also emphasized 'the great need for an endeavor on the part of all official agencies to extend the fullest cooperation to Nyaya Panchas'[77] and stressed also the need to accord them 'the status and courtesy due to them for the proper discharge of their functions'.[78] The Study Team was sympathetic to the general feeling voiced by *nyaya panchas*, some of whom felt that they were treated as if they were 'unwanted orphans'.[79]

Neither the Maharashtra nor the Rajasthan Committee Reports seriously attended to this aspect of the *Study Team Report*. These committees do not explain why the states found themselves unable to attend to the needs of NP, underscored by the painstaking 1962 *Report*. Was it because the states felt that it would be wrong in principle to interfere with the jurisdiction of village *panchayats*? Or was it because it was felt that the *sarpanchas* would make financial commitments to NP a bargaining counter to obtain more resources for *panchayats* themselves? Or because it was genuinely felt that funds required to meet the NP expenses would be too substantial (the number of NP could be quite large, e.g. Maharashtra has 3,446 NP, Rajasthan —as of 1966—had 1,370 NP)? Or was it felt that a greater degree of bureaucratization of the functioning of NP might result if additional resources—including permanent accommodation and secre-

[75] *Rajasthan Report* (1973: 42–3).
[76] *Study Team Report* (1962: 118–21).
[77] Id., 121.
[78] Id., 118.
[79] Id., 119.

tariat—were to be provided? Or was state inaction a compound of all those elements? Or due to just plain inertia? One could well understand that some *sarpanchas* of village *panchayats* would oppose the rise of parallel powerful institutions in the form of NP. But the inspecting magistrates and state departments might have looked at the matter primarily in terms of efficient performance of functions statutorily assigned to NP.

Both Maharashtra and Rajasthan subsequently abolished NP.[80] These actions do not conclusively indicate any general trend of the future of NP in their own states, or for India as a whole. There is still some persuasiveness in the urging of the *Law Commission Report*[81] and the *Study Team Report* that steps be taken to strengthen NP. Whether the existing workload accurately reflects the level of need for village based dispute institutions or points to structural impediments to their functioning, remains a matter for further investigation. The present state of affairs, in so far as one may generalize *at all*, illustrates some major difficulties of translating ideology into action, to which we will return in the last section of this paper.

Nature of Disputes and Quality of Justice

If available information on the organization and functioning of NP is meager, information on the subject matter of disputes and the quality of justice of NP processes and outcomes is virtually non-existent. Aside from the statistical breakdown of disputed matters into the categories of 'civil' and 'criminal,' neither the *Fourteenth Report of the Law Commission* nor the 1962 *Study Team Report* gives us information relating to those two dimensions. But to comprehend NP we need to know about the types of disputes that reach them, the nature of their adjudicatory processes, and the outcomes they

---

[80] The State of Maharashtra did abolish NP with the passage of the Bombay Village Panchayat (Amendment) Act, 1974, effective as of August 1, 1975. Cases pending before the NP were transferred to regular civil and criminal courts. Govt. of India, Ministry of Agriculture and Irrigation, Department of Rural Development. *Panchayati Raj at a Glance, Statistics,* 1975–6, 1977: 4, 20: 1; *Report on National Juridicare: Equal Justice — Social Justice Bhagwati Committee Report* (1978: 140–1).

Panchayati Raj at a Glance, Statistics (1975–6, 1977:4, 20–1); Bhagwati Committee Report (1978: 140–1). The State of Rajasthan also abolished NP in the Rajasthan Panchayat (Amendment) Ordinance of 24 September 1975, rectified by legislation passed in February 1976. New provisions to replace the old NP with new *nyaya up samitis* have apparently not been implemented. *See* Ministry of Agriculture and Irrigation, *supra.* According to scholarly informants in 1978 most village adjudications in Rajasthan are being heard by *gram panchayats.*

According to one official source, as of March 31, 1976, NP were in operation in only Gujarat, Himachal Pradesh, Jammu and Kashmir, Madhya Pradesh, Manipur Bihar, Tripura, Uttar Pradesh, West Bengal and the Andaman and Nicobar Islands. *See* Ministry of Agriculture and Irrigation, *supra.*

[81] *Law Commission Report* (1958: 923–5).

arrive at. An understanding of the interrelation of NP on the one hand and community dispute processing institutions on the other is inconceivable without a detailed grasp of the nature of disputes, sanctions and incentives entailed in the functioning of NP.[82]

The Study Team had before it the proposal that NP be vested with matrimonial jurisdiction. Its rejection of this proposal was based on the ground that both legal rules and the facts to which they would apply were extremely complex, and that the *nyaya panchas* could not be expected to have the requisite competence to deal with them.[83] This may well be true, as the diversity of personal laws is intimidating, even for specialists. On the other hand, the assumption that villagers will avail themselves of the court system for family law matters is unproven; indeed, available evidence shows lack of both knowledge and will to have recourse to the court system.[84] This means that non-state legal institutions will continue to meet this need, which is a major one. At the People's Court (*lok adalat*) at Rangpur, for example, the bulk of disputes coming before the court concern divorce, maintenance and custody. This is a non-governmental body, founded by a charismatic outsider, which has been functioning in a tribal area over a quarter of a century. The *lok adalat* settled 17,156 disputes in the period 1949–71; of these, 10,165 involved marital relations.[85] It is clear that the *lok adalat* system has displaced the NP system in the Rangpur area, in part because it has effectively met the needs of the communities concerned. Insofar as the eventual displacement of community dispute institutions is an aim of the NP system, it is undermined by the NP's lack of this vital jurisdiction.

Concern with the quality of justice administered by the NP system has been confined mainly to a theoretical and structural level. As against considerable attention (and space) given to the issues of the constitution of NP (election/nomination) or training programmes (frequently recommended but not yet implemented) for the *nyaya panchas*, existing literature does not probe the justice qualities of NP's functioning. We have no systematic data, for example, on 'impartiality' in hearing and disposing of cases, equality in implementation (regardless of socio-economic status of parties involved) or 'fairness' of individual decisions. We have reason to think that substantial timelags do occur in disposal of cases; post-judgment

[82] It is, of course, clear, given the jurisdiction of NP, that cases of ritual lapse or caste discipline would not directly come before NP, nor would disputes concerning matrimonial affairs, family arrangements in general, or land relations reach NP. A large area of dispute matters thus remains outside the jurisdiction of NP.

[83] *Study Team Report* (1962: 78–92).

[84] Chatterjee, Singh and Yadav (1971).

[85] Baxi (1976: 40).

conciliation does raise doubt (as documented by the Rajasthan study) concerning the justice of specific outcomes; and we know, broadly speaking, that the rate of appeals from NP decisions is not very high. We do not know whether this 'low' appeal rate is due to the appeal provisions themselves or due to differential party capabilities (in terms of knowledge and access) or due to party satisfaction with NP decisions. Nor do we know, though we do need to know, how far the levels of basic facilities made available to NP affect their capacity to provide 'justice' in terms of the variables mentioned here.

Discussion of how the NP should be constituted so as to maximize 'impartiality' has inevitably centred on the impact of 'factions' in villages upon the selection or election of *nyaya panchas*.[86] But the general conclusion on the issue (as to inevitability of factional politics in the context of PR institutions) relates only to the *structural* aspect of NP; not much attention has been devoted to how factionalism may affect the *functioning* of NP and the justice qualities of the system. A recent study in U.P. indicates that factions within villages can influence the NP substantially in favour of the powerful faction, at the expense of justice values.[87] The data are not conclusive even for the areas studied, but they suggest the justice qualities of the NP processes cannot be properly studied merely by examining the records or observing the processes of NP. The overall political systems, including factional groupings in the units concerned, must be carefully explored.

Concern for the justice qualities of the NP led the Law Commission in its *Fourteenth Report* to suggest short-term training programmes of *nyaya panchas*. The Study Team gave considerable attention to this aspect, noting that *panchas* needed training which would enable them to comprehend the values involved in the notion of 'acting judicially'. *Panchas* must be 'humble' as well as 'fearless'; they must learn to 'bear in mind that all are equal before the law and that the law is no respecter of persons'. They must 'conform to the principles of natural justice and must not only avoid bias, ill will or affection but must appear to have so avoided them'. *Nyaya Panchas* should also be trained to 'administer justice in such a way that respect for the law is maintained'; they must thus know well both the substance and the procedure of the law which they have to apply. For this (and related purposes) the Study Team prepared a comprehensive programme of training including coursework, preparation of a manual, periodic refresher courses, seminars, conferences, *panchayat* journals, radio programmes, etc. Judicial officers, lawyers, and law teachers

---

[86] *Law Commission Report* (1958: 912–19); *Study Team Report* (1962: 47–57).
[87] Singh (1972: 117–22). See also Hitchcock (1960); Nicholas & Mukhopadhyaya (1962); Robins (1962).

(among others) would supply the resources for training *panchas*.[88]

The Study Team noted that there was an overwhelming body of opinion in favour of imparting training to *nyaya panchas*, and yet few programmes of training were launched. Rajasthan trained several hundred *panchas* in 1962–4, but this number was but a small portion of the almost 25,000 *panchas* in the state (based on 20-25 *panchas* for each of 1,370 NP).[89] As recently as 1973, the state government of Manipur allocated Rs 20,000 for training of *nyaya panchas*.[90] Little information is available on the effectiveness of these training programmes. Bastedo noted the futility of Bihar's training efforts, reporting that the rapid turnover of *panchas* in the state and the slow pace of the training programme resulted in only one or two *panchas* in each bench having received any formal help.[91]

The *Study Team Report* suggested several reasons which may explain but not justify the difficulties Bastedo hints at. The number of persons to be trained reaches the hundreds of thousands, many barely literate. Financial problems of NP and the task of training *panchas* in a broadly elective context further compounds the problem. The *Study Team Report* still found that 'planned efforts' were necessary to combat the 'general apathy' toward properly training *panchas*.[92] It appears that apathy rather than planned effort is still the general reponse of the states towards the imaginative suggestions of the Study Team.

Conclusion

It may seem odd, after more than a decade of the era of democratic decentralization and of NPs to raise the question 'Why *Nyaya Panchayats*?' The *Maharashtra Report* urging the abolition of NP and the *Rajasthan Report* advocating reversal of separation from the administrative *panchayat* make this fundamental question of more than academic or historical interest. When we look at the purported justifications of the NP system, we find the same sort of ideological ambiguities as we found in relation to the PR system. These ambiguities may partially account for the present state of affairs in which the NP have been institutionalized, like *gram panchayats*, but in an unsystematic and uneven manner.

Like the village *panchayats*, NP have been perceived, despite sufficient historical evidence to the contrary, as reviving an important

---

[88] *Study Team Report* (1962: 65–72); *The Nyaya Panchayat Road to Justice* (1963). The latter is the manual for the *nyaya panchas*.

[89] *Rajasthan Administrative Reports* (1963–4, 1964–5).

[90] *Manipur Administrative Report* (1972–3: 13).

[91] Bastedo (1969: 217–18).

[92] *Study Team Report* (1962:65–71).

feature of traditional community life in India. Even when the intervening 'swing of the pendulum in favor of centralization' which occurred during the Moghul and British eras is acknowledged, administration of justice by villagers (and therefore decentralization) is perceived as a 'swing back' to the 'historic past and . . . deep rooted sentiment.'[93] The revivalist ideology is one important element accounting for the creation and maintenance of the NP system.

The fervour for 'democratic decentralization' and the institutionalization of the PR system contributed to the creation of the NP. As noted earlier, the logic of the PR system necessitated functional division of labour; it was felt that the wide range of functions to be performed by village *panchayats* should not include 'judicial' functions. Relief from the 'judicial' workload for these bodies was important if they were to perform developmental tasks more efficiently. To this consideration was added the rationale of the separation of the executive from the judiciary sanctified by Article 50 of the Constitution.

The value of access to state justice was another theme which influenced the creation of NPs. Administration of justice could be gotten 'at the door-step of the village' only through NP. When state justice was brought to the village this way, it would have a different complexion; it would be easy and cheap, less 'procedure ridden', more 'informal and flexible' and more community based.[94]

Nyaya Panchayats and village courts would (it was felt) also be carriers of the secular, egalitarian, modernistic, legal ideology, and thus assist in the desired transformation of society. The Law Commission in its *Fourteenth Report*, stresses the 'educative value' of NPs: just as village *panchayats* would educate the villager in the art of self-government, 'so would Nyaya Panchayats' train him in the art of doing 'justice between fellow citizens and instill in him a growing sense of fairness and responsibility'.[95] In this image, the NP constitute an aspect of the overall developmental process.

The NPs are also visualized as the lowest rungs of the state system of administration of justice. The Law Commission recognized NP (and village courts) as very useful devices to prevent court congestion at higher levels, and thought of NP primarily as petty courts. The jurisdictional range of NP, as presently structured, amply substantiates this role.

On the other hand, it is not clear whether NPs constitute primarily a sub-system of the state administration of justice system or a sub-system of the *panchayats* as organs of local self-government. At present, NPs continue to be both. In relation to constitution, finances, and

93 Id., 31.
94 Id., 35–6.
95 *Law Commission Report* (1958: 900–10).

secretariat, the NP constitute a sub-system of the *panchayats*; in their operation, however, they are a sub-system of the state apparatus of justice, both in terms of oversight and review and of relations with police.[96]

Finally (without being exhaustive), one may hypothesize that NPs, consciously or otherwise, represent an extension of the state legal system in rural areas. A desire for the eventual displacement of community dispute institutions and for the eventual spread of the ideology and values of the desired new order contributes to the motivation for the creation and maintenance of the NP. Extension of state law, of course, means extension of the formal polity at the expense of informal community processes. NP are supposed to combine some features of the state legal system (e.g. formal principles of organization and operation, hierarchy, oversight and control) with some prominent features of community dispute-handling processes (e.g. informality, flexibility). This makes NP institutions halfway between fully fledged judicial institutions of the state on the one hand and these 'judicial' institutions which flourish under the auspices of the communities themselves, without the aid of state law.

One can readily see the diversity and incompatibility of those various rationales for the creation of NP. Assessments of the efficacy of NP will, of course, depend on which rationale or objectives are preferred. Thus, if one were to view NP merely as a 'siphoning' device which prevents overburdening of 'higher' levels of the court system, the functioning of NP so far might seem satisfactory, as it was viewed by the *Law Commission* and the *Study Team Reports*. On the other hand, if one viewed NP in terms of their didactic/developmental roles, there would be much reason for dissatisfaction at their present state. Similarly, as components of the state judiciary NP continue to attract suggestions which will make them more accountable and more capable (through training programmes), though these do not seem imperative if NP are merely seen as 'siphoning' devices. On the other hand, greater inputs would be needed for disseminating the ideology and values of a constitutionally desired social order.

The diversity of objectives/rationales for the NP makes problematic both the status of NP and the role of the *nyaya panchas*. NPs have a relatively insecure and subordinate status *both* as judicial institutions and as institutions operating within the context of the PR system, without being an integral part of it. The indeterminacy of status also affects perceptions of role obligations; the *nyaya panchas* are not magistrates (except in some statutory context), nor are they

96 Bayley (1969: 385–408).

adjudicators in the traditional sense. They are *not* elders of caste *panchayats*; they do not necessarily enjoy a reputation for integrity and wisdom; nor are they necessarily members of dominant *jatis*, respected or feared and obeyed as such. Rather, especially when not directly elected, they are ultimately the nominees of the *sarpanch* or a distant official. However, the *nyaya panchas*, like community adjudicators, are to perform their tasks in the spirit of community service; unlike judicial officers they are not paid, their services being honorary. And yet they are to work on this basis *for* the state. Furthermore, unlike the community adjudicators, the powers of *nyaya panchas* are limited with overwhelming clarity; and the ideology inspiring their creation limits the range of sanctions to only those which are available in state law, and that, too, at a level of utmost modesty.

As members of the judiciary, the *nyaya panchas* are effectively isolated from the power hierarchy of the village societies; they cannot hold any formal position within village *panchayats*, *samitis* or *parishads*. On the other hand, because they are not full-time or tenured members of the judicial service, they are not full fledged members of the judiciary either.

In the discharge of their statutory functions, the *nyaya panchas* are to administer justice according to the law; but the law they are to administer requires basic training which they do not have. Nor are they to be assisted in the task by trained lawyers or lawmen. They have little option, in strict theory, except to follow the law, which at best they may only partially understand; when, in fact, they reach decisions in disregard of the law or on a basis other than the law (conciliating when not explicitly provided or deciding on solidary lines), they are liable to social and official criticism. The *nyaya panchas*, at any rate the conscientious among them, may thus be exposed to continuous role conflict.

The ambiguity of role spills over into the provision of facilities as well. Informality and flexibility do not go well with the bureaucratization entailed in adequate secretarial organization. Social distance, negating these features, may be reinforced if NP were to be provided a permanent office. Indeed, any visible symbol of status enhancement may be seen to move NP closer to the formal court systems, thus 'frustrating' some basic aims of the enterprise. As a consequence, we find ambivalence even on the provision of *minimum* facilities for the performance of adjudicatory tasks.

Our structural diagnosis is confirmed by the fragmentary data on workload which show the erosion of NP, once established. Although the evidence is indirect, it all points unmistakably to severe institutional attrition. This unhappy condition seems to reflect the basic ambivalence surrounding the very conception of NP. Earlier governmental policy never decided whether they were to be accessible local organs of official justice or community-based dispute institutions promoted by the state. The pathos of the NP is that they have achieved neither the impartiality of the regular courts (at their best) nor the intimacy, informality and ability to conciliate of traditional *panchayats* (at their best). Instead NP seem in large measure to have achieved a rather unpalatable combination of the mechanical formalism of the courts with the political malleability of traditional dispute processing.

Available data are insufficient to warrant a conclusive evaluation of NP. In the dismal aggregates may lie hidden some exemplary syntheses of intimacy and impartiality, conciliation and redress. We need more than data about structure and workload. We need to ascertain how NPs function and the needs and policies that are served and frustrated by their functioning. Only intensive comparative studies of the working of NP in their social context could provide a basis for a realistic assessment of the potentialities of NP as a means of providing access to justice.[97] Although administration of justice is a subject on which only states may legislate, the moribund condition of NP calls for a national policy to provide access to justice that is based upon a realistic appreciation of the dynamics of *panchayats* and of alternative means of providing access.

Since the Declaration of Emergency in June, 1975, India has entered a period of intensified reconsideration of various institutional arrangements for securing justice. There has been an enhanced concern with the implementation and effectiveness of various reforms. During Emergency Rule the Constitution was amended to reduce drastically the role of the higher judiciary in reviewing public policy. Many matters touching on crucial governmental policies (land reform, procurement of food grains, labour relations, etc.) were removed from the ambit of the regular courts and vested in special tribunals. At the same time, there has been a pronounced concern to increase the access of the poor to the regular courts, evidenced by intense activity to implement long dormant schemes for legal aid and symbolized by the addition to the Constitution of a Directive Principle enjoining the government to provide legal aid to the poor.

[97] Since the body of this paper was completed, the first book-length study of NP has been published: *Kushawaha* (1977); *Nyaya Panchayats in India: A Case Study of Varanasi District* (1977).

Kushawaha does not discuss the dispute process or justice aspects of NP adjudications and outcomes, but this useful work provides some figures on NP in three blocks of one district in U.P. Although the author's estimation is favourable, his data support this paper's conclusions on dwindling caseloads, protracted dispositions, inadequate financial and administrative support, lack of insulation from village politics, and general apathy in the operation of NP.

The renewed interest in access and legal aid has found expression in a variety of proposals for mobile legal aid units, rural law camps, lay advisory and grievance boards, and even 'barefoot judges' (i.e. law students in clinical programmes as adjudicators in villages). The upsurge of interest in making law accessible was not, however, noticeably marked by concern to change the character of the courts or to revivify the NPs.

The end of Emergency Rule and the restoration of judicial authority was accompanied by renewed interest in NP. The Bhagawati Committee's *Report on National Juridicare: Equal Justice — Social Justice*[98] recommends retention and reinvigoration of NP and exhorts the states to reconstitute them. In January 1978 Parliament commissioned a national committee chaired by Shri Asoka Mehta to investigate PR institutions throughout India and to report on the possibilities of further decentralization of local governmental, judicial and developmental bodies. A draft report available in May 1978 called for an improved NP system in every state. It thus appears that a new national policy on decentralized popular adjudicative bodies is emerging. It is a policy that points once again to the unanswered questions posed by the fate of the NP experiment.

[98] Draft Report: Chap. VI (1978). The proposals for NPs are only one part of the Report's expansive programme of legal aid, conciliation cells in the courts, public interest law, etc.

5

Indian Law as an Indigenous Conceptual System *

When I first became interested in Indian law more than twenty years ago, I encountered some quizzical reactions on the ground that since most of Indian law is palpably Western in origin, studying it could not reveal much about India. Nearly ten years ago, I wrote a paper expressing my feeling of wonder that independent India had not seen any serious attempts to dislodge the legal system imposed on it by the British.[1] Both intellectually and institutionally, there seemed an acceptance of the mid-twentieth century system as something fully Indian. There is nothing, for example, like the conflicted but very genuine desire to get rid of the English language as a medium of public life. This 1972 paper tries to explain this lack of interest in dismantling the legal system; the findings boil down to a widespread feeling among professionals, the urban elite, and some villagers that 'this is *our* system'.

This situation points to the question of what we mean by an indigenous conceptual system. Is it one that the indigenous people feel is theirs? Or do we have to look to some elite among the indigenes? If so, which elite do we look to? That is very much the question here, because in India there has been a displacement of an older elite, who were the carriers of an earlier indigenous tradition. This does not imply that the earlier legal complex was an exact counterpart of the law complex that developed during British rule, although each combined authoritative learning, governmental force, and local dispute institutions.[2] The strength of the modern law complex, I have argued, is that it introduced a new and powerful set of ties among these elements.[3]

The older legal tradition fit with other components of social life in ways recognizable to elites and to lay people alike — and to observers from afar. It elaborated such familiar notions as *dharma*

Reprinted from Social Science Research Council, *Items*, 32: 42–6 (Dec., 1978), pp. 42–6.

[1] Galanter (chap. 3).

[2] Rocher (1972) suggests that the British made a fundamental error in equating *dharmasastra* (Hindu systematic learning about duty) and its expositors with law and lawyers.

[3] Galanter (chap. 2).

and it was embodied in persons of religious and political eminence. What replaced it was an institutional-intellectual complex that does not fit neatly with the other institutions. In the new dispensation, there are discontinuities. But at the same time, law as an institution is more differentiated from everything else, so that its inconsistency with the internal content of other institutions (e.g. family, temples, castes) doesn't matter as much.[4]

Perhaps India has moved from a situation in which all the major intellectual-institutional complexes in the civilization were more or less compatible, based (in theory at least) on a common fund of ideas and worked out by a common fund of techniques. Now the institutional-intellectual complexes are bigger, more internally differentiated, more differentiated from each other, and more independent of one another. Thus, a complex like 'law' can exist with its own internal socialization of key actors, its own technical vocabulary, its own styles of thought, its own prestige hierarchy, its own system of social support, and its own system of concepts and meanings. It forms a large and partly autonomous subculture of its own. Is this the case with other institutional-intellectual complexes as well (e.g. government, education, art, journalism)?

When a society undergoes this kind of transformation, it seems to me that 'indigenous' takes on a new meaning. In the old dispensation, one could spot a foreign element because of its lack of fit with others. But in the new world, institutions don't flow into and reflect each other in the same way. It is a world of discontinuities and abrupt shifts between different parts of one's life experience. The indigenes are not always recognizable or accessible to one another. There may be great barriers between one indigenous phenomenon and another.

The law may loom as intrusive, alien, and external to many within Indian society, while others may find it comfortable, known, wholly familiar, and expressive of their values and outlook. Clearly, 'indigenous' cannot mean just that which is not of foreign origin. How far back do you go? There have been many grafts on the tree. The Indians have taken British law and have reworked it, imposed unique stylistic features on it.[5] The affinities with British or American law (there was a second wave of borrowing of American constitutionalism) are striking; the kinship is apparent, yet the Indian style is quite distinctive.

As George Gadbois[6] has pointed out, there must have been some special predilection among Indians underlying this extraordinary

---

[4] Cf. Mayhew (1971).
[5] Galanter (Chaps. 1 & 2).
[6] Gadbois (1977)

receptiveness. This eager and unfaltering embrace of Western law simply did not take place in Southeast Asia or in Japan (and I think this is true for Africa as well).

The attachment to Anglo-Indian law (and its American-style constitutional overlay) runs deep. Two major new pieces of evidence suggest the tenacity of these forms in late twentieth century India. First, there is the resistance to alterations in these institutional forms manifested in connection with the 1975–7 Emergency regime. Second, there is the disintegration of post-independence efforts to revive *nyaya panchayats* (village tribunals).

During the 1975–7 Emergency, there was a decisive break with the liberal constitutionalist vision of the legal order that was shared by the Bar, officials, the political elite, and a large part of the populace since before independence—an order in which independent courts vindicated private rights and curtailed official arbitrariness. The major contours of the Emergency regime are familiar by now: the elimination of civil liberties, particularly by control of the press and by a reduction in the autonomy of organized groups such as labour unions and the Bar; the elevation of the executive to decisive supremacy over Parliament: and especially a reduction of the power of the judiciary to decide public issues.

After a few gestures in the direction of repudiating the constitutional design, the government of Mrs Gandhi pursued a policy of removing the courts from politically sensitive areas, making them more subserviant politically but retaining, thus truncated, the rule of law. The new constitutional amendments eliminated the judiciary's power to pass on the propriety of constitutional amendments, severely curtailed the power of higher courts to adjudge statutes unconstitutional, and eliminated the jurisdiction of the ordinary courts over such politically sensitive areas as government employment, revenue, land reforms, labour disputes, and elections. Roughly, all of the politically volatile cases were withdrawn from the regular courts and put into tribunals that would, presumably, have been more politically responsive. The supervisory power of the regular courts over these tribunals was curtailed or eliminated. But with these significant amputations—and subject to some attempts to influence the judges by arbitrary transfers, etc.—the regular courts were left to function as before. The pattern that emerged was reminiscent of that the Spanish sociologist José Toharia[7] suggests is the classical authoritarian response to law. Referring to the final years of the Franco regime, he wrote.

> The situation of the ordinary courts under the present Spanish regime [is characterized] . . . on the one hand by the considerable degree of independence allowed to the judges (with a consequent lack of political

indoctrination and the exercise among them of a certain ideological diversity) and on the other by the sharply curtailed and relatively unimportant sphere of action afforded them.

The administration of justice is thus controlled by avoiding the political mobilization of individuals and institutions and by 'the systematic reduction of the area of competence of ordinary tribunals'.

In spite of these drastic incursions into the legal order, there were during the Emergency remarkable continuities. Most prominently, there was an adherence to superficial legal proprieties and an unwillingness to depart from the major institutional forms of Anglo-Indian law. Although there were a few official mutterings about people's courts, these did not amount to more than a scare. There seems to have been no serious consideration of replacing or transforming the familiar brand of Anglo-Indian courts either with 'people's courts' or with indigenous *panchayats*. Indeed, the government undertook new commitments to support a massive legal aid programme, presumably to increase access to and use of the legal system.[8] If there were outcroppings of a genuine desire to make legal institutions responsive to popular needs, there was little sense that these institutions would require alteration or transformation in order to accomplish this.

Although both technocratic and populist themes were invoked during the Emergency to constrain the autonomy of legal institutions, the new legal measures were in very much the same style that had previously animated courts, lawyers, and legislators. The Emergency measures involved no attempt to reshape legal forms, but only sought to truncate and contain them. Law remained formalist rather than instrumentalist, a matter of individual entitlements conferred from the top rather than one of facilitating group mobilization, a matter for professionals rather than for lay participants.

As soon as the Emergency ended there was an immediate concern to restore legal institutions to their former condition. Is this restoration only a temporary reversal of a long run trend toward the extrusion of a foreign element? In that light, one might see Mrs Gandhi as a precursor of a movement to develop truly Indian political and legal forms. But what is remarkable is that even during the Emergency there was no significant challenge to the basic form and style of legal institutions. There was griping from the government side

---

[7] Toharia (1975).

[8] While curtailing judicial power, the infamous 42nd. Amendment to the Indian Constitution added a new Directive Principle that the State shall 'provide free legal aid . . . to ensure that opportunities for securing justice are not denied to any citizen.' The justice to be secured is, by implication, that dispensed by the then current variety of laws, courts, and lawyers.

about the courts interfering with what were thought to be issues that should be left to Parliament and the Executive. But the basic shapes and style of courts as institutions and of the legal learning that they cultivate and dispense was never questioned. Both Mrs Gandhi and her opponents shared a basic acceptance of India's legal institutions.

The second body of data is that concerning the disintegration of the *nyaya panchayats* (village tribunals) promoted by the government after independence. In a recent assessment of this programme [see chap.4], Upendra Baxi and I found that the *nyaya panchayats* established in the 1950s and early 1960s are for the most part moribund. Where these tribunals have not been abolished by state governments dismayed at their workings, they have low case loads and enjoy little public or official regard. Although lack of government support undoubtedly made an important contribution to their demise, it is clear that they never attracted significant support from the villagers in whose name they were established. It is not clear whether they withered away because they lacked the qualities of the traditional indigenous tribunals or because they displayed them all too well. Most likely, it was because they represented an unappetizing combination of the formality of official law with the political malleability of village tribunals. Nevertheless, there now appears to be a new wave of interest in these institutions.

Recent proposals to promote *nyaya panchayats* tend to visualize them as agents for the efficacious delivery of official justice rather than as means to revive indigenous justice. Thus, in a report[9] that views itself as a radical critique of Indian legal arrangements, Justice Krishna Iyer speaks glowingly of *nyaya panchayats*, citing Lenin in their favour, and commends them as a promising counterpart to 'the system of justices of peace in the United Kingdom and People's Courts in the Soviet Union. . . .' These examples he reads as establishing that 'laymen may well be entrusted with dispensing legal justice provided certain safeguards are written into the scheme.'[10] It is clear that the 'legal justice' to be dispensed is the law of the land and not that of the villagers or their spiritual advisors. *Nyaya panchayats* are commended as inexpensive, accessible, expeditious, and suitable to preside over conciliatory hearings. The Report recommends that they be staffed by superannuated judges and retired advocates. *Panchayat* justice is seen as part of a larger scheme of legal aid and as providing public access to the courts.

Charged with proposing concrete measures to secure access to justice for the poor, the Bhagwati Report elaborates this discussion

[9] Government of India (1973).
[10] For an illuminating critique, see Baxi (1975).

of *panchayats*. After the obligatory nod to Mohandas Gandhi, the Report observes that

> ...there can be little doubt that our legal and judicial system is not adequate to meet the needs of the new society which is emerging in our country. It is not effective to provide a solution to the new problems which are coming up and presenting a challenge to contemporary society. It is not sufficiently responsive to the new norms and values which are replacing the old and it does not reflect properly and adequately the new approach which characterizes the true purpose and function of law.[11]

Thus, the legal system needs to be changed to become a more effective instrument for delivering justice to the poor and disadvantaged and to do this it is necessary to consider ways of making justice cheap, expeditious, and accessible. But again, this is justice of the current variety.

The Report continues with an eloquent characterization of the 'file-fed . . . dehumanized' and procrastinating character of the existing system. It comes out for a system of 'law and justice at the *panchayat* level with a conciliatory methodology.' An idealized account of pre-British *panchayat* justice and lengthy citation of Gandhi are juxtaposed with Lenin's commendation of popular participation and a former law minister's equation of *panchayat* justice with people's courts.

Nyaya panchayats are endorsed on the ground that they would remove many of the defects of the British system of administration of justice since they would be manned by people with knowledge of local customs and habits, attitudes and values, familiar with the ways of living and thought of the parties before them. '[T]he poor would feel that the authority which is administering justice to them is their own and not that of an alien system which they neither understant nor trust.' Yet the proposed *panchayats* do not depart from current notions of law. The *panchayat* is to be comprised of a chairman, the *panchayat* judge, 'having knowledge of law,' and two respected lay members who would be drawn from a select panel for each case by the *panchayat* judge. The lay members would be given rudimentary legal training. The tribunal would proceed informally, would emphasize conciliation; there would be no professional lawyers, etc. Legal questions would be decided by the judge and his decisions could be reviewed by the District Judge. What seems to be proposed is an informal, conciliatory, nonadversarial, small claims court with some lay participation.

These reports indicate that even those Indian jurists who are most critical of the present system don't visualize an 'indigenous' alterna-

11 Government of India (1978: chap. 6).

tive to it—at least if we take 'indigenous' to mean a return to earlier forms. I think it is fair to say that at the level of discourse about policy, there is an almost total obliteration of indigenous elements.[12] However, if one looks at the social organization of modern Indian law, I think it is possible to identify elements that seem to be the extension or projection of traditional patterns.

For example, when I recently attempted to trace in detail the way a body of legal doctrine developed in contemporary India, I was surprised to see how equivocal it was and how that equivocation was so solidly institutionalized. Doctrines of authority and precedent and the hierarchic organization of courts led me to expect a body of fixed hierarchically-established doctrine. But the pressures of case load, turnover of judges, the system of sitting in small benches, the limitations of legal research, and other factors combined to transform the law into something very different. What purported to be a pyramidal hierarchy establishing fixed doctrine turned out to be a loose collegium presiding over an open-textured body of learning within which conflicting tendencies could be accommodated and elaborated.[13]

This suggests that the pattern of continuities and discontinuities is more complex than we might infer from the overt and dramatic displacement of pre-British law. What was displaced was, after all, not a seamless monolith but a complex comprising a distinctive body of elevated legal learning, linked indirectly and unevenly to governmental practice and to the self-regulatory activities of villages, guilds, castes, and other groupings. Hindu legal learning contemplated broad delegation of regulatory functions to these groups, an expectation that was if anything exceded in practice.[14] 'Law' in the indigenous style included not only regulation that was local and comprehensible, but elements that were more external and opaque to the locals. The imposition of British law introduced both new legal learning and new techniques for impressing this learning on the various lesser regulatory systems. Much of the subsequent history of law in India may be seen as the interplay between the nationalization of legal activity and the continuing drive for self-regulation in many sectors of social life: local interests and understandings found new channels of expression and

[12] The indigenous content of family law, the last redoubt of traditional law in the modern system, has been eroded by a series of post-independence reforms. A brief account may be found in Galanter (1978). J.D.M. Derrett (1978), the leading student of these matters, contends that the traditional mould of Hindu marriage law was shattered by 1976 legislation.

[13] The example is elaborated in Galanter (1984).

[14] The actual situation was rendered complicated beyond elaboration here by the presence of Islamic as well as Hindu legal traditions.

the new national institutions learned to accommodate diverse sorts of local regulatory activity.

The tension between authoritative higher law and local law-ways was not introduced with British law, but was a constituent part of the earlier indigenous system.[15] Perhaps modes of combining local and 'external' elements have survived the drastic changes of content. Perhaps new and distinctively Indian ways of accommodating these tensions have arisen. Or perhaps the Indian experience exhibits universal features of the accommodation of diverse normative orderings in the legal systems of complex heterogeneous societies. Certainly neither the presence of a multiplicity of normative orders nor the gap between local law-ways and the most authoritative legal doctrine are phenomena unique to India. But the sociology of law has had great difficulty in addressing these phenomena— in part because it has shadowed a tradition of normative learning which insistently obscured the visibility and centrality of these features by branding them as marginal, transient, or pathological.[16]

Hindu law; on the other hand, openly embraced normative diversity; legal learning was attuned to a multiplicity of legitimate group norms. It may well be that the older Indian learning has something to tell us about the patterns that prevail in India today. For eample, indigenous notions of collegial authority or of deference to local decision makers may help us to understand patterns of regulation in contemporary India. And by extension they may help us to develop concepts for exploring the way in which the official, national legal systems in modern societies accommodate the variety of normative orderings which, in spite of the law's pretensions to hierarchic control, coexist with it. The relation of official law to the other normative orders may, I surmise, provide the key to a deeper understanding of law in modern societies.[17] By learning from India's very different way of visualizing them, we can hope to illuminate these relations.

[15] Kidder (1978) attempts to separate the foreignness and the 'external' character of British law in India.

[16] Galanter (1974b).

[17] Galanter (chap. 10).

III

Legal Conceptions
of the Social Structure

6

Group Membership and Group Preferences in India *

The Constitution of India authorizes the government to provide special benefits and preferences to previously disadvantaged sections of the population[1] Reserved posts in government, reserved seats in legislatures and on local political bodies, reserved places in public educational institutions and an array of preferences and welfare measures have been made available to Scheduled Castes ('untouchables') and Scheduled Tribes and, to a lesser extent, to Backward Classes.[2]

* Reprinted from *Journal of Asian and African Studies*, 2:91–124 (1967).

[1] Governmental discrimination on grounds of race, religion and caste is prohibited by Art. 15(1) (all references are to the Constitution of India, 1950, as amended.). More specifically, discrimination is prohibited by Art. 16(2) (in regard to state employment); Art. 23(2) (in regard to compulsory public service); Art. 29(2) (in regard to state-run and state-aided educational institutions); Art. 325 (in regard to electoral rolls).

It is a 'Directive Principle of State Policy' that: 'The State shall promote with special care the educational and economic interests of the weaker sections of the people, and, in particular, of the Scheduled Castes and the Scheduled Tribes, and shall protect them from social injustice and all forms of exploitation.' Art. 46. Consonant with this directive, Art. 15 and Art. 29(2) are qualified by Art. 15(4), which provides that the State may make '. . . any special provision for the advancement of any socially and educationally backward classes of citizens or for the Scheduled Castes and the Scheduled Tribes.' This proviso was added by the Constitution (First Amendment) Act, 1951, §2, after the Supreme Court held that a state had no power to reserve seats in educational institutions for members of backward communities. *State of Madras v. Champakan Dorairajan*, A.I.R. 1951 S.C. 226. Art. 16 is qualified in Art. 16(4) to permit the State to make 'any provision for the reservation of appointments or posts in favour of any backward class of citizens which, in the opinion of the State, is not adequately represented in the services under the State.' Cf. Art. 335.

In the sequel, unless the context otherwise requires, the word 'reservations' is used to refer to reserved posts in government service authorized by Art. 16(4); the word 'benefits' is used to refer to all special provisions authorized by Art. 15(4); the word 'preferences' is used as a general term including both of the above.

[2] Arts. 330 and 332 specifically provide reserved seats in Parliament and Legislature for Scheduled Castes and Scheduled Tribes. Backward Classes do not enjoy this preference. Art. 334 provided that such reservations should expire ten years after the commencement of the Constitution. They were extended for another ten years by the Constitution (8th. Amendment) Act, 1959.

The term 'backward classes' is sometimes used in the broader sense of including the former two groups as well. In the sequel it is used in the narrower sense.

With membership in these groups a qualification for preferment of various kinds, it is not surprising that disputes have arisen concerning such membership. In a number of cases over the past fifteen years, the courts have had to pass on the question of whether a person was in fact a member of such a preferred group. The cases raise many puzzling questions: is membership in a caste or tribe to be determined solely by birth, or by allegiance, or by the opinion of its members or of the general neighbourhood? Does one lose (or gain) caste by conversion? By excommunication? By assimilation? Does one lose tribal membership by claiming or achieving caste status? Who is a communicant of a particular religion? Those born into it? Those who have been converted to it? Those who adhere to its precepts? What is the effect of unorthodoxy? Of excommunication? Are the tests used for the application of personal law appropriate in the area of preferences?

Prior to independence and to a diminishing extent since, the courts have faced problems of determining membership in social groups for the purpose of applying appropriate rules of personal law in such fields as marriage, divorce, and inheritance. The recent judicial treatment of group affiliation represents an adaptation of this older jurisprudence to a new purpose. More notably, it reveals some of the assumptions, explicit and implicit, about the structure of Indian society, which guide the courts. It is proposed to examine these assumptions from the view-point of their consonance with the principles of the Indian Constitution and with empirical knowledge of Indian social organization.

Where a nation's fundamental law envisages a far-reaching reconstruction of society, the judiciary are inevitably engaged in the delicate task of mediating between social actualities and the avowed goals of the polity. They are both authoritative interpreters of these goals and assessors of the changing actuality in which these are to be realized. These are not wholly separate undertakings, for the interpretation of goals and values takes on content and colour from one's picture of what is; and the perception of the latter tends to be informed by an awareness of one's goals. What one imagines to be desirable and attainable and what one imagines to be real are interdependent and interpenetrated. In examining judicial methods of solving problems of group membership we can discern competing views of Indian society and divergent proposals for the implementation of the principles of the Indian Constitution.

For description of the various schemes of preference employed see the annual reports of the Commissioner for Scheduled Castes and Scheduled Tribes (Delhi, 1951). For a critical analysis, see Dushkin (1961).

Caste and Sect

The Constitution does not itself define the groups which may receive preferences nor does it provide standards by which such groups are to be designated. It provides for the initial designation of Scheduled Castes and Scheduled Tribes by Presidential order[3] with subsequent modification only by act of Parliament.[4] The wider and more inclusive group of Backward Classes is not only undefined in the Constitution, but no such method or agency for their determination is provided.[5] Unlike Scheduled Castes and Tribes, Backward Classes may be designated by state and local as well as central government and by administrative as well as legislative authorities.[6] Although there is a growing tendency to substitute economic criteria of backwardness[7], for the most part the Backward Classes designated by the states have been caste groups.[8]

In order to qualify for preferences, one must be a member of a listed caste or tribe. The tests for determining membership were given extensive consideration by the Supreme Court in *Chatturbhuj Vithaldas Jasani v. Moreshwar Parashram*. An Election Tribunal had rejected the nomination papers for a reserved seat submitted by a Mahar who had joined the Mahanubhava Panth, a Hindu sect which

[3] Art. 341(1) empowers the President to specify, after consulting the Governor of a State, those 'castes or races or tribes or parts of or groups within castes, races or tribes which shall for purposes of this Constitution be deemed to be Scheduled Castes in relation to that State'. See Constitution (Scheduled Castes) Order, 1950. Art. 342(2) provides that the President may similarly specify 'tribes and tribal communities or parts of or groups within tribes or tribal communities' to be the Scheduled Tribes. See Constitution (Scheduled Tribes) Order, 1950.

[4] Art. 341(2), Art. 342(2). The list of Scheduled Castes was revised by Parliament in 1956. Scheduled Castes and Scheduled Tribes (Amendment) Act, 1956.

[5] Art. 340 provided for the establishment of a Backward Classes Commission to be appointed by the President to investigate the conditions of the Backward Classes and recommend measures for improvement. This investigative commission, established in 1953, was directed to determine criteria for designating Backward Classes but its report did not meet a favourable response. For a discussion of difficulties in designating Backward Classes, see Galanter (1961b).

[6] *Kesava Iyengar v. State of Mysore*, A.I.R. 1956 Mys. 20; *Ramakrishna Singh v. State of Mysore*, A.I.R. 1960 Mys. 338; Cf. *Venkataramana v. Madras* (1951), S.C.J. 318.

[7] See *Report of the Commission for Scheduled Castes and Scheduled Tribes*, 1958–9, p. 12.

[8] The use of caste or communal groups as units that may be designated Backward Classes has been consistently upheld by the courts. *Balaji v. State of Mysore*, A.I.R. 1961 Mys. 220. But it is now clear that the low social standing of a class is in itself neither a necessary nor a sufficient criterion of 'social backwardness' to justify its designation as a Backward Class. *Balaji v. State of Mysore, supra*; *Chitralekha v. State of Mysore*, A.I.R. 1964 S.C. 1832. The latter case also throws some doubt on the permissability of using caste groups as the units which are designated as Backward Classes. But cf. *State of Kerala v. Jacob Mathew*, I.L.R. 1964 (2) Ker.53. In any event, caste groups may still be used in designating the Scheduled Castes.

repudiated the multiplicity of gods and the caste system. Reversing the Tribunal, the Supreme Court held that the candidate remained a Mahar and was thus entitled to stand for the seat reserved for Scheduled Castes. To determine whether adherence to this sect made the candidate cease to be a Mahar, the Court specified three factors to be considered: '(1) the reactions of the old body, (2) the intentions of the individual himself and (3) the rules of the new order.'[9]

The candidate was admitted to all Mahar caste functions and had been allowed to marry within the community. He twice married Mahar girls, neither of whom were Panth members at the time of the marriage. He always identified himself as a Mahar. The Panth, in spite of its doctrinal repudiation of caste, had not penalized him for his adherence to the caste.

The Supreme Court concluded that 'conversion to this sect imports little beyond an intellectual acceptance of certain ideological tenets and does not alter the convert's caste status.'[10] It is clear that the primary consideration was the second test—i.e. the intentions of the convert himself—intention not in the sense of mere declaration but as evidenced by a consistent course of conduct and dealings. The Court applied the 'broad underlying principle' of *Abraham v. Abraham*, decided a century before by the Privy Council in determining the law of inheritance applicable to a Hindu convert to Christianity: [h]e may renounce the old law by which he was bound, as he has renounced his old religion, or, if he thinks fit, he may abide by the old law, notwithstanding he has renounced his religion.'[11] Applying this principle, the Supreme Court found that 'if the individual . . . desires and intends to retain his old social and political ties' and if the old order is tolerant of the new faith and does not expel the convert, the conversion has no effect. 'On the other hand, if the convert has shown by his conduct and dealings that his break from the old order is so complete and final that he no longer regards himself as a member of the old body and there is no reconversion and readmittance to the old fold . . . [he cannot] claim temporal privileges and political advantages which are special to the old order.'[12]

---

[9] [1954] S.C.R. 817, 838. This case is hereafter referred to as *Jasani*.

[10] Id. at 840. The Panth is a devotional sect, founded in the eleventh century, which not only eschewed caste and polydeism, but also challenged the validity of the Vedas, image worship, and the system of *asramas* (stages of life). On the role of this sect in Mahar tradition, see R. Miller (1966).

[11] Quoted from reporter's note to *Abraham v. Abraham*, 9 M.I.A. 199 at 196 (1963). Cf. pp. 242–4. The rule was subsequently overturned by the Indian Succession Act, 1865, (now Indian Succession Act, 1925, s.58). But the courts are still divided over whether the Hindu rule of survivorship is applicable to Christian families who continue to be joint after conversion.

[12] [1954] S.C.R. at 838.

Although the test is primarily one of the convert's intention or behaviour this intention must be confirmed by acceptance by the old group. The inclusion of this additional test is important because as the court says, 'the only modification here is that it is not only his choice which must be taken into account but also the views of the body whose religious tenets he has renounced, because here the right we are considering is the right of the old body, the right conferred on it as a special privilege to send a member of its own fold to Parliament.'[13]

The third test, 'the rules of the new order', is of minor significance. Since it is the legal and political rights of the old body that are being considered 'the views of the new faith hardly matter.'[14] 'The new body is free to ostracise and outcaste the convert from its fold if he does not adhere to its tenets but it can hardly claim the right to interfere in matters which concern the political rights of the old body, when neither the old body nor the convert is seeking legal or political favours from the new body as opposed to purely spiritual advantages.'[15] If this test has to be taken into account at all, it is only as indirect evidence of the intentions and conduct of the convert. For example, continued acceptance by a new group which was notoriously intolerant of the retention of the old ties might well evince an intention to break with the old group. But here the Court found it 'evident that present day Mahanubhavas admit to their fold persons who elect to retain their old caste customs.'[16]

In *Shyamsundar v. Shankar Deo Udgir*,[17] the principles of the *Jasani* case were applied to decide whether a candidate for a reserved seat had lost his membership in the Samgar caste by joining the Arya Samaj, a Hindu sect which rejects idolatry and ascription of caste by birth.[18] He had been accepted for membership in a local Arya Samaj organization, had paid membership dues, had married a girl of Sonar caste in accordance with Arya Samaj rites and had reported himself as an 'Arya' in the 1951 census. The Mysore High Court, citing the *Jasani* case, announced that there would be no deprivation of caste unless there was either (1) expulsion by the old caste or

13 Id. at 839.

14 [1954] S.C.R. at 838.

15 Id.

16 Id. at 840. For a situation in which the rules of the new order played a more prominent role, see *Rhagava Dass v. Sarju Bayamina*, A.I.R. 1942 Mad. 413, where by joining the Byragi: sect a person ceased to belong to his original caste.

17 A.I.R. 1960 Mys. 27.

18 On the Arya Samaj, see Lala Lajpat Rai (1915). The caste referred to as 'Samgar' here is evidently the same as that listed as 'Samagara' in census of India, Paper No.2 (1960): Scheduled Castes and Scheduled Tribes Arranged in Alphabetical Order, p. 84. The Sonar caste, mentioned below, are traditionally goldsmiths and of a higher social standing, usually associated with Vaishya status. See note 26 below.

(2) intentional abandonment or renunciation by the convert. There being no evidence of expulsion or ostracism by the old caste, the question was whether there had been a break from the old order 'so complete and final that . . . he no longer regarded himself as a member of the Samgar caste.'[19] The Court found that his activities evinced that he regarded himself as a Samgar, as did his testimony that he believed in idolatry and in texts repudiated by the Samajists. The Court found no evidence that he could not have married the Sonar girl 'in the ordinary way' and thus the marriage was not inconsistent with his membership in the caste, nor was the census report since the Court refused to accept 'Arya' as equivalent to Arya Samajist. Almost as an afterthought, the Court notes that the Samaj did not expel him for departure from their tenets. Such expulsion would only have reinforced the Court's conclusion, where the absence of it (if there ever were expulsions) might indicate an acceptance inconsistent with his remaining in the caste. The test that emerges, somewhat inchoately, is that so long as the person identifies himself with the old caste and is accepted by the caste—no matter if he is accepted by the new group or not—he remains a member of the caste for purposes of qualification for nomination to a reserved seat.

Tribe

In *Kartik Oraon v. David Munzni*,[20] the challenged candidate was a member of an Oraon family that had been Christian since his grandfather's time. His eligibility to stand for a seat reserved for Scheduled Tribes was challenged on the ground that as a Christian he was no longer an Oraon, since he had abandoned the animistic faith, did not follow the manners and customs of the tribe, and had no affinity of interests or aspirations with the tribal people. The High Court found the candidate's active participation in the civic life of the tribe belied the charge of lack of common interests and aspirations. As to manners and customs, there was evidence that although Christian tribals did omit certain observances of tribal religion, they retained such practices as exogamy based on totemistic lineage, certain harvest rites, ceremonial eating of first fruits, birth and marriage observances, and style of writing surnames. Even if he omitted to observe certain festivals or observed some in a manner different than other tribals, the Court concluded, the 'most important thing . . . is that the non-Christian tribals treat the converted Oraons as

---

[19] A.I.R. 1960 Mys. at 32.
[20] A.I.R. 1964 Pat. 201. On the Oraons see Roy, (1928).

tribals, calling them 'Christian Oraons'.'[21] Christian and non-Christian Oraons intermarry and their descendents are treated as full members of the tribe. Christian Orans are invited to feasts and participate in them. Applying the *Jasani* tests, which it found 'fully applicable', the Court concluded that conversion did not extinguish membership in the tribe.[22]

In *Wilson Reade v. C. S. Booth*[23] an Election Commission had rejected nomination papers for a Scheduled Tribe seat from a candidate whose father was English and whose mother was a Khasi. In pre-independence days he had accepted for himself and his children (his wife was a Khasi) privileges restricted to Anglo-Indians. But he was accepted as a Khasi by the tribespeople, the group being matrilineal and anyone born of a Khasi mother being regarded as a member of the tribe; he had followed 'the customs and the way of life of the tribe', was treated by them as one of themselves and had been active in Khasi, politics. The Assam High Court found that even though he was an Anglo-Indian within the Constitutional definition,[24] this did not prevent him from being a member of this tribe or some other community. Whether he was in fact a Khasi depended not on purity of blood but on his conduct and on the acceptance of the community.

The situation in the *Wilson Reade* case is the reverse of that in the *Jasani*, *Shyamsundar* and *Kartik Oraon* cases. Here, it is the 'new' group that is the politically privileged group rather than the 'old' one. The question is not whether the new identification precludes the old but whether the old one precludes the new. In *Wilson Reade* as in the other cases, the Court looked to the views of the privileged group to confirm the individual's claimed membership. It was the Khasis who where entitled to special representation and it was their acceptance of him that was determinative. Neither birth nor the

---

[21] Id. at 203.

[22] A similar approach was applied by an Election Tribunal which had to decide whether a Konda Dora who had converted to Christianity at the age of ten 'for the purpose of his education' was a member of the Tribe and thus qualified to stand for a reserved seat. There was no evidence that the convert had been excommunicated by the Tribe. Finding that converts observe the same customs and habits, intermarry and are treated as members of the tribe, the Tribunal held that 'mere acceptance of Christianity is not sufficient to make him cease to be a member'. *Gadipalli Parayya v. Boyina Rajayya*, XII E.L.R. 83 (1956).

[23] A.I.R. 1958 Ass. 128.

[24] Art. 366(2) defines an Anglo-Indian as 'a person whose father or any of whose other male progenitors in the male line is or was of European descent but who is domiciled within the territory of India and is or was born within such territory of parents habitually resident therein. . . .' Some preferences for Anglo-Indians on a declining scale for a ten-year period were provided by the Constitution. Arts, 331, 333, 336, 337. On the Anglo-Indian community, see Grimshaw (1959). On this tribe, consult P. Gordon (1914).

possibility that the Anglo-Indian community might also have accepted him was considered relevant. Thus all of these cases permit overlapping and multiple group affiliations. The possibility that an individual might be accepted by a second group is not taken to automatically remove him from the first.

Tribe and Caste

The empirical approach of the *Jasani* case has not been applied in the problem of tribals attaining caste status. The case of *V. V. Giri v. D. Suri Dora*[25] arose out of an election to a seat in Parliament reserved for a member of a Scheduled Tribe. The candidate was born a Moka Dora and his family had described itself as such in all documents from 1885 to 1888. Since that time they had described themselves as Kshatriyas.[26] There was evidence that the family had adopted Kshatriya customs, celebrated marriages in Kshatriya style, was connected by marriage to Kshatriya families, employed Brahmin priests and wore the sacred thread in the manner of Kshatriyas. His election was challenged on the ground that he was no longer a Moka Dora and was therefore ineligible to stand for the seat.

Applying the tests set forth in the *Jasani* case the Election Tribunal concluded that the candidate was no longer a Moka Dora, finding that 'he has expressed unequivocal intention of drifting away from the clan, has totally given up feeling himself to be a member of the Moka Dora tribe and considers himself a Kshatriya'.[27] Apparently the candidate's family was one of a number of families of Mokasadars or large landowners who, according to the Tribunal, 'would not like to be called Moka Doras but considered themselves Kshatriyas'.[28] The Tribunal found support for its finding in the observation that 'persons of [this] type . . . who have drifted away from their old clan and renounced the tribal customs and manners and chosen to adopt the prevailing practices of the higher caste [sic] of the Hindu community could not be entrusted with the task of representing the genuine grievances' of the tribal communities; to do so would amount to a denial of the benefit of special representation conferred on the tribals by the Constitution.[29] Since the Tribunal

[25] A.I.R. 1959 S.C. 1318. Hereafter referred to as the *Dora* case.
[26] Classical Hindu legal and social theory divides society into four great *varnas* (literally 'colours') or classes: Brahmins (priests and scholars); Kshatriyas (rulers and warriors); Vaishyas (merchants and agriculturists); Sudras (menials). *Varna* distinctions are influential but bear no direct relationship to existing social divisions. See Srinivas (1962: chap. 3).
[27] XV E.L.R. 1 at 38 (1957).
[28] Id.
[29] Id.

had found against the candidate on the question of his intention to remain a Moka Dora, it was not necessary to go into the question of acceptance by either the tribe or the Kshatriyas.

So far the case, although reaching the opposite result, proceeded along the lines laid down in the *Jasani* case. However, the case took a radically different turn when it reached the High Court of Andhra Pradesh[30] which addressed itself not to the question of whether he had remained a Moka Dora but to the quite distinct question of whether he had become a Kshatriya. Starting with the principle that caste is a matter of birth rather than choice and that higher caste cannot be gained,[31] the Court conceded that 'it is possible that a member of a Scheduled Tribe may in course of time adopt certain customs and practices in vogue among the Hindus, but in order to bring them within the fold of Hinduism it would take generations. Even if they came within the fold of Hinduism, [a] question would arise whether they have formed a separate sect among themselves, or [whether] they would belong to the . . . [Sudras] or to the twice-born class.'[32] Having thus indicated the severe limitations of possible mobility, the Court proceeded to lay down as requirements for proving that such movement had taken place its version of the tests employed in the *Jasani* case. These tests — intention, reaction of the old group, rules of the new group — are used in the *Jasani* case to test whether the individual has remained in his old group. But here the Court uses them as tests of assimilation into the new group. They become in effect a set of binding requirements, which must all be satisfied in order to prove a case of successful mobility.

The High Court found no evidence of the reaction of the old tribe (which, by the *Jasani* approach, would have been irrelevant had he failed on the test of intention) and 'no evidence as regards the reaction of the new fold, except that some of the Kshatriyas recognize appellant as a Kshatriya. We can understand this if this had been the result of generations, but the acceptance of the appellant as a Kshatriya by one of two families would not . . . be sufficient.'[33] Having thus failed to attain Kshatriya status, the High Court

---

[30] A.I.R. 1958 A.P. 724.

[31] The cases cited by the High Court in support of this point, id. at 735, are easily distinguishable. *Sahdeonarain v. Kusumkumari*, A.I.R. 1923 P.C. 21, and *Chunku Manjhi v. Bhabani Majhan*, A.I.R. 1946 Pat. 218, are concerned with whether tribals are governed by Hindu personal law. *Maharajah of Kolhapur v. Sundaram Iyer*, A.I.R. 1925 Mad. 497 is concerned with *varna* status for purposes of finding the applicable rule of inheritance.

[32] A.I.R. 1958 A.P. at 735. The 'twice-born' are the three higher *varnas* — Brahmins, Kshatriyas, and Vaishyas — who have supposedly undergone a second or intellectual birth upon investiture with the sacred thread. The Sudra *varna* is only once-born.

[33] Id. at 736.

assumed that he therefore remained a Moka Dora and found him eligible for nomination to the reserved seat.

The Supreme Court, rather than reasserting the *Jasani* tests and disengaging them from the High Court's theories about caste mobility, took a third tack. Where the Election Tribunal had addressed itself primarily to the 'intention' test laid down by *Jasani*, and the High Court had insisted that all three factors mentioned in *Jasani* were required to prove mobility, the Supreme Court fixed its attention only on the third — and originally least important — of the *Jasani* tests: the reactions of the new group.

The Supreme Court found the evidence insufficient to demonstrate that the candidate was a Kshatriya, since 'the caste status of a person in this context would necessarily have to be determined in the light of the recognition received by him from the members of the caste into which he seeks an entry'.[34] Finding no evidence of such recognition, the Court said 'unilateral acts cannot be easily taken to prove that the claim for the higher status which the said acts purport to make is established.'[35] The Court concluded that the candidate had not become a Kshatriya and had therefore remained a Moka Dora, eligible for the reserved seat.[36]

In spite of the similarity in outcome, the course of reasoning taken by the Supreme Court here is in sharp contrast to that in the cases discussed earlier. The *Dora* case agrees with its predecessors that neither birth nor mere intention is determinative of group membership; the conduct of the individual and the attitudes of the groups must be considered. But which groups? And their attitudes about what? In the cases discussed above, when the question was whether X had, by joining a new group B, ceased to be a member of privileged group A, it was the group A whose reactions were consulted — the Mahars in the *Jasani* case and the Samgars in the *Shyamsundar* case. But when the question was whether a person had become a member of a privileged group B, then the views not of the old group A but only of group B were pertinent — the Khasis in the *Wilson Reade* case. It was irrelevant what the Anglo-Indians might have thought of Reade, just as the views of the Manahubhava Panth and the Arya Samaj received only a passing glance. Had the courts in these cases

[34] A.I.R. 1959 S.C. 1318 at 1327.

[35] Id.

[36] Another election tribunal dealing with a similar Mokhasadar family reached the same conclusion by anticipating the approach followed by the Supreme Court. The candidate remained a tribal since 'there was no evidence that the Kshatriya community as a whole recognized him as belonging to their class.' *Gadipalli Parayya v. Boyina Rajayya*, XII E.L.R. 93 (1956). One may wonder who are 'the Kshatriya community as a whole' — whether this is defined in terms of the locality, the district, the state or all-India?

seriously considered acceptance by the non-privileged group as incompatible with membership in the privileged group, the cases would most probably have had different outcomes. Following on these lines one would have expected the Supreme Court in the *Dora* case to address itself to the views of the Moka Doras. But like the High Court below, they considered only the views of Kshatriyas.

True, the *Jasani* case did mention 'the rules of the new order' as one of the factors to be considered. But it is clear that this was not only the least important factor, but was intended to mean the rules of the new order respecting the retention by X of his membership in the old group. It was not the views of the Bs as to his membership in the Bs that counted, but their views as to his membership in the As. In the *Dora* case when the Supreme Court consults the attitudes of the new group, it is on the question of X's membership in the Bs.[37]

Since the Court never discusses the question of his membership in the Moka Doras, one can only gather that there is implicit in the Court's view a logical incompatibility between membership in the two groups. Had his Kshatriya status been upheld, he would *ipso facto* not have been a member of the Moka Doras for the purpose of standing for the reserved seat. Faced with the question of whether

[37] The practical effect of these divergent approaches can be easily seen in tabular form. Let us imagine that X, a member of privileged Group A, has somehow aspired to membership in Group B.

	Test I Did X intend to remain an A?	Test II Did the A's accept X as an A?	Test III Did the B's accept X as a B?	Is X an A? *Jasani* approach	*Dora* approach
1.	yes	yes	yes	yes	no
2.	yes	yes	no	yes	yes
3.	yes	no	yes	no	no
4.	yes	no	no	no	yes
5.	no	yes	yes	no	no
6.	no	yes	no	no	yes
7.	no	no	yes	no	no
8.	no	no	no	no	yes

This table merely restates the requirement of *Jasani* that to be an A (for the purpose of filling a reserved seat), X must fulfill both tests I and II. This occurs only in lines 1 and 2. It is clear that according to the *Dora* Court the 'yes' on test III in line 1 would make the answer 'no'. And in lines 4 and 6 and 8 the *Dora* method would make him an A by virtue of his not being a B (i.e., failing test III) whereas the *Jasani* method would find him not an A because of failing either test I (line 6), test II (line 4) or both tests (line 8).

It should be noted that using the approach of the High Court in the *Dora* case the answer would be yes in every case except for line 7 and then only upon the additional condition that the As accepted X as a B.

X remained an A, the Court addressed itself to the question of whether he had become a B. But this course of reasoning is only plausible if it is assumed first, that the two memberships exhaust the possibilities and second, that they are mutually exclusive.

In assuming that they exhausted the possibilities the Court seems to deny the possibility that the candidate's family had, although failing in some sense to become Kshatriyas, so separated themselves from the Tribe as to lose acceptance as members. Such an intermediate possibility was considered by the Election Tribunal and to some extent by the High Court.[38] Such splitting off is one of the classic and well documented methods by which new castes are formed.[39]

The *Jasani* line of cases had allowed overlapping and multiple affiliations. It was possible to be simultaneously a Mahar and a member of the Mahanubhava Panth; a Samgar and an Arya Samajist, a Christian and an Oraon, an Anglo-Indian and a Khasi. Why is it not possible to be both a Moka Dora and a Kshatriya? The Court does not indicate the source of its notion that these affiliations are mutually exclusive. But it seems that this incompatibility is felt because Indian society is visualized as consisting of groups with unique corporate ranks in some definite rank ordering. Thus membership in one such group entails occupying such a rank and is inconsistent with membership in another group, which would mean simultaneously holding a lower rank in the same system of ranks. The Court refers to the claim here as one for 'higher status,' which presumably cannot be achieved without giving up membership in the group with lower status.

In part this notion of Indian society as consisting of mutually exclusive groups ranked in a definite and unique order is a carryover from the area of personal law. The courts have long applied to members of different religious communions their respective laws in matters of marriage, divorce, inheritance, succession and religious endowments. Since in Hindu law there were some differences in the rules applicable to the three higher *varnas* on the one hand and Sudras on the other,[40] the courts had from time to time to determine which rules were applicable to particular persons or groups. For this purpose Hindu society was visualized as if it consisted of four ranked compartments — the lowest being residual — and any of the actually existing caste groups would be assigned, if need be, to one

---

[38] XVI E.L.R. at 38; A.I.R. 1958 A.P. at 735.

[39] See e.g., *Muthasami Mudaliar v. Masilamani*, I.L.R. 33 Mad. 342 (1919); Hutton (1961: 50ff).

[40] These differences are concisely summarized by Derrett (1958: 383–5).

of these theoretical compartments.[41] Since the Constitution, the courts have continued to make such *varna* assignments when necessary.[42] Now that the various Hindu Code Acts of 1955–6 have eliminated almost all of the instances in which *varna* might make a difference in applicable law[43] the courts can look forward to the day when they will no longer be faced with the task of making these imponderable and often fictitious *varna* identifications.

In personal law cases the question was not whether an individual was or was not a member of some existing social group, but whether he should be assigned the status of one or the other *varna*.[44] Ordinarily the individual was indisputably a member of some actual caste or group and the proceedings took the form of determining the *varna*

41 In order to make such assignments, the courts evolved various tests: lists of diagnostic customs (see e.g., *Gopal v. Hanmant*, I.L.R. 3 Bom. 278 (1879) where the tests of Sudra status are widow remarriage and admission of illegitimate sons to dine and marry within the caste and to inheritance), or alternatively, tests of reputation (see e.g. *Subrao v. Radha*, I.L.R. 52 Bom. 497 (1928) where it is held that *varna* depends on the consciousness of the caste as to its status and the acceptance of this estimate by other castes).

According to the theory of *varna* Hinduism consisted of the four *varnas* and every caste group could be assigned to one of these; caste and *varna* were co-extensive with Hinduism. But departures from the symmetry of this scheme are found in many instances where courts modified it to account for the actualities of the situations before them. Thus it is possible to have *varna* standing without belonging to a caste group. *Sunder Devi v. Jheboo Lal*, A.I.R. 1957 All.215 (convert to Hinduism); *Upoma Kuchain v. Bholaram* I.L.R. 15 Cal. 708 (1888) (daughter of outcaste); cf. *Ratansi v. Administrator General*, A.I.R. 1928 Mad. 1279. Caste and *varna* may apply to persons who are not strictly Hindus, *Inder Singh v. Sadhan Singh*, I.L.R. (1944) 1 Cal. 233 (Sikh Brahmins). Caste groups have been recognized which have no *varna* nor are Hindu in any sense. *Abdul Kadir v. Dharma* I.L.R. 20 Bom. 190 (1895). Again, members of the same caste may hold different *varna* statuses. *Subrao v. Radha*, above.

42 After the advent of the Constitution the administration of separate personal law to the respective religious communities was challenged as discriminatory and *ultra vires* Art. 15. Although Art. 44 directs the eventual elimination of separate personal laws, the continuing validity of disparate rules of personal law and the power of the State to create new rules applicable to particular religious communities has been upheld. E.g., *State of Bombay v. Narasu Appa*, A.I.R. 1952 Bom. 84. The assignment of a community to a *varna* has been held not to constitute a deprivation of rights to equality before the law, nor is it religious discrimination. *Sangannagonda v. Kallangonda*, A.I.R. 1960 Mys. 147.

43 i.e. the Hindu Marriage Act of 1955, the Hindu Succession Act of 1956, the Hindu Minority and Guardianship Act of 1956 and the Hindu Adoptions and Maintainance Act of 1956. Derrett (1958) suggests that the only instances in which *varna* might continue to have effect are succession to *sanyasis* and determination of the maximum age for adoption.

44 The judicial treatment of the relation between *varna* and caste was plagued by confusion, engendered in part by the use of 'caste' to refer both to the four great classes or *varnas* into which Hindu society is theoretically divided by the Sanskrit law-books and to the multitude of existing endogamous groups or *jatis*. In the sequel caste is used only in the latter sense.

of this group. It was assumed for this purpose that all caste groups could be assigned to one or the other of the *varnas*. Since the purpose of determining *varna* was to ascertain the appropriate rule of law, and since *varnas* clearly stood in a ranked order, the whole object of the proceeding was to arriave at a unique determination of status.

The question before the courts in these election cases is quite different in kind. It is whether in fact a person is for a particular purpose to be considered a member of some existing group. There is no necessary relationship between membership in such a group and holding of *varna* status.[45] Apparently the candidate in the *Dora* case — and his family and possibly the whole group of Mokasadars — claimed to be Kshatriyas. It is unclear whether they were merely asserting Kshatriya *varna* status or whether they were claiming membership in some particular endogamous group of Kshatriya families. It would seem possible to achieve *varna* status without necessarily becoming effective members of a caste with that status. In any event, it is unclear whether attainment of the *varna* status claimed in the *Dora* case would have been felt by the tribe to be incompatible with continued membership. Whatever theoretical incompatibility there may be in belonging to two *varnas*, it is not impossible to be accepted as a member of two actual social groups.

But *varna* theory is not the sole source of the Supreme Court's notion of mutually exclusive group membership. This notion is supported by its picture of Indian society. The Court indicates that it bears in mind 'the recognized features of the hierarchial social structure prevailing amongst the Hindus' and the 'inflexible and exclusive nature of the caste system'.[46] The Court is, of course, only giving its view of the conditions that obtain; it expresses the hope 'that this position will change, and in course of time the cherished ideal of a casteless society . . . will be attained'.[47] Nevertheless, in its anxiety not to be 'unrealistic and utopian', the Court seems unnecessarily to give currency to the view that all groups in Indian society are ranked in some unique and definite order.[48] Thus the Court

---

[45] Cf. *Mulai v. Lal Dan Bahadur Singh*, IX E.L.R. 9 (1952) where an Election Tribunal found that the proclamation of a former prince had not transformed the Gonds of Rewa state into Kshatriyas. 'But even if the Gonds ... could be deemed to be Kshatriyas, they would not cease to be members of Scheduled Tribes.' Even 'their aristocratic sub-division known as Raj Gonds, still continue to be Gonds and ... belong to the Scheduled Tribes'.

[46] A.I.R. 1959 S.C. at 1327.

[47] Id.

[48] This tendency to picture Indian society as series of graded corporate ranks seems congenial to the courts, perhaps because of the felt necessity of having some conceptual means for reducing the immense variety of Indian society to terms which could be applied without extensive investigation in each individual instance. Cf. Marriott's (1959: 104) suggestion that urban and educated Indians (and foreigners) tend to con-

is impelled to formulate a general standard for assigning standing
in such a rank ordering. Noting that whatever may have been the
case in ancient times 'status came to be based on birth alone . . .,'
the Court says 'it is well known that a person who belongs by birth
to a depressed class or tribe would find it very difficult, if not impossible, to attain the status of a higher caste amongst the Hindus by
virtue of his volition, education, culture and status'.[49] Thus Hindu
society is not only hierarchic but inflexible as well. If the Court meant
literally that caste status was determined by birth alone, it would
of course be redundant to consider the views of any group in order
to determine it. But the Court apparently means that birth is determinative in the first instance and that this can be varied — but not
by an individual's conduct but only by the unanimous recognition
of his claims by members of the higher group. J. L. Kapur, J., dissenting, vigorously rejected the primacy of birth, put forward another
general theory of assignment of rank. Holding that caste varies as
a consequence of the *gunas*, *karma* and *subhavana* and is dependent
on actions, he found that the candidate had 'by his actions raised
himself to the position of a Kshatriya and he was no longer a member
of the Scheduled . . . Tribe. . . .'[50]

Either of these general theories may prove embarrassing. Acceptance of the dissenting judge's theory that caste (or *varna*) status may
be gained by entirely individual action would expose the courts to
a torrent of litigation in which they would be faced with the necessity
of setting up legal tests for caste standing and assigning it to individuals and communities.[51] The majority's acceptance of birth as the
primary determinant of group membership avoids this difficulty
and is no doubt accurate in the overwhelming majority of instances.
But by making membership in the old group dependent on failure
to achieve the purported membership in the new group, the majority's theory may disincline the courts from giving legal recognition
to existing patterns of mobility, which ordinarily involve a period
of conflicting claims and overlapping identifications. Successful

---

ceive of caste in terms of criteria which constitute or imply a scale of Hindu ritual values
rather than according to the structure of interaction among various groups.

[49] A.I.R. 1959 S.C. at 1327.

[50] Id. at 1331. *Gunas*, etc. means roughly material nature, deeds and temperament.

[51] See *Sankaran Namboodri v. Madhavan*, A.I.R. 1955 Mad. 579, where one section
of a family of Embrandiri Brahmins changed its name and claimed that they were
Nambudiri Brahmins and therefore governed by the special personal law of the latter.
The Court was willing to concede that they had voluntarily effected a partition in
accordance with Nambudiri law, which would help establish their status as Nambu-
diris for purposes of personal law in the future, but held that the change of name was
a mere unilateral declaration of no effect on their status in the absence of evidence of
renunciation by the whole family of their present status or of their acceptance by the
Nambudiri community as a whole.

separation from an old group may be overlooked, with the result of imposing on a privileged group a candidate who is not an accepted member of it. More generally, existing channels of mobility may be discredited. For if acceptance by a new group removes one from the old, the Hinduization of tribals and the formation of new sects would be accompanied by the danger of disqualification for receipt of preferences. Since the system of preferences is designed to increase flexibility and mobility within Indian society, there seems little reason to make abandonment of older and slower methods of mobility a condition for the utilization of the new ones.

The dilemma posed by these opposing theories can be solved by eschewing any general theory of assignment and deciding questions of group membership by the pragmatic approach of the *Jasani* case. Since this permits multiple identifications, the courts would not pose for themselves the kind of either/or puzzles that have no satisfactory answer. They would be concerned only with whether for the purpose of the particular measure the individual concerned ought to be counted a member of the privileged group.[52] As far as it is compatible with the particular legislative policy, membership would depend entirely on the voluntary affiliation of the individual —as confirmed, where necessary, by voluntary acceptance by the group's members. Consonant with the constitutional principles of freedom of association and the autonomy of social groups, state imposition of standards of membership would be minimized,[53] and judicial determination of the relative status of groups would be eliminated.

[52] In *Janilal v. Jabarshingh*, A.I.R. 1957 Nag. 87 the Court found that 'a person can be a Hindu and also be deemed a member of an aboriginal tribe' for the purposes of the provisions of the Central Provinces Land Alienation Act which outlawed conditional sale provisions in mortgages made by listed tribes.

[53] It has been argued with some persuasiveness that the Constitution withdraws all governmental power to determine whether an individual is a member of a particular caste. 'Can a secular government force a citizen to belong or not to belong to a particular caste?' queries the Chairman of the Backward Classes Commission in considering whether census clerks may put down the caste of an individual according to their conception of it or whether the individual's conception is determinative. The Chairman contends that if caste is to be a voluntary affiliation, government must refrain from assigning it. Report of the Backward Classes Commission, 1956, Vol. I. p. xviii. But once caste and tribe have been accepted as appropriate units for the distribution of preferences, some governmental determinations, in order to prevent abuses, are unavoidable. It would seem possible to have more or less objective standards, even for determining a voluntary affiliation and to refrain from any assignments other than the determination of whether or not an individual is within the group to which he claims to belong.

Religion

The Indian Constitution is openly and determinedly secular. Religious discrimination on the part of the State is forbidden. Freedom of religion is guaranteed. The courts have been vigilant in invalidating governmental measures framed along religious lines.[54] Nevertheless, in some instances religion has been made a qualification for preferential treatment. The President's Order specifying Scheduled Castes provided that 'no person professing a religion different from Hinduism shall be deemed a member of a Scheduled Caste'.[55] In 1956 this was broadened to include Sikhs.[56]

Who meets this religious qualification? The legal definition of Hinduism, developed for the purpose of applying appropriate personal law, was neither a measure of religious belief nor a description of social behaviour as much as a civil status describing everyone subjected to the application of 'Hindu law' in the areas reserved for personal law.[57] Heterodox practice, lack of belief, active support of non-Hindu religious groups, expulsion by a group within Hin-

[54] *State of Rajasthan v. Pratap Singh*, A.I.R. 1960 S.C. 1208; *Nain Sukh Das v. State of U.P.*, A.I.R. 1953 S.C. 384; *State of Jammu and Kashmir v. Jagar Nath*, A.I.R. 1958 J. & K. 14. On Indian secularism generally, see D. Smith (1963); Galanter (1965). For a discussion of the problem treated in this section, see Smith (1963) 322 ff.

[55] Constitution (Scheduled Castes) Order, 1950, Para. 3. An exception was included for Sikh members of four of the thirty-four Scheduled Castes listed for the Punjab. Cf. the Government of India (Scheduled Caste) Order, 1936, para. 3, which provided that 'No Indian Christian shall be deemed a member of a Scheduled Caste.' The Constitution (Scheduled Tribes) Order, 1950, contains no religious provision.

[56] The Scheduled Castes and Scheduled Tribes Orders (Amendment) Act, 1956 (63 of 1956), para. 3. The main provision now reads: 'No person who professes a religion different from the Hindu or Sikh religion shall be deemed a member of a Scheduled Caste.' The inclusion of Sikhs has an interesting history going back to dissatisfaction with their exclusion from the Scheduled Castes under the 1936 Order. In the Constituent Assembly Sikhs gave up their demands for political safeguards for minorities in return for assurances that their backward classes would be recognized and listed as Scheduled Castes (there was some opposition to this inclusion among the Scheduled Castes). See the speech of Sardar Hukum Singh at IX *Constituent Assembly Debates* 235, remarks of Sardar V. J. Patel, id. at 247. Sardar Hukum Singh moved an amendment to this purpose in the Constituent Assembly but it was withdrawn with the understanding that the Sikhs would be considered in the making of the lists of Scheduled Castes. VIII C.A.D. 552 ff.

[57] Or, more accurately, all who would be subject to Hindu law in the absence of proved special custom or of a contingency such as marriage under the Special Marriage Act (III of 1872, now 43 of 1954). But persons who are not Hindus may attract the application of Hindu law. These include some Christians (see e.g. *Chinnaswamy v. Anthonyswamy*, A.I.R. 1961 Ker. 161. Cf. note 11, above). Tribals may be subject to Hindu law although they are not Hindus. See *Mira Devi v. Aman Kumari*, A.I.R. 1962 M.P. 212. Hindu personal law was applied to some Muslim groups until the passage of the Muslim Personal Law (Shariat) Application Act (XXXVI of 1937).

duism—none of these removed one from the Hindu category, which included all who did not openly renounce it or explicitly accept a hostile religion. The individual could venture as far as he wished over any doctrinal or behavioural borders; the gates would not shut behind him if he did not explicitly adhere to another communion.[58] The same negative definition remains today for purposes of application of personal law. In *Chandrasekhara Mudaliar v. Kulandaivelu Mudaliar*, the Supreme Court had to decide on the validity of a consent to adoption by one who disavowed belief in the religious efficacy of adoption, in Hindu rituals and scriptures, in '*atma*' (soul) and salvation. The Court found that 'the fact that he does not believe in such things does not make him any less a Hindu. . . .He was born a Hindu and continues to be one until he takes to another religion . . . whatever may be his personal predilections or views on Hindu religion and its rituals.'[59] One who has converted may readily effect a re-conversion to Hinduism.[60]

[58] *Bhagwan Koer v. Bose*, 30 I.A. 249 (1903); *Ratansi D. Morarji v. Admr. General of Madras*, A.I.R. 1928 Mad. 1279, 1283. The concept of 'Muslim' is treated somewhat differently. There are expressions of the same negative (no conversion) test. See *Bhagwan Bakhsh Singh v. Dribijai Singh*, I.L.R. 6 Luck. 487 (1931). Generally it is held that adherence to some minimum of beliefs (the unity of God, the mission of Mohammed as his prophet, and the authority of the Koran) is necessary and sufficient to make one a Muslim. *Narantakath v. Parakkal*, I.L.R. 45 Mad. 986 (1922); *Jiwan Khan v. Habib*, I.L.R. 14 Lah. 518 (1933). Repudiation of these beliefs, even without conversion to another religion, makes one not a Muslim. *Resham Bibi v. Khuda Bakhsh*, I.L.R. 19 Lah. 277 (1938). Compare the test for Sikhs, note 3.

[59] A.I.R. 1963 S.C. 185, 200. See also the definition of 'Hindu' set out in the Hindu Marriage Act, 1955, Sec. 2, and discussion in Derrett (1963; Sec. 17–20). A similar latitudinarianism may be observed in the tests for whether a tribe is sufficiently Hinduized to attract the application of Hindu law. Orthodoxy is unnecessary; it is sufficient that the tribe acknowledge themselves as Hindus and adopt some Hindu social usages, notwithstanding retention of non-Hindu usages. *Chunku Manjhi v. Bhabani Majhan*, A.I.R. 1946 Pat. 219.

[60] No proof of formal abandonment of his new religion is necessary for the convert to effect a successful reconversion to Hinduism. While a mere declaration is not sufficient to restore him to Hinduism, acceptance by a Hindu community with whatformalities it deems proper—even none at all—is sufficient. *Durgaprasada Rao v. Sundarasanswami*, A.I.R. 1940 Mad. 513; *Gurusami Nadar v. Irulappa Konar*, A.I.R. 1934 Mad. 630. However, Cf. *Marthamma v. Munuswami*, A.I.R. 1951 Mad. 888, 890, where the primary test is the 'intention' of the re-convert; the court says 'the religious persuasion of a man now-a-days depends on his 'subjective preference' for any religion.' Cf. the declaration test discussed in note 63, below.

For purposes of at least certain preferences, re-converts to Hinduism who were born in Scheduled Castes are deemed members of the Scheduled Castes. But those born in another religion (e.g., whose fathers were converts) are not treated as members of Scheduled Castes 'whatever may be their original family connections'. *Report of the Commissioner for Scheduled Castes and Scheduled Tribes*, 1953, p. 132. In the personal law cases, acceptance by the community was a measure of one's success in re-entering Hinduism; here, Hindu birth is the condition of enjoying privileges devolving on the community.

In the post-Constitution cases involving preferences the same latitudinarian conception of Hinduism has been carried over from the area of personal law, To 'profess' Hinduism, merely means to be a Hindu by birth or conversion.[61] Unorthodox practice or lack of personal belief in its tenets does not mean lack of profession for this purpose — one may eat beef and deny the authority of the Vedas. In effect the test seems to amount to a willingness to refrain from calling oneself something else. Thus where the election to a reserved seat of an active supporter of Dr Ambedker's Buddhist movement was challenged on the ground that he was not a Hindu, the Court found that 'it has to be established that the person concerned has publicly entered a religion different from the Hindu . . . religion'. Mere declarations falling short of this would not be sufficient.[62] The candidate had supported the movement for mass conversion by serving on the reception committee, editing a newspaper supporting the movement, and attending a rally where an oath, 'I abandon the Hindu religion and accept the Buddha religion' was administered by Dr Ambedkar. When those who wished to convert were asked to stand, the candidate stood. But there was no evidence that he did in fact take the oath; the Court held that in the absence of such a declaration, he remained a Hindu. [63]

The same test of public declaration was recently upheld by the Supreme Court in *Punjabrao v. D. P. Meshram,* where there were eye-witnesses that the candidate did convert to Buddhism at a mass ceremony. In spite of their political enmity to the candidate, the Court finds the testimony of these witnesses credible in view of the fact that the candidate was a prominent follower of Dr Ambedkar and 'it would be highly improbable that he would have remained aloof' from the Buddhist movement.[64] This direct evidence was

---

[61] *Michael v. Venkateswaran,* A.I.R. 1952 Mad. 474.

[62] *Karwadi v. Shambarkar,* A.I.R. 1958 Bom. 296, 297. On the Buddhist movement (or neo-Buddhist as it is sometimes called) see Zelliot (1966); Keer (1962); Isaacs (1965); Miller (1966). More than 3 million persons, mostly Mahars from Maharashtra, have become Buddhists since the movement began in 1956. An unknown but large number have refrained from, or concealed, conversions in order to remain eligible for preferences.

[63] A.I.R. 1958 Bom. at 299. The vagaries of the declaration test are illustrated in *Rattan Singh v. Devinder Singh,* VII E.L.R. 234 (1953), XI E.L.R. 67 (1955), where the candidate had at various times described himself as a Mazhabi Sikh, a Harijan Hindu, a Balmiki, an a Balmiki Hindu. The Tribunal, holding that the minimum qualification for being a Sikh is willingness to declare 'I solemnly affirm that I believe in the ten gurus and that I have no other religion', found him to be a Balmiki Hindu in 1953 and a Mazhabi Sikh in 1955. Any objective evidence was rigorously excluded since 'the question of . . . religion . . . is a matter of personnal faith and cannot be the subject of any evidence of a third party'.

[64] III (1965) *Maharashtra Law Journal* 162, 164. The witnesses, who were also Buddhist converts, were members of a rival faction within Dr. Ambedkar's Repub-

corroborated by a later written declaration and by certain acts and omissions by the candidate. The 'strongest circumstance' corroborating the evidence of the conversion is that as chairman of a local committee, the candidate presided over the conversion of a small Shiva temple into a Buddhist temple. '[H]owever great the admiration or regard that a Hindu may have for Lord Buddha, he would shudder at the idea of desecrating a Shiva Linga in this manner or ever of converting what was once a Shiva Temple into a Buddhist Temple.'[65] Secondly, the candidate willingly married his daughters to Buddhists, upon which the court reflects that 'if he were a Hindu ... it is unlikely that he would have reconciled himself with the idea of giving his daughters in marriage to non-Hindus, more particularly when the bridegrooms' side insisted on following the Buddhist ritual.'[66] Third, the wedding announcements substituted a picture of the Buddha and an invocation of his blessing for the usual picture and invocation of the *Kuladaivata* (household deity). Rejecting the candidate's contention that he was treating Buddha as 'the 11th [sic] incarnation', the Court observed that he was not of that 'sophisticated' class that have discarded altogether the picture and the blessing, and 'had [he] considered himself to be a Hindu, he would have followed the usual practice'.[67]

All of the above indicate to the Court that a conversion did in fact occur. The test of conversion is public declaration. Entry into another religion must be so public 'that it would be known to those whom it may interest.'[68] Therefore if one publicly declares that he has ceased to belong to his old religion and has embraced a new one, he will be accepted as professing the new one. There is no religious test for membership in the new group beyond the declaration. 'In the face of such an open declaration it would be idle to enquire further whether the conversion ... was efficacious.'[69]

---

lican Party. Id. at 163. Since this was written the case has been reported at A.I.R. 1965 S.C. 1179.

[65] Id. at 168.

[66] Id. at 166.

[67] Id. at 166. The incarnations (*avataras*) of Visnu are commonly accounted to number ten, Buddha being the last to have appeared. However, in general Buddha receives little attention from Vaishnavite Hindus. Apparently the candidate here either had a different list or was unaware that Buddha was already included.

[68] Id. at 169.

[69] Id. at 169. The formalities of conversion are to be tested by the new religion itself, not by external standards. In the case of these Buddhists it appeared that officiation by a *bhikku* (monk) was not required and recitation of three vows and five precepts was considered sufficient. This finding, along with the non-inclusion of Buddhists among Hindus, was gratifying to most Buddhists, who felt that the decision confirmed the efficacy of their conversions as well as the correctness of their conviction that they are no longer Hindus.

Once it is established that the candidate has 'professed' Buddhism, the question arises whether this is 'a religion different from the Hindu . . . religion' within the meaning of the Scheduled Castes Order. Converts to Christianity and Islam are, of course, non-Hindus.[70] But Buddhists, Jains, and Sikhs are treated as Hindus for many purposes.[71] Hindu is an equivocal term, sometimes used with reference to adherents of more or less 'orthodox' Vedic and Brahmanical communions and at other times used to embrace the full array of heterodox sects, including Sikhs, Jains, and Buddhists. Yet Jains and Buddhists are considered non-Hindus for purposes of preferences, and Sikhs are mentioned separately.[72] The candidate in *Punjabrao* argued that 'Hindu' in the Scheduled Caste order should be given a broad meaning that would include Buddhists, in line much other official usage. The Supreme Court finds this clearly negatived by the language of the Order: the separate mention of Sikhs would otherwise be redundant. This reading of the Order clearly accords with legislative intention to exclude the major 'heterodox' communions. But the Court's observation that 'the word 'Hindu' is used in the narrower sense of the orthodox Hindu religion which recognizes castes and contains injunctions based on caste distinctions',[73] is troubling. It comes close to suggesting a positive definition of Hinduism which goes beyond merely excluding Sikhs, Jains, and Buddhists to the point of throwing doubt on the inclusion of those sects which in one fashion or another repudiate caste distinctions in doctrine or practice.

70 *Michael v. Venkateswaran*, A.I.R. 1952 Mad. 474. But one may remain a Hindu even while belonging to a religious sect that includes non-Hindus among its members, *State of Bombay v. Yagna Sastri Purushadasji*, 61 *Bom. L. Reporter* 700 (1960).

71 These groups are Hindus for purposes of personal law. See, e.g. Hindu Marriage Act,, 1955, 2. They are specifically included among Hindus for purposes of authorizing the State to open Hindu temples to Untouchables. See the explanation II to Art. 25 (2) of the Constitution. But their separateness is recognized in some contexts, e.g., Jains are not of 'the same religion' as Hindus for purposes of current temple-entry legislation. *State v. Puranchand*, A.I.R. 1958 M.P. 352; *Devarajiah v. Padmanna*, A.I.R. 1958 Mys. 84. Differences in regard to the organization or religious trusts may be recognized by having a Hindu religious trusts law which covers Jains and Buddhists but not Sikhs. *Moti Das v. S. P. Sahi*, A.I.R. 1959 S.C. 942.

72 They were excluded until 1956. *Gurmukh Singh v. Union of India*, A.I.R. 1952 Pun. 143; *Rattan Singh v. Devinder Singh*, note 63, above.

73 III *Maharashtra Law Journal* at 169. The Supreme Court has more recently enunciated views of Hinduism which seem to conflict with its description here. In *Sastri Yagnapurushadasji v. Muldas Bhundardas Vaishya* (Civil Appeal No. 517 of 1964, recently reported at A.I.R. 1966 S.C. 1119) it was held that the Swaminarayan sect was, in spite of its protestations, part of Hinduism and subject to Bombay's temple entry act. The Supreme Court observed that the sectarians' belief in caste pollution 'is founded on superstition, ignorance and complete misunderstanding of the true teachings of the Hindu religion. . . .'

This is particularly discomfiting because the Court, while inquiring what it means to 'profess' a religion 'different from the Hindu', reverses the question to examine what it means to profess Hinduism. It reads the Order as requiring that to be treated as a member of a Scheduled Caste 'a person . . . must be one who professes either Hindu or Sikh religion'.[74] And it notes, 'the word 'profess' in the Presidential Order appears to have been used in the sense of open declaration or practice by a person of the Hindu [or the Sikh] religion.'[75]

The Court is obviously not attempting to establish a positive religious test of eligibility for preferences. But the unfortunate juxtaposition of these dicta on Hinduism and its profession with the evidentiary examination of conventional Hindu practice and attitude with regard to intermarriage, invitations, and respect for idols comes uncomfortably close to suggesting that these Hindu proprieties are requirements for preferences. Not only converts, but anti-caste sectarians, atheists, and militant iconoclasts would feel the pinch of such a test. It would not only be extremely difficult to apply, but it would no doubt violate the understandings of many Hindus, impede innovation and change within Hinduism, and present serious problems of religious discrimination and freedom of religion.[76]

In spite of the Court's remarks, it is clear that the Scheduled Caste Order itself does not establish or sanction such a positive religious test. It merely requires that he does not profess a different religion. Professing Hinduism and professing a non-Hindu religion may be mutually exclusive categories, but they are not exhaustive.[77] The question is whether one professes a non-Hindu religion; the profession and practice of Hinduism need be considered, as they are in *Punjabrao*, only for their evidentiary value in this inquiry.

It is clear then that Buddhists are excluded by the Presidential Order. The Supreme Court merely notes that when a person 'has ceased to be a Hindu he cannot derive any benefit from that Order'.[78] Persistent efforts by Buddhists to be treated on a parity with Scheduled Castes have achieved only partial success. Exclusion of Buddhists is usually justified on the ground that conversion to

[74] III *Maharashtra Law Journal* at 168.

[75] Id. at 169.

[76] While the Ordeer's religious distinction itself is open to these objections, it might conceivably correspond to some difference in conditions. Any relevant differences between those who observe Hindu proprieties and those who don't is even more implausible.

[77] One wonders whether someone who repudiated Hinduism, or an avowed atheist, would be held to 'profess a religion other than Hinduism.' Cf. *Chandrasekhara Mudaliar v. Kulandaivelu Mudaliar*, A.I.R. 1963 S.C. 185.

[78] III *Maharashtra Law Journal* at 169.

Buddhism operates as loss of caste. 'As Buddhism is different from the Hindu religion, any person belonging to a Scheduled Caste ceases to be so if he changes his religion. He is not, therefore, entitled to the facilities provided under the Constitution specifically for the Scheduled Castes.'[79] The state of Maharashtra, where most of the Buddhists are concentrated, restored to them those of the benefits enjoyed by the Scheduled Castes which are conferred by the state, but the central government and the other states have remained adamant in their unwillingness to include the Buddhists.[80]

In *Punjabrao v. Meshram* the Supreme Court never reaches the question of whether the 'Hinduism' test for recipients of preferences infringes the constitutional ban on religious discrimination by the State.[81] The constitutional challenge has been raised in several earlier cases, involving two sorts of factual situations. First, instances

[79] *Report of the Commissioner of Scheduled Castes and Scheduled Tribes*, 1957–8, Vol. I, p. 25. This ruling is based squarely on the 'Hinduism' requirement of the President's Order. See the statement of Pandit Pant, *Times of India*, Aug. 21, 1957, p. 12, col. 3. Compare the observation of the Supreme Court that 'Hindu' for purposes of the Scheduled Caste Order includes that religion which 'recognizes caste and injunction based on caste distinctions.' *Punjabrao v. D. P. Meshram*, III *Maharashtra Law Journal* 162 at 169 (1965).

[80] The major benefits outside the purview of the state government are reservations in legislative bodies, post-matriculation scholarships and reservations in central government employment. Most Buddhists oppose the continuation of legislative reservations, so the dispute has centred on the latter benefits. A bill that would have restored them to Buddhists was defeated in the Lok Sabha. *New York Times*, Aug. 30, 1961, p. 2, col. 6. The Central government, recognizing that conversion was of itself unlikely to improve the condition of the converts, recommended that the state governments accord neo-Buddhists the concessions available to the Backward Classes. Such preferences, less in scope and quantity than those for Scheduled Castes, have been granted in some states, but others have withdrawn all preferences. *Report of the Commissioner for Scheduled Castes and Scheduled Tribes*, 1957–8, Vol. I, p. 25; Vol. II, p. 60. The treatment of Christians of Scheduled Caste background is comparable. See Smith (1963: 323). A vigorous campaign to secure favoured treatment from the central government was mounted by the Republican Party in late 1964. In April, 1966, an affirmative response by the central government was anticipated in Buddhist circles.

[81] The failure to reach the constitutional issue was due to the preoccupations of the Meshram's counsel in the High Court. The Election Tribunal had decided the factual question of a conversion against the candidate, much as the Supreme Court did later. On appeal to the High Court, Meshram's counsel never raised the question of the constitutionality of the religious classification. Instead, counsel were eager to argue that Buddhism was in fact not a religion different from Hinduism. The High Court held that this was a factual matter which should have been pleaded and proved by evidence and refused to allow the argument to go forward. Assessing the evidence, the High Court found no evidence of authority to effectuate a conversion to Buddhism and a predominant political rather than a religious motive. The Court concluded that the ceremony was a protest against caste and demonstrated petitioner's sympathy with Buddhism but was not tantamount to the requisite profession of a non-Hindu religion. On Appeal, the Supreme Court re-assessed the evidence once more and reinstated the view of the election Tribunal that the conversion was effectuated.

in which a group straddles the Hindu/non-Hindu border and includes among its members both Hindus and non-Hindus; second,
in instances of conversions by individual members of a privileged
Hindu group.

In *Gurmukh Singh v. Union of India*[82] a Bawaria Sikh protested his
exclusion from the Scheduled Castes, in which the Presidential Order
had included Hindu Bawarias. A full bench of the Punjab High
Court found that Art. 341 empowered the President to select those
'parts of castes' which he felt should be included and that he could
select those parts on the basis of religion without violating Art. 15(1),
the ban on religious discrimination, since Art. 15(4), which authorized preferential treatment for the backward, operated as an
exception to the prohibitions of 15(1). The Court conceded that
Scheduled Castes were to be designated on the basis of their backwardness. But, finding that the Constitution vested in the President
the entire power to make such determinations, the Court refused
to review his order by considering whether the Sikh Bawarias were
in fact sufficiently backward to be included.

In *Michael v. Venkataswaran* the religious requirement was upheld
against a Paraiyan convert to Christianity who wished to stand for
a reserved seat. Even if there are cases in which both the convert
and his caste fellows consider him as still being a member of the caste,
the Court found, 'the general rule, is [that] conversion operates as
an expulsion from the caste . . . a convert ceases to have any caste.'[83]
The Presidential Order, according to the Court, proceeds on this
assumption and takes note of a few exceptions.[84] The Court declined
to sit in judgment on the President's determination that similar
exceptional conditions do not prevail in other instances. Thus the
Presidential Order was upheld not because of an absence of judicial
power to review it but because of its accuracy in the general run
of cases.

In *In re Thomas* another bench of the Madras Court considered
a convert case which did not involve the Presidential Order. The
Madras government had extended school-fee concession to converts
from Scheduled Castes 'provided . . . that the conversion was of
the . . . student or of his parent. . . .' A Christian student whose
grandfather had converted could not, it was held, complain of
discrimination on grounds of religion. The basic criterion was not
religion but caste, and converts did not belong to the Scheduled
Castes. By conversion they had 'ceased to belong to any caste because

[82] A.I.R. 1952 Pun. 143.
[83] A.I.R. 1952 Mad. 474, 478.
[84] i.e. the four Sikh groups listed in the 1950 Order. See note 55 above.

the Christian religion does not recognize a system of castes'.[85] The concessions to recent converts were merely an indulgence and the State could determine the extent of this indulgence.

These cases indicate the three distinct grounds on which the religious disqualification had been upheld: first, the inappropriateness of reviewing the President's Order; second, the constitutional authorization for using religious criteria in the designation of beneficiaries; third, the theory that no religious criteria are used since non-Hindus have no caste.

Is the Scheduled Castes Order unreviewable? Executive action, even in pursuance of expressly granted and exclusive constitutional powers, is not immune from judicial review for conformity with constitutional guarantees of fundamental rights. The special status of the Order as an act of the President is no bar (if it ever was),[86] for since 1956 the religious requirement has been promulgated by Parliament. In any event, the posture of the *Gurmukh Singh* and *Michael* cases was one of judicial restraint rather than judicial powerlessness.[87] Is such restraint appropriate? Since those cases were decided, judicial power to review the government's designation of beneficiaries of preferences has been firmly established. In recent cases the courts have subjected the standards used by the government to designate Backward Classes to close and detailed scrutiny.[88] There is no indication that the power of the President and Parliament

---

[85] A.I.R. 1953 Mad. 21, 22. The exclusion of Buddhists from the preferences for Scheduled Castes has been similarly justified by the notion that 'Buddhism [does] not recognize castes'. Statement of Mr. B. N. Datar in the Rajya Sabha, reported in *Times of India*, Aug. 27, 1957, p. 10, col. 1.

[86] Arts. 12, 13. In *Karkare v. Shevde*, A.I.R. 1952 Nag. 330, the Court found that the immunity conferred on President and State Governors by Art. 361 'does not place the actions of the Governor purporting to be done in pursuance of the Constitution beyond the scrutiny of the Courts. . . . Unless there is a provision excluding a particular matter from the purview of the Courts [as in Arts. 122, 212, 263, 329(a)] it is for the Courts to examine how far any act done in pursuance of the Constitution is in conformity with it'. But cf. *Biman Chandra v. Governor*, A.I.R. 1952 Cal. 799, holding that Art. 361 removes the acts of a governor from judicial review unless there is evidence of dishonesty or bad faith.

[87] Art. 12 provides that for purposes of the Fundamental Rights provisions of the Constitution, 'the State' includes the Government and Parliament of India and the Government and the Legislature of each of the State and all local or other authorities. . . .' In *Gurmukh Singh* the Court concedes that the President is included and that therefore his action is governed by the requirements of the chapter on Fundamental Rights.

[88] *Balaji v. State of Mysore*, A.I.R. 1963 S.C. 649; *Chitralekha v. State of Mysore*, A.I.R. 1964 S.C. 1823; *Ramakrishna Singh v. State of Mysore*, A.I.R. 1960 Mys. 338; *S. A. Partha v. State of Mysore*, A.I.R. 1961 Mys. 220. These were foreshadowed by *Venkataramana v. State of Madras* [1951] S.C.J. 318.

to delineate Scheduled Castes stands on a different footing than that of government to name Backward Classes.[89]

Article 15(4) permits special provisions in favour of Scheduled Castes notwithstanding the provisions of Art. 15(1), which bans discrimination by government on grounds of caste or religion. Does Art. 15(4) authorize the use of religious criteria in selecting the Scheduled Castes? Caste and religion are normally forbidden bases of classification.[90] Constitutional authorization for special provisions to disadvantaged groups has been held to authorize the use of such otherwise forbidden classifications.[91] However their use is still subject to the standards applicable to any governmental classification. The classification 'must be founded upon an intelligible differentia which distinguishes persons or things that have been grouped together from those left out . . . and . . . the differentia must have a rational relationship to the object sought to be achieved'.[92] In selecting Backward Classes, caste criteria may be used only to the extent that they are useful in identifying social backwardness and cannot be the sole or dominant test for that purpose.[93] Although the constitutionality of the use of religion as a criterion for selecting Backward Classes has never been explicitly rejected, the courts have shown a pronounced tendency to reject its application in practice.[94] Non-Hindus cannot be made the

---

[89] See Art. 12, note 87 above. The Hinduism requirement is an expression of the power conferred by Art. 341 to select 'castes, races or tribes or *parts of or groups* within castes, races or tribes.' (Emphasis supplied.) Is the power of the President and of Parliament (which has exclusive power to modify the list) subject to review on the question of whether the delineation of 'parts' or 'groups' is reasonably related to the object of the classification? There is no indication that this power is exempt from such review. See Art. 13(2) and (3)a.

[90] On the religious classification, see cases note 54. For striking down of caste classifications see *Sanghar Umar v. State*, A.I.R. 1952 Saur. 124; *Bhopalsingh v. State*, A.I.R. 1958 Raj. 41; *State of Madras v. Champakam Dorairajan*, [1951] S.C.J. 313.

[91] The power to designate 'classes' in Art. 15(4) and Art. 16(4) operates as an exception to the prohibition on the use of all grounds of classification in Articles 15(1) and 16(2) respectively. See *Ramakrishna Singh v. State of Mysore*, op. cit., note 88, at 349.

[92] A.I.R. 1960 Mys. at 346–48, A.I.R. 1961 Mys. at 229. These cases represent the application to the field of preferences of the general standards for the constitutionality of classifications, firmly established by numerous rulings of the Supreme Court, See e.g., *Budhan Chaudry v. State of Bihar*, A.I.R. 1955 S.C. 191: *Bidi Supply Co. v. Union of India*, A.I.R. 1956 S.C. 479.

[93] *Balaji v. State of Mysore*, A.I.R. 1963 S.C. 649 at 658; *Chitralekha v. State of Mysore*, A.I.R. 1964 S.C. 1823 at 1833.

[94] In *Venkataramana v. State of Madras* [1951] S.C.J. 318, the Supreme Court rejected the inclusion of Muslims and Christians as Backward Classes. *State of Jammu and Kashmir v. Jagar Nath*, A.I.R. 1958 J&K 14, a cabinet order authorizing appointment of Muslims to certain posts 'to remove the communal disparity' was held void. reserved seats for Muslims and Indian Christians on the Madras Corporation Council were held invalid in *A. R. V. Achar v. State of Madras*, Writ Petition No. 568, High Court

recipients of preferences on the basis of purely religious classification in the absence of other evidence of backwardness. One wonders why they can be excluded from preferences on a purely religious basis in spite of the presence of such evidence.[95]

The religious requirement is an expression of the power which Art. 341 confers on the President and Parliament to determine which 'caste, race or tribe or *part of or group within* any caste, race or tribe' shall be included in the list of Scheduled Castes.[96] If the standards of reasonable classification are to be applied the crucial consideration would be whether the division of the group into Hindu and non-Hindu corresponds to some difference in conditions, resources, or the incidence of disabilities so that the division is rationally related to the object of the preferences. Existing precedents would not seem to foreclose such an approach, since the cases upholding religious tests were all decided before judicial review was firmly established in the preference area and before the *Jasani* case and others had developed an empirical approach to questions of group membership.[97]

Is the third ground—that acceptance of a non-Hindu religion operates as loss of caste—more substantial? The existence of caste among non-Hindu groups in India is well known and has long been recognized by the judiciary.[98] Does an individual convert's accep-

---

at Madras, Aug. 25, 1952, aff'd on other grounds A.I.R. 1954 Mad. 563. In *Ramakrishna Singh v. State of Mysore*, note 88, the Court invalidated a scheme of educational reservations under which Sikhs, Jains, Muslims, and Indian Christians were among the beneficiaries. In *Balaji v. State of Mysore*, A.I.R. 1963 S.C. 649, Muslims were rejected as a Backward Class. But cf. *Kesava Iyengar v. State of Mysore*, note 6 (religious groups allowed without discussion. This case is now seriously discredited on other grounds); *State of Kerala v. Jacob Mathew*, I.L.R. 1964 (2) Ker. 53 (Muslims and Latin Catholics included among Backward Classes without discussion) rev'g on other ground, A.I.R. 1964 Ker. 39 (these groups disallowed).

95 Perhaps the State's power to define Backward Classes is not as broad as the President's to define Sscheduled Castes. And perhaps the relevance of religion to untouchability, notwithstanding its inappropriateness in marginal cases, is more apparent than to defining backwardness. But it cannot be inferred that this exempts the power from review in those cases where the use of such standards is challenged as inappropriate.

96 Italics supplied.

97 *Punjabrao v. D. P. Meshram*, note 64, clearly does not foreclose such an approach since it decides only the disputed fact of whether there was a conversion and never reaches any of the constitutional arguments against the religious classification.

98 *Abdul Kadir v. Dharma*, I.L.R. 20 Bom. 190 (1895) (Muslims); *Inder Singh v. Sadhan Singh* 1944 (1) Cal. 233: *Gurumukh Singh v. Union of India*, A.I.R. 1952 Pun. 143 (Sikhs). See also the *Report of the Backward Classes Commission*, Vol. I, p. 28 ff. In *Michael Pillai v. Barthe*, A.I.R. 1917 Mad. 431, a claim of high caste Christians for restoration of a wall separating them from low-caste Christians in church was rejected on the ground that such a claim for precedence could not be enforced among Christians, since it was based on Hindu notions of pollution.

tance of Christianity, Islam, or Buddhism invariably evidence a loss of membership in the caste group to which he belonged at the time of conversion? This is a double faced question of both fact and law — a question about his observable interactions with others and about his legal status. There is evidence that in some cases at least the convert continues to regard himself and to be regarded by others as a member of the old caste.[99] Why in such cases do the courts insist that the act of conversion has as a matter of law deprived him of membership? In other instances of conversion by members of privileged groups, the courts have addressed themselves to the factual side of the question.[100] Thus in the *Jasani* and *Shyamsundar* cases the courts addressed themselves to whether the adherence to sects within Hinduism had actually removed the person from his caste. A similar empirical approach has been applied to conversions among Scheduled Tribes. One wonders why the courts have forsaken this empirical approach when dealing with religious classification among the Scheduled Castes. Why, rather than look to the facts of the individual case have they chosen to apply this theoretical rule about loss of caste? As the Supreme Court observed in the *Jasani* case, 'conversion . . . imports a complex composite composed of many ingredients. Religious beliefs, spiritual experience and emotion and intellectual conviction mingle with more material considerations such as severance of family and social ties and the casting off or retention of old customs and observances. The exact proportions of the mixture vary from person to person.'[101] It is surprising that the courts in cases like *Michael* and *Thomas* have accepted a picture of conversion which corresponds more closely to missionary aspirations than to observable consequences.

This acceptance reflects the continued force of a view of caste groups which sees them as units in an overarching sacral order of Hinduism. To abandon the whole is to abandon the part. Accord-

---

[99] The reports are replete with cases in which converts have lived so indistinguishably with their caste-fellows that the courts retrospectively infer a tacit reconversion without either formal abjuration of the new religion or formal expiation and readmittance to Hinduism. *Durgaprasada Rao v. Sundarasanaswami*, A.I.R. 1940 Mad. 513; *Gurusami Nadar v. Irulappa Konar*, A.I.R. 1942 Mad. 193. The 'indulgence' extended by the State in the *Thomas* case, A.I.R. 1952 Mad. 474 seems to reflect an awareness that recent converts, if not effective members of their old castes, are at least subject to similar disabilities. Cf. the Kerala fee concessions in education for Christian converts from the Backward Classes. Report of the Commissioner for Scheduled Castes and Scheduled Tribes 1960–1, p. 320. And cf. *Muthusami Mudaliar v. Masilamani*, I.L.R. 33 Mad. 342 (1909), where Christian wives were accepted as members of a Hindu caste.

[100] In the *Jasani* case the Supreme Court undertook to determine 'the social and political consequences of such conversion . . . in a common-sense practical way rather than on theoretical grounds' (1954) S.C.J. at 325.

[101] [1954] S.C.J. at 326.

ing to this view, which prevailed prior to the Constitution, caste groups were visualized as occupying a unique place in an integrated but differentiated religious order in which the different parts enjoyed rights and duties, privileges and disabilities as determined by their position in this order.[102] From this view of caste derived the long-standing reluctance of the courts to give legal effect to caste standing among non-Hindu communities.[103] But the Courts have always recognized castes among non-Hindus when the claim was not an assertion of standing in this sacral order, but was a claim regarding the corporate autonomy or internal regulation of the group. For the latter purposes 'caste' comprised 'any well-defined native community governed for certain internal poses by its own rules and regulations' and was thus not confined to Hindus.[104]

In the preference area the question of non-Hindus having caste arises in two sorts of factual situations: first, those involving a caste group or a section of a caste made up of members who are non-Hindus; second, those involving individual converts.[105] In the first type, there is little dispute that such persons as, e.g. the Sikh Bawarias in the *Gurmukh Singh* case are, in fact, members of a caste group of the kind that has always been recognized to exist among non-Hindus. There is evidence that some non-Hindu castes or parts of castes are in circumstances equivalent to those of the Hindu Untouchables.[106] To refuse to recognize such caste membership in such cases implies that the 'caste' to which the Court is addressing itself is not caste in the sense of a body of persons bound by social ties, but caste in the sense of a body of persons which occupies a given place in the ritual order of Hinduism.

In the case of individual converts, the question facing the court would seem to be whether the individual's acceptance of Christianity, Islam, or Buddhism evidence a loss of membership in the caste group to which he belonged at the time of conversion.[107] There is evidence

102 See Galanter (chap. 7).

103 See e.g., *Michael Pillai v. Barthe*, A.I.R. 1917 Mad. 431.

104 *Abdul Kadir v. Dharma*, I.L.R. 20 Bom. 190 (1895) Cf. *Yusuf Beg v. Maliq*, A.I.R. 1927 Mad. 397.

105 In the case of the converts to Buddhism, though there is a paucity of data, there is little doubt that those who do not remain closely tied to their old castes tend to form a 'community' in the Indian sense. They not only share religious tenets and practices, but marry and socialize chiefly with one another.

106 See e.g. the observations of the Commissioner regarding Christian and Muslim sweepers, who do not receive special treatment. *Report of the Commissioner for Scheduled Castes and Scheduled Tribes*, 1960–1, p. 316.

107 One of the paradoxes of the religious test as presently applied is that an individual convert from Hinduism to Sikhism or vice versa automatically remains within the Scheduled Castes without any consideration of his intention to remain in the caste, continued acceptance by his caste fellows, or change in his condition. He is accounted a member of the caste even though, according to the loss of caste by conversion theory, he has renounced it.

that in at least some cases of conversion the convert continues to regard himself as a member of the old caste and to be so regarded by others.[108] The unwillingness of the courts to credit such membership derives not from an examination of the facts but from an identification of castes as components in the sacral order of Hinduism. Caste and Hinduism are regarded as coterminous; when Hinduism is abandoned, so is caste membership.

This view of caste is at variance with the constitutional and statutory 'disestablishment' of the sacral view of caste since Indian independence. *Varna* distinctions have been abolished in the personal law; untouchability has been abolished; disabilities have been prohibited; all support for claims of precedence based on caste standing has been withdrawn.[109] Caste groups are, of course, free to entertain any views they wish concerning their station within (or without) Hindu society. But they may no longer call on the government or the courts to confirm and enforce these views. The law recognizes caste groups as corporate entities, bound by manifold ties of association. The nature of these ties, religious and social, varies from group in accordance with its own internal order. But the Constitution bars governmental recognition of hierarchical or ering, the apportioning of sacramental honour among caste groups, or the imposition of religious tests for recognition of a group as a caste.[110]

Since preferences for Scheduled Castes are meant to alleviate the disadvantages historically associated with low standing in the Hindu sacral order, an attenuated recognition of this hierarchical order is utilized for the purpose of deciding who is to receive these preferences. Scheduled Castes are identified in part by considerations of *varna*, pollution, and ritual distance.[111] But it is clear that these otherwise discredited legal categories are appropriate only to the extent that they serve to identify the needy and deserving. They are appropriate only as convenient descriptions of practical conditions, not as prescriptive requirements.

Conversion, both to sects within Hinduism and to religions outside Hinduism, has traditionally been and continues to be one of the

---

108 See note 99.

109 Art. 17, Art. 15(2). Untouchability Offences Act, 1955 (XXII of 1955) and kindred state legislation, the Hindu Marriages Validity Act, 1949 (XXI of 1948) and the Hindu Code Acts (see note 43).

110 It is open to a caste group itself to impose religious requirements for membership. See *Devchand Totaram v. Ghaneshyam*, A.I.R. 1935 Bom. 361. Cf. note 121.

111 In fact, attempts to identify the Scheduled Castes in terms of criteria of pollution, impurity, and disabilities proved inadequate to isolate the groups which local administrators felt deserving of inclusion. Additional criteria of poverty and illiteracy had to be added. Thus the Scheduled Castes were determined by an uneven mixture of ritual and socioeconomic criteria. See Dushkin (1961).

common expedients of those at the bottom of the caste hierarchy seeking to improve their position.[112] Historically, the disadvantages associated with their low position have often followed converts across religious lines; they continue to do so in at least some cases in the present. However effective or ineffective a means of escape conversion has proved to be in a particular case should be evaluated in empirical terms. To deem conversion to non-Hindu communions an automatic disqualification for aid violates the constitutional command of equal treatment for different religions. It also restricts freedom of religion, which would seem to require that government refrain from administering its welfare schemes so as to put a heavy price-tag on its exercise.

The Hindu requirement seems to reflect a hostility toward conversions which is anachronistic.[113] There is little reason to expect large-scale conversions to non-Indian religions as this point in Indian history, and anxiety inspired by the threat of separate electorates is similarly out of date.[114] The test is certainly not designed to bolster Hindu orthodoxy since one can remain a Hindu while embracing

---

112 Titus (1959); Pickett (1933); Hazari (1951); Keer (1962); Isaacs (1965).

113 Cf. the disabilities imposed upon converts from Hinduism in the various Hindu Code acts. See e.g., *Sundarambal v. Suppiah*, A.I.R. 1963 Mad. 260. Derrett (1958: 393) remarks that it is 'strangely inconsistent with the claim to be a secular state' that a Hindu who changes his religion is liable to be divorced by his wife, may forfeit an existing claim to maintenance, may lose the right to give his child in adoption, may lose the right to be the guardian of his own issue and may have his issue deprived of the right of inheritance from unconverted relations. See also D. Smith (1963: chap. 6).

114 By separate electorates is meant the representation of religious (or other) minorities by legislators elected from an electorate composed only of members of that minority. The provision and extent of such representation was an extremely troublesome political issue in India during the last forty years prior to independence. The Government of India Act, 1909, gave separate electorates to Muslims; the Government of India Acts of 1919 and 1935 provided separate electorates for Muslims, Sikhs, Indian Christians, and other groups. Proposals to give separate electorates to the Scheduled Castes under the 1935 Act, were withdrawn after adamant resistance by M. K. Gandhi. See Ambedkar (1946). They were given instead reserved seats in the legislature—i.e., only they could be candidates for these seats, but the whole general electorate chose among the candidates. The existence and extent of separate electorates was a constant source of dispute between Hindus and Muslims. See. Coupland (1944); Dalal (1940). Concern about separate electorates gave rise to extreme sensitivity to respective population figures of minority groups.

The Constituent Assembly definitively rejected special safeguards for religious minorities. See 12 Constituent Assembly Debates 299. Article 325 outlaws separate electorates for Parliament and the State legislatures. The Supreme Court has indicated that Art. 15 prohibits communal electorates in local bodies. *Nain Sukh Das v. State of U.P.*, A.I.R. 1953 S.C. 384. The reservations for Scheduled Castes and Scheduled Tribes are not separate electorates, but are reservations of seats whose occupants are voted on by the general electorate. Candidates must belong to the privileged group, but the entire electorate participates in choosing among candidates with such qualifications.

the most heterodox beliefs and practices. Not only is the test inconsistent with the avowed lack of religious discrimination but, ironically, it seems to impose on Hinduism the notion of a hard and fast line between creeds and communions. This is neither historically nor philosophically a Hindu notion, but is more consonant with the exclusivist creeds of the West, which require the convert to abjure his previous faith. It is rather surprising that Indian jurisprudence should give currency to this notion rather than to more cosmopolitian views that are more congenial to the Indian tradition of religious tolerance.[115]

Preferences are designed to remedy certain conditions — particularly to offset or dispel the disadvantages created by the imposition of social disabilities and the lack of economic and educational resources. The usefulness of the theory that converts are casteless in describing these conditions is questionable. Where the preference is one that devolves on the members of the group as individuals, surely the appropriate question is whether the disabilities and disadvantages have, in the particular case, been effectively dispelled by the conversion. Where the preference is a political one that devolves on the group corporately, the appropriate question would be whether the group still accepts the convert as a member and is willing to have him as a representative. In neither instance would it be necessary to decide whether he is a member of the group for all purposes or whether he is a Hindu.

Conclusion

The judicial treatment of the group membership problem displays two divergent tendencies, each with some support from the Supreme Court. The first, which might be called the pragmatic or empirical approach is represented in the *Jasani*, *Shyamsundar*, *Kartik Oraon*, and *Wilson Reade* cases. The second, which might be called the formal or fictional approach is represented in the *Dora* case and in the cases dealing with the 'Hinduism' test for Scheduled Castes.

Both approaches recognize the 'compartmental' nature of Indian society. But the empirical approach is willing to give recognition to the areas of blurring and overlap that are found within it. The

---

[115] For an example of judicial expression of Hindu cosmopolitanism, see *Kolandei v. Gnanavarum*, A.I.R. 1944 Mad. 156, where a gift by the manager of a Hindu joint family to a Roman Catholic church was upheld as binding on his heirs, since 'the very idea that pious and charitable purposes should . . . mean objects that can be said to promote a particular faith, creed, or dogma is foreign' to the ideals of Hinduism. Id. at 157. Cf. Sastri, 663. While such cosmopolitan tolerance should not be imposed on groups which find it uncongenial, certainly the courts should not impose narrower views on those groups which are willing to accept members of different religions.

fictional approach, emphasizing theoretical symmetry, tends to picture the society as one of mutually exclusive and hierarchically ranked compartments; where in fact individuals straddle compartments, the Court sees its task as assigning them to one or the other. The pragmatic approach does not share this notion of resolving the ambiguity into a single identification but is congenial to multiple and overlapping affiliations; it addresses itself to whether, in the light of the policy of the particular legislation involved, the individual can be a member of the group concerned.

The fictional approach concentrates on the theoretical consequences of certain acts — one who attains caste status loses his tribal affiliation, one who declares himself a member of a non-Hindu religion loses caste membership. The pragmatic approach pays less attention to such theoretical incompatibility and gives greater weight to the facts of intention and acceptance. Thus sect members can retain their caste memberships, an Anglo-Indian can become a tribal and a convert can remain a tribal.

Is one of these approaches preferable or constitutionally incumbent upon the courts? The pragmatic approach is clearly more consonant with empirical knowledge of Indian society. It avoids the encumbrance of theories of caste which reflects perceptions of Indian society current among observers a generation or more ago. Modern students of Indian society have modified older notions of the caste system which emphasized *varna* and inflexibility and have achieved a new understanding of its complexity which includes its local variability, the ambiguity of caste-ranking, the existence of mobility, and the limitations of *varna* theory. The Courts have long recognized the deficiencies of the older notions.[116] Now that Courts are no longer hampered by the necessity of giving legal effect to a picture of society consisting of the four *varnas*,[117] they may appropriately employ the new perspectives in an attempt to confront the actualities of Indian society in order to implement the principles of the Constitution.

It is submitted that the empirical approach is also to be preferred because of its consonance with the ideals and principles of the Constitution, while the fictional approach contravenes these and should be discarded. The Constitution sets forth a general programme for the reconstruction of Indian society. In spite of its length, it is suprisingly undetailed in its treatment of the institution of caste and the existing group structure of Indian society. But it clearly

---

116 On the limitations of the *varna* model, see note 41. For instructive discussion of the complexity of the caste-system, see *Muthusami v. Masilamani*, note 99.

117 The enactment of the various Hindu Code acts in 1955–6 virtually eliminates *varna* as a distinct legal status is still permissible for limited purposes (see notes 42 and 43), the Constitution surely provides a mandate to confine it within the narrowest limits.

sets out to secure to individuals equality of status and opportunity, to abolish invidious distinctions among groups, to protect the integrity of a variety of groups — religious, linguistic, and cultural, to give free play to voluntary associations, the widest freedom of association to the individual, and generally the widest personal freedom consonant with the public good. Without pursuing all of these in detail, it is clear that the following general principles are consistently in evidence; (1) a commitment to the replacement of ascribed status by voluntary affiliations; (2) an emphasis on the integrity and autonomy of groups within society; (3) a withdrawal of governmental recognition of rank ordering among groups.

In all of these, the pragmatic approach seems more congenial to the constitutional design. Its emphasis on the actual conduct of the individual and the actual acceptance by members of the group gives greater play to the voluntary principle. The control this gives the group over determination of its own membership seems implicit in the recognition of the integrity of the group. Finally it avoids the necessity of giving official recognition to the ranking of groups.

The fictional approach, on the other hand, severely limits the voluntary principle and the autonomy of the group by giving conduct unintended consequences on theoretical grounds and by determining the question of membership without consulting the views of the relevant groups. It gives, if unwillingly, legal effectiveness to the notions of rank order among groups and mutual exclusiveness among them. The appearance and persistence of such an approach at this time should be regarded as an anachronism and one exprects that before long the Supreme Court will refine and extend the empirical approach it pioneered in the *Jasani* case.

A pragmatic approach, sensitive to the egalitarian and welfare objectives of the policy of protective discrimination, is particularly appropriate in view of the welcome trend-away from exclusive reliance on communal membership in selecting the recipients of preferences.[118] This development, notably encouraged by the courts, represents a determination that preferences are to be directed to those most needy and are not to be permitted to ossify into communal quotas. Eventually the replacement of caste criteria by economic and educational tests would eliminate litigation about group membership in the preference area. To the extent that group criteria continue to be used, it is appropriate that the courts visualize groups and their boundaries in the light of the constitutional principles of equality and voluntarism.

There are no automatic answers to practical questions which must be decided in the specific social setting and in the context of

[118] See cases cited note 8.

particular statutory policies. In such cases the courts must undertake the difficult task of assessing the social actualities and construing legislative policy in the light of constitutional principles. In doing so, the courts have an opportunity to demonstrate how principles of equality and voluntarism can be implemented in a plural society with a social structure that is mainly traditional and where status is mainly ascribed. By the thoughtful and coherent treatment of these problems of group membership and group preferences, Indian jurisprudence can contribute much-needed guidance both to new nations attempting to construct a viable plural society and to older nations which have been unable to resolve the problems of diversity. By such guidance the courts may play a crucial role in assuring that the world-wide transformation from traditional to modern society takes place in conformity to the principles of freedom and equality.

APPENDIX

Group Integrity and Excommunication

However, the mere avoidance of fictional tests which violate the understanding of those involved does not solve all problems in this area. While the principles of voluntary affiliation and group auto-nomy are compatible in most of these membership cases, it is easy to imagine a situation where conflict between them is inevitable — that of excommunication or expulsion. What effect should excom-munication or expulsion from the group have on eligibility for preferential treatment? The Constitution leaves unchanged the power of a caste group to excommunicate those who offend it.[119] In religious denominations — and these include at least some caste groups[120] — this right of excommunication is constitutionally pro-tected insofar as it is used for the enforcement of religious discipline.[121]

In the *Jasani* case we saw that the test for retention of membership in the privileged group was primarily one of the convert's intention. However it was necessary that this intention be confirmed by

[119] Section 9 of the Civil Procedure Code forbids the courts to take cognizance of 'caste questions'. It is still a good defence to a criminal action for defamation to assert the privilege of communicating news of an excommunication to one's caste fellows. *Panduram v. Biswambar*, A.I.R., 1958 Or. 259; *Varadiah v. Parthasarathy*, I.L.R. (1964) (2) Mad. 417. Cf. *Hadibandhu v. Banamali*, A.I.R. 1960 Or. 33. The power to excom-municate is subject to two important restrictions. It is an offence to undertake any disciplinary action directed toward the enforcement of 'untouchability.' Untoucha-bility Offences Act, § 7. It is a corrupt election practice to threaten a candidate or elector with excommunication. Representation of the People Act, 1951, § 123(2).

[120] See, e.g., *Sri Venkataramana Devaru v. State of Mysore*, A.I.R. 1958 S.C. 255.

[121] *Saifuddin Saheb v. State of Bombay*, A.I.R. 1962 S.C. 863.

acceptance by the old group. The inclusion of this additional test is important because, as the Court says, 'the only modification here is that it is not only his choice which must be taken into account, but also the views of the body whose religious tenets he has renounced, because here the right we are considering is the right of the old body, the right to send one of its own fold to Parliament.'[122] Since the general electorate choose among the candidates standing for the reserved seat, the only way that representation of the privilegd group can be assured is by seeing that only those accepted by it are permitted to stand for these seats. But if the preference were not in the political area, if it did not devolve on the Scheduled Castes collectively but on their members individually, the primary test of intention and conduct of the convert would apparently be conclusive and it would not be necessary to confirm this by acceptability to the privileged group. It seems doubtful whether ostracism or excommunication by the old group would effectively deprive one of membership in the group for the purpose of qualifying for benefits which were not directed to the group collectively. For such benefits as education and housing do not devolve on the group in any corporate or collective capacity; membership in the group is merely a convenient device for identification of deserving beneficiaries. In such a case the relevant question would be whether the expulsion was associated with any conditions which effectively dispelled the disabilities or other backwardness which had caused the group's members to be singled out for preferential treatment. Presumably, if the individual still suffers these disadvantages he would remain a member of the privileged group in the sense of being a legitimate recipient of these preferences. So while the group's opinion of the membership of the individual should be decisive in the case of preferences which devolve on the group collectively, where the preferences are distributive (i.e. devolve on the members of the group individually), the state may legitimately intervene to protect the individual from group power. The appropriate question is whether his condition has changed from that which caused members of the group to be given preferences.

Is membership merely a convenient device for identifying deserving beneficiaries? It might be argued that preferences are directed to

[122] (1959) S.C.R. at 839. Separate electorates along caste or religious lines are barred by Art. 325. Since seats are not reserved for particular Scheduled Castes but for the Scheduled Castes as a whole, one might question the relevance of looking specifically to his own community to see whether he was accepted. E.g., if the privileged group consists of A's, B's, and C's and X—a member of A—converts to sect S, the question of whether X is a member of the Scheduled Castes might equally depend not only on whether the A's accept him as an A but whether the B's and C's do as well or instead.

members of particular groups not only to benefit the individual members directly, but to utilize the ties of kinship, loyalty, and co-operation among them to share and multiply these benefits. But while group cohesion is an instrument of welfare policy, it is not necessarily an objective of that policy or a necessary condition for its effectiveness. Preferences which include among their objectives the promotion of the cohesion and integrity of the group must be distinguished from those where group cohesion is used only as an instrument. It is clear that preferences for the Scheduled Castes and Backward Classes are for the sole purpose of ameliorating the deprivations of the lowly; they are not designed to promote the cohesion and integrity of the groups that make up these categories. On the other hand, the preferences for Schedule Tribes include among their purposes the promotion of the cultural integrity and autonomy of the tribal groups. While expulsion should not outweigh the right of the individual to enjoy preferences among the Scheduled Castes and Backward Classes, in the case of the tribes the group's view of the individual's membership must be given greater weight.

Some caste groups among the Scheduled Castes or Backward Classes might regard themselves as religious denominations and invoke their constitutionally protected right to excommunicate for the purpose of upholding religious discipline.[123] However, since preferences to these groups are for the purpose of ameliorating their condition rather than promoting their religious unity, such expulsion should not outweigh the right of the expelled individual to share in these preferences — at least so far as his enjoyment of them does not detract from the religious freedom of the group.

Finally, we might speculate about the situation where an individual aspired to become a member of the preferred group. Entry to the group, like departure from it, involves three factors which are relevant for the purpose of preferences: the intention and conduct of the individual; acceptance of him by the group to which he aspires to belong; and an affinity of conditions and resources with those of members of that group. As in the departure cases, the test of intention and conduct is primary. But how should the other factors be weighed? (1) Where intention is accompanied by both acceptance and by affinity of conditions there is no difficulty in deeming the aspirant a member of the group. (2) Similarly, where intention is confirmed by neither group acceptance nor by affinity of conditions, there is no reason to include him under any scheme of preferences. The other possibilities are more troublesome: (3) When intention is coupled with group acceptance but there is no substantial affinity

---

[123] On the loss of civil rights consequent upon religious excommunication, see *Saifuddin Saheb v. State of Bombay*, note 121.

of condition, presumably acceptance should outweigh affinity of condition in dealing with those benefits like political representation for the Scheduled Tribes, where preservation of the cultural integrity of the group is an aim of policy. But among Scheduled Castes and Backward Classes (other than in the case of collective benefits) consideration of affinity of condition should be determinative. (4) When intention is coupled with affinity of condition, but there is no acceptance by the group, the positions are reversed. Here, the individual can certainly be excluded from enjoyment of political benefits. Affinity of condition should be enough to secure other benefits among the Scheduled Castes and the Backward Classes, but among the Tribes, lack of acceptance should be conclusive and outweigh affinity of condition, even for these distributive benefits.

7

Changing Legal Conceptions of Caste*

It is a commonplace that the Constitution of India envisages a new order as to both the place of caste in Indian life and the role of law in regulating it. There is a clear commitment to eliminate inequality of status and invidious treatment, and to have a society in which government takes minimal account of ascriptive ties. Beyond this the Constitution is undetailed and in some respects unclear about the posture of the legal system with respect to role of the caste group in Indian life.

In this paper I hope to elucidate some features of the relation between caste and law by considering the ways in which the law conceives of the caste group. The legal characterization of caste is of some interest as a reflection of the views of an important and influential group in Indian society. Second, it is of some historical importance, since there is evidence that the legal system is a powerful disseminator of notions about the various groups in Indian society and may affect their self-image and the image others have of them.[1] Finally, it is of great practical importance, for, on the one hand, the Indian government is committed to the abolition of certain features of caste[2] and the legal system places restrictions on the powers of caste groups, on governmental recognition of caste, and on claims that can be made in the name of caste standing; on the other hand, religious groupings and voluntary associations enjoy constitutional protections, and government is committed to allowing them free play within broad limits.[3] The extent to which caste groups are legally characterized as religious or voluntary groups may have considerable practical import. And, to the extent that these legal notions influence behaviour, the legal characterization of the caste

* Reprinted from *Structure and Change in Indian Society*, Milton Singer and Bernard S. Cohn, eds., Viking Fund Publications in Anthropology, No. 47. Copyright 1968 by the Wenner-Gren Foundation for Anthropological Research, Inc., New York.

An earlier version of this paper was prepared for a seminar on Religion and Politics in South Asia held in Colombo in July, 1964, and has appeared in D.E. Smith (ed.), *South Asian Politics and Religion* (Princeton, N.J.: Princeton University Press, 1966).

1 For example, McCormack (1963) suggests that the notion of a unitary Lingayat group with a single distinctive culture appeared as a result of the application of the Anglo-Hindu law and British judicial administration.

2 Constitution of India, Arts. 15(1), 15(2), 16(2), 17, 23, 29(2).

3 Constitution of India, Arts. 19, 25, 26, 29, 30.

group may be an influential factor in the contempoary reform and reorganization of Hinduism and of Indian society.

In order to describe the judicial conceptualization of the caste group, I propose to use four models. These models represent different ways of visualizing caste groups and their mutual relations.[4] All of them are employed by courts in dealing with concrete issues. Often the judicial response to an issue may employ more than one of these models or approaches. It is probably unnecessary to emphasize that the 'models' discussed here exist in the eye of the present beholder. Courts do not often speak explicitly in these terms. It is my contention that the models are demonstrably but implicitly present in the work of the courts (and legislatures). The models point to different ways of visualizing the caste group and prove useful in describing recent changes in the legal view of caste.

The first model sees the caste group as a component in an overarching sacral order of Hindu society. Hindu society is seen as a differentiated but integrated order in which the different parts may enjoy different rights, duties, privileges and disabilities; these are determined by the position of the caste group in relation to the whole. We may call this the *sacral* view of caste.

In contrast to this is what we might call the *sectarian* view, which sees the caste as an isolable religious community distinguished from others by idiosyncratic doctrine, ritual, or culture.[5] It is a self-contained religious unit, disassociated from any larger religious order. The rights and duties of the group and its members follow from its own characteristics, not from its place in a larger order. Where the sacral view visualizes castes as occupying the various rooms, shrines, courtyards, and outbuildings of the great labyrinthine temple of Hinduism (to each of which is attached special prerogatives and disabilities), the sectarian view visualizes castes as a series of separate chapels under independent management. In the sacral view, the rights and duties of a caste (or its members) can be determined by its relation to the whole (or at least to its surroundings); in the sectarian view, they can be described by reference to its own internal order. It is the difference between a ward in a great and dense city, and a self-sufficient small town.

---

[4] This paper does not address itself to the question of the geographical spread and boundaries of the units that are called castes; for example, whether they are village groups, local endogamous groups, regional networks, whether they contain endogamous subgroups, etc. My general impression is that the courts have made little distinction between these kinds of 'castes'. I suspect that there may be some differences in the kind of groups involved in different sorts of cases, a suspicion that I hope to confirm upon a recheck of the cases. I have followed the designation and spelling of the courts in referring to castes, tribes, and sects.

[5] In employing the term 'sectarian' it is necessary to resist both the connotation that such groups are associated with a distinctive and precise doctrine.

Both these views characterize caste in terms of religious factors.[6] The second pair of views is secular, not in the sense that they omit religion entirely, but because they do not give it a central place. The first of this second pair might be called the *associational* view of caste. Here, the caste is seen as an autonomous association with its own internal order and rule-making powers, but characterized neither by a fixed place in some larger religious order nor by distinctive and idiosyncratic religious beliefs or practices. It is a body with its own principles of affiliation and its own internal order. These may be in some respects like those of a corporation, a club, a dissenting church (in English law), or some other voluntary association, but they render the caste a form of association *sui generis*. The nature of the tie is not characterized solely or conclusively by religious fellowship. The bonds of association may include religious ones, but the religious tie is only one among a constellation of affinities. As is the case with the sacral view of the caste group, the associational view avoids characterization in terms of specific religious factors. And in common with the sectarian view, it does not identify the caste by its standing in a differentiated religious order of society. The sacral view regards the caste group in terms of its relation to the larger body of Hinduism; the sectarian view sees it in terms of its own religious distinctiveness; the associational view defines caste in terms of its associational bonds, which may include religious features, although they are not conclusive in identifying or characterizing it.

Recently, we find evidence of a fourth view of caste, which I call the *organic*. Here, the caste group is seen as occupying a particular place in a social order made up of many such groups. This place is determined by a certain level of resources and attainments relative to other groups in the society. As does the sacral model, this view characterizes the caste group in terms of its relative standing in a larger whole. But this organic view sees the standing of a caste as determined, not by its possession of Hindu ritual values as does the sacral view, but by its share of mundane accomplishments and resources. It does not take religious factors as the sole or primary determinant of the nature of the caste group; in this respect it resembles the associational view. But, this organic view, in contradic-

---

6 Western and Indian writers alike disagree about the 'religious' character of caste groups. To some they are the very units of Hinduism; to others they are 'purely social', with only an accidental relation to Hinduism. I shall not attempt any reconciliation of different views of 'religion', I have not attempted to define 'religion' or 'religious', but have been content to call 'religious' those characterizations of caste that involve such rubrics as religion, worship, sacred, denomination, Hinduism, impurity, pollution, etc. References to castes in terms of clubs, associations, corporations, economic and educational level, political influence etc. I have deemed nonreligious.

tinction to both the associational and the sectarian views, does not see the caste as an isolated or idiosyncratic entity.

The models may be schematized as on the following chart:

		Characterization of the group in terms of its position in the larger society		
		Yes	No	
Conclusiveness of religious factors in characterizing the group	Yes	SACRAL View	SECTARIAN View	Religious Views
	No	ORGANIC View	ASSOCIATIONAL View	Secular Views
		Holistic Views	Segmental Views	

In order to trace the changing judicial conceptualization of caste, I propose to examine several kinds of cases in which caste comes before the courts. The matters I have chosen are (1) the administration of 'personal law;' (2) the recognition of claims for precedence and for the imposition of disabilities; (3) the recognition of castes as autonomous self-governing groups. After briefly suggesting the judicial characterization of caste that prevailed in each of these fields in the latter days of British rule[7] in India, I shall attempt to trace developments since independence to show the emerging judicial view of caste.[8]

The Old Regime

Personal Law

The Hindu law applied by the courts in matters of 'personal law'[9]

[7] By this I mean the period since the consolidation of the modern legal system, which can be dated about 1860.

[8] The developments described here are at the higher and more authoritative levels of the legal system. In describing the development and application of doctrine by legislatures and higher courts, it is not intended to imply any one-to-one correspondence between the pronouncements of these higher authorities and the day-to-day operations of magistrates, officials, and lawyers, and much less the lay public. In the long run, however, the higher courts' pronouncements not only tend to reflect what the officials and the public are doing but are uniquely influential, first, by disseminating influential 'official' conceptions of caste which have an impact on the caste system and, second, by deflecting behaviour towards conformity with the doctrines they promulgate.

[9] Under the legal system which the British established in India, all persons were subject to the same law in criminal, civil, and commercial matters. However, a group

did not address itself to the multitude of caste groups, but recognized only the four varnas (and occasionally the intermediate classes of classical Hindu legal theory).[10] This law contained a number of instances in which different rules were to be applied to members of different varnas — in most cases one rule for the three twice-born varnas and a different rule for the Shudras. The most notable of these differences were in the law of succession, the law of adoption and, the law of marriage.[11] With limited exceptions, marriages and adoptions among members of different varnas were not valid at all. In order to apply these rules which differed according to varna, it was necessary for the courts to determine which castes and individuals were included within which varna. The assignment of standing in the four-varna system to actual castes presented an opportunity for eliciting legal recognition of the ceremonial status of the group and certification of its claims for higher status.

The courts developed several kinds of tests to determine the varna standing of particular castes. One was the listing of certain diagnostic customs, for example, admission of illegitimate sons to commensality and marriage within the group and the prevalence of second marriages for widows, marked the group as Shudras.[12] Another line of cases developed an alternative approach of testing the varna standing of a caste group by its own consciousness of its status and by the acceptance of this self-estimate by other castes in the locality.[13] These tests involve reliance on widespread conventional notions of purity and pollution; they emphasize orthodox and prestigious practice rather than refinements of doctrine or ritual.[14] These notions of

---

of matters that might roughly be described as 'family law' — marriage and divorce, adoption, joint family, guardianship, minority, legitimacy, inheritance and succession, and religious endowments — was set aside and left subject to the laws of the various religious communities. The applicable law in these fields was 'personal' rather than territorial. In these family and religious matters, Hindus were ruled by Dharma-shastra — not by the ancient texts as such, but by the texts as interpreted by the commentators accepted in the locality.

[10] The judicial treatment of the relation between varna and caste was plagued by confusion, engendered in part by the use of 'caste' to refer both to the four great classes or varnas into which Hindu society is theoretically divided by the Sanskrit lawbooks and to the multitude of existing endogamous groups or *jātis*. In the sequel, unless the context indicates otherwise, caste is used only in the latter sense.

[11] These differences are concisely summarized by Derrett (1958).

[12] See, for example, *Gopal v. Hanmant*, I.L.R. 3 Bom. 273 (1879).

[13] See, for example, *Subrao v. Radha*, I.L.R. 52 Bom. 497 (1928).

[14] Mere performance of ceremonies associated with higher castes will not elevate lower classes to that station, though 'where caste is doubtful, the performance of Vedic or Puranic ritual may be important evidence as to caste. . . .'. *Maharajah of Kolhapur Sundaram Ayyar*, A.I.R. 1925 Mad. 497 at 553.

differential purity are used to assign castes to their proper varna.[15] It is assumed that the castes are components of the varnas, which in turn comprise Hinduism. It is assumed that all groups within Hinduism are subsumed under one or another varna. Although there are some instances of judicial departures from the symmetry of this scheme,[16] generally the picture of Hinduism found in the administration of personal law is one which regards caste and varna as coextensive with Hinduism. Castes, therefore, have certain religious characteristics; they occupy their respective places in an overarching sacral order of ranks which embraces all groups within Hinduism. Positions in this order could be assigned by certain widely shared notions about the relative standing implied by certain practices.

The textual law recognized varnas but not castes. But the textual law was to be modified by prevailing custom; the doctrine that 'clear proof of usage will outweigh the written text of the law' was early accepted as part of the Hindu law.[17] The custom proved might be that of a family, a locality, or a caste. The latter was most commonly pleaded. William McCormack suggests that 'most of the alleged 'caste customs' which came before the courts were in fact regional customs'.[18] The strategy of pleading caste customs tended to associate the caste group with a distinctive set of customs, providing models for patterning social behaviour, serving to maintain caste identity in new settings and facilitating the transformation of local caste groups into wider congregations.[19] Thus the caste group

---

[15] On pollution as a differentiating and integrative factor in caste systems, see Orenstein (1965) and Harper (1964).

[16] Thus it is possible to have varna standing without belonging to a caste group. *Sunder Devi v. Jheboo Lal*, A.I.R. 1957 All. 215 (convert to Hinduism); *Upoma Kuchain v. Bholaram*, I.L.R. 15 Cal. 708 (1888) (daughter of outcaste); cf. *Ratansi v. Administrator General*, A.I.R. 1928 Mad. 1279 (convert to Hinduism). For some purposes at least, Hindu caste groups may fall outside of or below the four *varnas. Sankaralinga Nadan v. Raja Rajeswara Nadan*, 35 I.A. 176 (1908). Possibly one can be a Hindu without caste or *varna*. See *Ratansi v. Administrator General, supra*. Caste and *varna* may apply to persons who are not strictly Hindus, *Inder Singh v. Sadhan Singh*, I.L.R. (1944), 1 Cal. 233 (Sikh Brahmans). Caste groups have been recognized for some purposes which have no varna nor are Hindu in any sense (*Abdul Kadir v. Dharma*, I.L.R. 20 Bom. 190 (1895). Again members of the same caste may hold different varna statuses. *Subrao v. Radha*, I.L.R. 52 Bom. 497 (1928).

[17] *Collector of Madura v. Moottoo Ramalinga Sathupathy*, 12 M.I.A. 397 at 436 (1868). However, the application of stringent common law requirements for proving a valid custom makes it difficult to prove variation from the rules of the lawbooks and had the effect, it appears, of extending the rules of the classical lawbooks to sections of the population which had previously been strangers to them. The British period then was marked by an attrition of local customary law at the expense of the written and refined law of the texts (Galanter, 1964a).

[18] McCormack (1966:30).

[19] McCormack (1966).

was seen as a carrier of a distinctive set of cultural traits. When these traits were employed in assessing varna standing they implied the sacral view of caste; when employed to vary the textual law they implied a view of the caste group as a corporate body culturally distinct from its neighbours.

Precedence and Disabilities

Prior to British rule, some Indian regimes had actively enforced the privileges and disabilities of various caste groups. Indeed, such enforcement of the caste order is urged by Hindu legal tradition as the prime duty of the Hindu king. During the latter part of the British period the prerogatives and dignities of castes received only limited support by active governmental sanctions. This limited support was undertaken on the basis of upholding customary rights, but these rights were often conceptualized in terms of the religious chatacteristics of caste groups.

With respect to the use of religious premises, caste groups did enjoy the support of the courts in upholding their claims for preference and exclusiveness. Courts granted injunctions to restrain members of particular castes from entering temples — even temples that were publicly supported and dedicated to the entire Hindu community.[20] Damages were awarded for purificatory ceremonies necessitated by the pollution caused by the presence of lower castes; such pollution was actionable as a trespass to the person of the higher caste worshippers.[21] It was a criminal offence for a member of an excluded caste knowingly to pollute a temple by his presence.[22] These rights to exclusiveness were vindicated by the courts not only where the interlopers were 'Untouchables' but also against such 'touchables' as Palshe Brahmans and Lingayats, whose presence in the particular temple was polluting.

In these cases the courts were giving effect to the notion of an overarching, differentiated Hindu ritual order in which the various castes were assigned, by text or by custom, certain prerogatives and disabilities to be measured by concepts of varna, of pollution, and of required ceremonial distance. Thus, in *Anandrav Bhikaji Phadke v. Shankar Daji Charya* the court upheld the right of Chitpavan Brahmans to exclude Palshe Brahmans from worshipping at a temple, on the ground that such an exclusive right 'is one which the

[20] *Anandrav Bhikiji Phadke v. Shankar Daji Charya*, I.L.R. 7 Bom. 323 (1883); *Sankaralinga Nadan v. Raja Rajeswara Dorai*, 35 I.A.C. 176 (1908); *Chathunni v. Appukuttan*, A.I.R. 1945 Mad. 232.

[21] See cases cited, note 18 *supra*. Cf. *S.K. Wodeyar v. Ganapati*, A.I.R. 1935 Bom. 371, where damages were awarded although the parties agreed there should be no finding on the question of pollution.

[22] *Atmaram v. King-emperor*, A.I.R. 1924 Nag. 121.

Courts must guard, as otherwise all high-caste Hindus would hold their sanctuaries and perform their worship, only so far as those of the lower castes chose to allow them'.23

In 1908 the Privy Council upheld the exclusion of Shanars from a temple and granted damages for its purification after a careful scrutiny of their social standing. Finding that 'their position in general social estimation appears to have been just above that of Pallas, Pariahs, and Chucklies (who are on all hands regarded as unclean and prohibited from the use of Hindu temples) and below that of the Vellalas, Maravars, and other cultivating castes usually classed as Shudras, and admittedly free to worship in the Hindu temples', the Judicial Committee of the Privy Council concluded that the presence of Shanars was repugnant to the 'religious principles of the Hindu worship of Shiva' as well as to the sentiments and customs of the caste Hindu worshippers.24 As late as 1945, Nair users of a public temple were granted damages for pollution for the purificatory ceremonies necessitated by Ezhuvas' bathing in tanks.25 These exclusionary rights were supported by criminal as well as by civil sanctions. Untouchable Mahars who entered the enclosure of a village idol were convicted on the ground that 'where custom . . . ordains that an untouchable, whose very touch is in the opinion of devout Hindus pollution, should not enter the enclosure surrounding the shrine of any Hindu god, . . .' such entry is a defilement in violation of Section 295 of the Penal Code.26

While Hinduism is seen as a unified and overarching order, it is also seen as differentiated. The religious obligations and prerogatives of groups differ according to their standing in this whole. Where Brahmans tore the sacred thread from the neck of an Ahir who had lately taken to wearing it, the Court ruled that, since he was a Shudra, the wearing of it was not 'part of his religion' *vis-à-vis* other Hindus. To them it was an assertion of a claim to higher rank. Therefore the injury was not to his religious susceptibilities — an offence — but only to his dignity.27 Had it been torn by non-Hindus, it might have been an insult to his religion itself.

In these cases the courts clearly express their notion of a rank ordering of all Hindu groups in a scheme of articulated prerogatives and disabilities. One looks to the position of the caste in the whole — its position on the scale relative to the other groups — to ascertain

---

23 7 Bom. 323 at 329. On the origins of this dispute, see Joshi (1913).
24 35 I.A.C. 176 at 182.
25 A.I.R. 1945 Mad. 232.
26 A.I.R. 1924 Nag. 121.
27 *Sheo Shankar v. Emperor*, A.I.R. 1940 Oudh 348. For another instance in the same neighborhood and time of a lower caste adopting (and relinquishing) the sacred thread, see Sinha (1960).

its rights. This approach did not always work to the disadvantage of the excluded class. In *Gopala v. Subramania*, members of the Elaivaniyar community obtained a declaration of their right to enter the outer hall of the temple and an injunction restraining other worshippers from ejecting them. The court declared that each group enjoyed a prima facie right to enter that part of the temple assigned their caste (that is, varna) by the Agamas (texts on use of temples), that these texts authorized the entry of Shudras in this part of the temple, and that the plaintiffs were 'at least Sudras'. Their right could be overcome only by proof of a special custom of exclusion.[28] Similarly, where Moothans were convicted for defiling a temple by entering the part open to 'non-Brahmans', the court reserved the conviction on the ground that Moothans are Shudras, no lower or more polluting than the Nairs, who were allowed to enter the temple.[29]

Again we see the notion of a single articulated Hindu community in which there are authoritative opinions (supplied by custom and accepted texts) which determine the respective rights of its component groups. The effect of this conception of the overarching Hindu order is revealed clearly in the case of *Michael Pillai v. Barthe*. Here a group of Roman Catholic Pillais and Mudalis sued for an injunction to require the Bishop of Trichinopoly to re-erect a wall separating their part of the church from that entered by 'low-caste Christians' and to declare plaintiffs' exclusive right to perform services at the altar. The court characterized the claim as one for 'a right of freedom from contact which can have but one origin . . . that of pollution',[30] but refused to recognize pollution as either a spiritual or a temporal injury among Christians. Nor could Christians constitute 'castes' with rights based on their respective purity. Not being Hindus, plaintiffs 'cannot . . . invoke the authority of accepted sacerdotal texts for perpetuating the distinction between touchables and untouchables during a particular life solely by reason of birth'.[31] When individuals have placed themselves by conversion outside the sacral order of Hinduism, their caste groupings are not invested with those rights which follow only upon their occupying a place in that order.[32]

Exclusionary practices did not enjoy the same judicial support in regard to 'secular' public facilities such as schools, wells, and roads. The courts declared that no right could be maintained to exclude

28 A.I.R. 1914 Mad. 363.

29 *Kutti Chami Moothan v. Rama Pattar*, A.I.R. 1919 Mad. 755.

30 A.I.R. 1917 Mad. 431 at 433.

31 Id. at 442.

32 For a similar unwillingness to enforce exclusion in regard to Parsis, since there was no defilement, see *Saklat v. Bella*, 53 I.A. 42 at 56–7 (P.C., 1925).

other castes or sects from the use of streets and roads.[33] The situation is more complicated regarding the use of water sources. The Lahore Court held other users had no right to prevent Chamars from drawing water from a public well.[34] However, other courts conceded that a right to exclude might be upheld if a custom of exclusive use by higher castes could be proved. Such customs were in fact difficult to prove. In *Marriappa v. Vaithilinga,* Shanars obtained an order allowing them to use a large tank on the ground that no custom of exclusion was proved (a right of exclusion was upheld in regard to one well in the dispute where such a custom was proved). What is interesting for our purpose is that even in denying the exclusionary claims of the higher groups, the court reveals an implicit view of an integrated Hindu community with graded rights. The absence of a custom of exclusion from the large tank, as distinguished from the well, is indicated by textual passages to the effect that precautions for impurity may be less intense in a body of water of this size.[35] Again, in *N. D. Vaidya v. B. R. Ambedkar,* the court found it unproved that there was any longstanding custom of exclusion. Textual provisions indicating that no elaborate precautions against pollution are required in a tank of that size rendered it 'doubtful whether any attempt would have been made to secure exclusive use of the water until such time as the tank came to be surrounded by houses of caste Hindus.'[36]

In dealing with exclusionary rights, the courts tried to confine themselves to claims involving civil or property rights as opposed to claims merely for standing or social acceptance. Thus the courts refused to penalize such defiance of customary disabilities as failure to dismount from a wedding palanquin or failure to concede another caste an exclusive right to ceremonial deference.[37] The prevailing notion was that social and religious matters did not give rise to legal rights unless the right was the sort of thing that could be possessed and made use of. Thus we find gradation from the temple cases, where there was ready enforcement of exclusionary rights, to water sources, where it seems enforcement might be forthcoming if difficult technical requirements were met, to customs in no way connected with the use of specific property, where there was no enforcement at all.[38] But where government intervened, it upheld custom and

[33] For example, *Sadogopachariar v. Rama Rao,* I.L.R. 26 Mad. 376, aff'd 35 I.A. 93.
[34] *Kazan Chand v. Emperor,* A.I.R. 1926 Lah. 683.
[35] 1913 M.W.N. 247.
[36] A.I.R. 1938 Bom. 146 at 148.
[37] *Jasnami v. Emperor,* A.I.R. 1936 All 534; *Govinda Amrita v. Emperor,* A.I.R. 1942 Nag. 45.
[38] While there was no support for these usages at the high court level, there is evidence of widespread local acquiescence in and enforcement of such practices. See,

this custom was evaluated and rationalized by the courts in terms of notions of ceremonial purity and pollution—existing in different degrees among different groups of Hindus.

It should be emphasized that prescriptive rights and disabilities received their greatest governmental support not from direct judicial enforcement but from the recognition of caste autonomy, that is, from the refusal of courts to interfere with the right of the caste group to apply sanctions against those who defied its usages or contested its claims. Members of a caste could be outcasted, and outsiders could be boycotted for violations of customary privileges and disabilities.[39] The broad sphere of autonomy enjoyed by caste groups permitted effective enforcement of their claims without resort to the courts and with immunity from governmental interference.

Caste Autonomy

Castes were early recognized as juridical entities with the right to sue and be sued, to sue on behalf of their members, and to acquire, hold, and manage property. More important for our purposes here, the caste was recognized as a group having the power to make rules for itself and to constitute tribunals to enforce these rules.[40] While caste power was limited by the official courts, which had exclusive jurisdiction over many matters (for example, criminal law), on most matters the caste could make, modify, and revoke its rules. The majority, or the established authorities within the caste, could not be overruled by the civil courts on these 'caste questions'. Caste questions were said to include all matters affecting the internal autonomy and social relations of a caste.[41] A caste then might make whatever

---

for example, the actions of the local officials described in *Kazan Chand v. Emperor*, A.I.R. 1926 Lah. 683, A.I.R. 1927 Lah. 430; *Jasnami v. Emperor*, A.I.R. 1936 All. 534; *Govinda Amrita v. Emperor*, A.I.R. 1942 Nag. 45.

39 Assertion of caste superiority by members of one caste over another and withdrawal of social intercouse does not amount to criminal defamation. *Venkata Reddi* (Mad. High Court, 1885), reported in I Weir 575. Cf. *Salar Mannaji Row v. C. Herojee Row*, I Weir 614 (Mad. High Court, 1887); *Babulal v. Tundilal*, 33 Crim. L. J. 835 (Nag., 1932). Refusal of villagers to have social intercourse with members of an unpopular sect or allow them to use a well is not criminal annoyance or nuisance. *Ramditta v. Kirpa Singh*, 1883 P. R. (Criminal) 3.

40 For detailed analysis and references in the area of caste autonomy, see Kikani (1912); Ramakrishna Aiyer (1918); Mulla (1901). The only legislation directly impinging on caste autonomy was the Caste Disabilities Removal Act (Act XXI of 1850, also known as the Freedom of Religion Act), which provided that there was to be no forfeiture of civil or property rights 'by reason of renouncing or, having been excluded from the communion of, any religion, or being deprived of caste. . . .'

41 But they did not include the economic interests of the group, where these conflicted with the property right of a member. See *Pothuraj Setty v. Padda Poliah*, 1939 (1) Mad. L.J. 116.

rules it wished about these matters. It might forbid the wearing of European clothing, departure from customary headdress, crossing the sea; it might prohibit intercourse with members who participated in widow remarriage; it might excommunicate those who failed to observe customary avoidance of lower castes.[42]

The right to have a fellow caste member accept one's food, gifts, or invitations; the right to receive invitations from him; the right to have precedence in leading one's bullock in a procession — in all these cases of dignity, acceptance, or precedence within the caste, the civil courts would not entertain a suit. Again, claims to leadership of a caste, claims to a caste office, claims to enjoy privileges and honours by virtue of such office, and claims to officiate as priest were held to be caste questions. Even if the dispute resulted in the expulsion of one person or faction, the courts would take no cognizance in such cases. Unwarranted attribution of loss of caste was defamatory. But so long as publication was not more extensive than necessary to effect the purpose of informing the caste,[43] announcement of a duly pronounced sentence of excommunication to other caste members was privileged, that is, immune from a claim for defamation as tending to protect the public good.[44]

But the courts were willing to take jurisdiction where they found that the claim was not merely for social acceptance or dignities but involved enforceable civil or property rights, which included rights in caste property, the right to offices with pecuniary emoluments, and the right to reputation. Even here, the courts were wary about the extent of intervention and set up standards that emphasized procedural rather than substantive supervision. The courts would entertain claims only: (1) that the decision of a caste tribunal had not been arrived at *bona fides*; (2) that the decision was taken under a mistaken belief; (3) that the decision was actually contrary to the

[42] *Sri Sukratendar Thritha Swami of Kashi Mutt v. Prabhu*, A.I.R., 1923 Mad. 587. Cf. *Khamani v. Emperor*, A.I.R. 1926 All. 306.

[43] See *Queen v. Sankara*, I.L.R. 6 Mad. 381 (1883), where the use of a postcard to inform the excommunicant was found to be a 'wanton excess of privilege'. Cf. *Thiagaraya v. Krishnasami*, I.L.R. 15 Mad. 214 (1892) (circulation of accusation against Brahman to persons of all castes in bazaar not privileged).

[44] See the Exceptions to Section 499 of the Indian Penal Code (Act XLV of 1960). Generally courts were well disposed toward caste tribunals and paid high regard to the beneficient qualities of caste discipline. See, for example, *Empress v. Ramanand*, I.L.R. 3 All. 664 at 667 (1881) ('No court would wish to interfere with those domestic rules and laws which regulate and control the relations between the members of a caste. On the contrary, the tendency would rather be to countenance and protect them.'); *Umed Singh v. Emperor*, A.I.R. 1924 All. 299 at 301 ('So long as caste . . . [remains] one of the fundamental characteristics of social life in India, any attempt to minimize, ignore or brush to one side existing regulations, existing sanctions or respect for existing decisions must be regarded as contrary to the public good.')

rules or usage of the caste; or (4) that it was contrary to natural justice. The last was the most important of these rules. Violations of natural justice included omission of proper notice to the accused and the denial of an opportunity to the accused to be heard and to defend himself.

Here we have a judicial view of caste more congenial to the secterian or associational models than to the sacral one. Castes are seen as independent bodies with their own internal order, and the rights and duties of individual members follow from this order. This order is not determined by the position of the caste in an overarching order of Hindu society. Although analogies are sometimes drawn from such associations as clubs,[45] corporations, partnerships, or dissenting churches, the courts never subsume the caste group under any of these. It is a group *sui generis*.[46] Although some courts speak of the caste as a voluntary organization in the sense that one can leave it, it was generally conceded that 'the caste is a social combination, the members of which are enlisted by birth, not by enrollment'.[47]

Is the caste group a 'religious body'? We have seen that the courts refused to take cognizance of suits for mere 'religious honours' or to enforce obligations they regarded as purely religious. The caste group was recognized as a proper forum for settling these religious questions. The caste was recognized as a corporate body with the right to prescribe and enforce its own religious doctrine, ritual, and leadership.[48] But in many cases it could not be characterized solely by its religious attributes. 'The caste is not a religious body, though its usages, like all other Hindu usages, are based upon religious feelings. In religious matters, strictly so called, the members of the caste are governed by their religious preceptors. In social matters they lay down their own laws.[49] Thus the caste unit was not solely religious in its concerns and nature. It was mixed — partly civil and partly religious.[50] Or as a Madras court summed it up, 'a caste is a combination of a number of persons governed by a body of usages which differentiate them from others. The usages may refer to social or religious observances, to drink, food, ceremonies, pollution,

---

[45] See *Appaya v. Padappa*, I.L.R. 23 Bom. 112 (1898).

[46] 'The Hindu caste is an unique aggregation so wholly unknown to the English law that English decisions, concerning English corporations and partnerships tend rather to confusion than to guidance upon matters relating to caste.' *Jeethabhai Narsey Chapsey Kooverji*, I.L.R. 15 Bom. 599 at 611 (1891).

[47] *Raghunath v. Javardhan*, I.L.R. 15 Bom. 599 (1891).

[48] See, for example, *Devchand Totaram v. Ghaneshyam*, A.I.R. 1935 Bom. 361. (Jurisdiction of caste includes outcasting of members for adherence to sub-sect said to be outside Vedic religion.)

[49] *Raghunath v. Javardhan*, I.L.R. 15 Bom. 599 (1891).

[50] *Haroon v. Haji Adam*, 11 Bom. L. Reporter 1267.

occupation, or marriage.'[51] The caste group, then, is not wholly or solely to be characterized by religion, either in doctrine or in practice.

It is here that we find a departure from the characterization of caste in terms of an overarching sacral order of Hinduism. Castes are autonomous units with internal government and characterized partly by religious and partly by nonreligious usages. In contradistinction to the personal law[52] and to what was held in the cases involving precedence and disabilities where castes were allocated differential religious honour because of their place in the wider Hindu scheme, here the castes are treated as autonomous and self-sufficient entities whose order proceeds from internal organs.

This detachment from the context of the wider Hindu society comes out clearly in the treatment of non-Hindu groups under the heading of caste autonomy. Here we find that the autonomous caste group is recognized not only among Hindus but also among Muslims, Jews, Sikhs, Jains and Christians.[53] In this context caste groups are not subsumed under the varnas; they are treated as a special kind of group. Claims of rights and powers which derive from a place in a larger Hindu order are not recognised among non-Hindu groups. But where the rights and powers derive from the internal order, customary and deliberative, of the group as an autonomous entity, they are recognized among all religions, not only among Hindus.

The New Dispensation

The Constitution sets forth a general programme for the reconstruction of Indian society.[54] In spite of its length, it is undetailed in its treatment of the institution of caste and of the existing group structure of Indian society. But it clearly sets out to secure to indi-

---

[51] *Muthusami v. Masilamani*, I.L.R. 33 Mad. 352 (1909).

[52] The personal law inclined away from the sacral view toward a view more like that found in the caste autonomy area in the recognition of castes as units whose customs, where proved, would serve to vary the law of the textbooks.

[53] See, for example, *Abdul Kadir v. Dharma*, I.L.R. 20 Bom. 190 (1895), where the court observed that 'caste' comprised 'any well-defined native community governed for certain internal purposes by its own rules and regulations,' and was thus not confined to Hindus. Cf. *Yusef Beg Sahib v. Maliq Md. Syed Sahib*, A.I.R. 1927 Mad. 397, where the court rejected the notion that the word 'caste' was confined to Hindus and found that it 'refers to any class who keep themselves socially distinct or inherit exclusive privileges.'

[54] This new dispensation did not arrive on the scene suddenly. It represents the culmination of more than half a century of increasing anti-caste sentiment among reformers, the gradual acceptance by politicians of the need for reform of caste, and a variety of provincial anti-disabilities and temple-entry legislation, and the growing conviction that caste is inimical to democracy and progress and should play a restricted role in the new India.

viduals equality of status and opportunity,[55] to abolish invidious distinctions among groups,[56] to protect the integrity of a variety of groups — religious, linguistic, and cultural,[57] to give free play to voluntary associations,[58] the widest freedom of association to the individual,[59] and generally the widest personal freedom consonant with the public good.[60] Without pursuing all these in detail, it is clear that the following general principles are consistently in evidence: (1) a commitment to the replacement of ascribed status by voluntary affiliations; (2) an emphasis on the integrity and autonomy of groups within society; (3) a withdrawal of governmental recognition of rank ordering among groups. In order to see how the new constitutional scheme has affected the judicial view of caste, we shall trace recent developments in the areas previously discussed and in some new problem areas that have emerged since independence.

Personal Law

The Constitution contains a commitment to replace the system of separate personal laws with a 'uniform civil code'.[61] In spite of its strictures against discrimination on the ground of religion, the Constitution has been interpreted to permit the continuing application of their respective personal laws to Hindus and Muslims. The continuing validity of disparate rules of personal law and the power of the state to create new rules applicable to particular communities has been upheld.[62] Within the Hindu law itself, the constitutional ban on caste discrimination has not been read as abolishing differences in personal law between Hindus of different castes. Although legal enforcement of disabilities against lower castes was sometimes rationalized in varna terms, the use of varna distinctions in the personal law is not included within the constitutional abolition of untouchability.[63] However, the Hindu Code Acts[64] of 1955–6 have

55 Preamble, Articles 14, 15, 16, 17, 18, 23, 46.

56 Articles 14–17, 25–30.

57 Articles 25–30, 347, 350A, 350B.

58 Articles 19(1)c, 25, 26, 30.

59 Id.

60 See generally, Parts III and IV of the Constitution.

61 Article 44.

62 For example, *State of Bombay v. Narasu Appa Mali*, A.I.R. 1952 Bom. 84.

63 The assignment of a community to a *varna* has been held not to constitute a deprivation of rights to equality before the law, nor is it religious discrimination. *Sangannagonda v. Kallangonda*, A.I.R. 1960 Mys. 147. The classification of the offspring of a Shudra and his Brahman concubine as a *cāṇḍāla*, the lowest of Untouchables in the traditional scheme, did not strike the court as unconstitutional in *Bachubhai v. Bal Dhanlaxmi*, A.I.R. 1961 Guj. 141.

64 That is, the Hindu Marriage Act of 1955, the Hindu Succession Act of 1956, the Hindu Minority and Guardianship Act of 1956, and the Hindu Adoptions and Maintenance Act of 1956.

largely abandoned the shastric basis of Hindu law and established a more or less uniform law for Hindus of all regions and castes. The new law creates the hitherto unknown capacity to marry and adopt across varna lines and, with a few minor exceptions, eliminates all of the distinctions along varna lines embodied in the old law.[65] Varna has virtually been eliminated as an operative legal concept, although for the present the courts are required to apply it to transactions covered by the older law. In addition, the new legislation severely curtails but does not eliminate the opportunities for invoking caste custom in order to vary the generally applicable Hindu law.[66] Where sanctioned by custom, caste tribunals may still dissolve marriages.[67] Where caste is still relevant to the operation of personal law, its significance is as a vessel of custom and as a body with its own rules and tribunal rather than as the holder of a position in a larger system. Here, the sacral view of caste has given way entirely to a view of castes as autonomous corporate units.[68]

Precedence and Disabilities

The Preamble to the Constitution resolves 'to secure to all of its citizens. . . . Equality of status and opportunity'. Accordingly, it confers on all its citizens a fundamental right to be free of discriminatin by the state on the ground of caste. The Constitution not only forbids caste discrimination by the government; it goes on to outlaw invidious treatment on the basis of caste by private citizens as well. Article 15(2) prohibits discrimination by private persons in regard to use of facilities and accommodations open to the public, such

[65] Derrett (1958) suggests that the only instances in which varna might continue to have effect are succession to sannyasis and determination of the maximum age for adoption. The former is an instance of a varna rule left intact (Shudras cannot become sannyasis); the possiblity of variation by caste custom also remains intact. See Derrett (1963: Sec. 592). The latter is an instance where custom may vary the law, and custom here is based at least in part on varna distinctions. See Derrett (1963b: Sec. 159ff.).

[66] Derrett (1963b: Sec. 13) lists as matters remaining open to customary variation the prohibited degrees for marriage, the maximum age of adoption, the right to dissolution of marriage by a caste tribunal, right to be a sannyasi, right to maintenance out of impartible estates, and other joint family rights.

[67] For example, *Premenbhai v. Chanoolal*, A.I.R. 1963 M.P. 57.

[68] The caste autonomy recognized here is of a limited kind. Castes carry and can administer established customs, possibly they can abrogate an old custom, but they cannot deliberately create new customary law. A caste cannot attract to itself a different body of personal law by changes of name, etc. *Sankaran v. Madhavan*, A.I.R. 1955 Mad. 579. Nor is it possible to form a caste voluntarily which enjoys even this limited prerogative, since it derives from the ancient character of the custom rather than the legislative power of the caste. *Deivani Achi v. Chidambaram Chettiar*, A.I.R. 1954 Mad. 657. Nor can traditional caste powers be transferred to a composite group made up of members of several castes. *Ellappa v. Ellappa*, A.I.R. 1950 Mad. 409.

as wells, tanks, shops, and restaurants.[69] Under these provisions, there is no longer any governmental power to make discriminations among citizens on caste lines.[70] Nor may government enforce any customary right to exclude certain castes from a public facility. [71] Article 17 provides:

'Untouchability' is abolished and its practice in any form is forbidden. The enforcement of any disability arising out of 'Untouchability' shall be an offence punishable in accordance with law.

The guarantee of freedom of religion is explicitly qualified to permit temple-entry legislation. [72]

The Untouchability (Offences) Act (UOA) of 1955 [73] outlaws the imposition of disabilities 'on grounds of untouchability' in regard to, *inter alia*, entrace and worship at temples, access to shops and restaurants, the practice of occupations and trades, use of water sources, places of public resort and accommodation, public conveyances, hospitals, educational institutions, construction and occupation of residential premises, holding of religious ceremonies and processions, and use of jewelry and finery. Enforcement of disabilities is made a crime, punishable by fine or imprisonment, and the power of civil courts to recognize any custom, usage, or right which would result in the enforcement of any disability is withdrawn.

In order to gauge the scope of Article 17 and the UOA, it is necessary to determine the meaning of 'untouchability'. Although it is yet unclear in detail, judicial construction so far provides some guidelines. The 'untouchability' forbidden by the Constitution does not include every instance in which one person is treated as ritually unclean and polluting. It does not include such temporary and expiable states of uncleanliness as that suffered by women in childbirth, mourners, etc.[74] It does not include attribution of impurity

---

[69] See also Articles 28(3) and 29(2), which forbid discrimination in private educational institutions.

[70] See, for example, *State of Madras v. Champakam Dorairajan* [1951] S.C.J. 313; *Sanghar Umar v. State*, A.I.R. 1952 Saur. 124. Caste cannot be recognized for electoral purposes. The Constitution rules out caste-wise electorates for parliament and state legislatures. Art. 325. Communal electorates in local bodies are unconstitutional. *Nain Sukh Das v. State of U.P.*, A.I.R. 1953, S.C. 384; nor can caste be used as a criterion in delimiting territorial constituencies (by excluding from a ward 'houses of Rajputs in the east of the village'). *Bhopal Singh v. State*, A.I.R. 1958 Raj. 41.

[71] *Aramugha Konar v. Narayana Asari*, A.I.R. 1958 Mad. 282.

[72] Article 25(2)b. When the Constitution was enacted, customary exclusion of lower castes from temples and secular facilities, previously recognized and to some extent India. For a survey of this provincial legislation and its continuing efficacy, see Galanter (1961a).

[73] Act XXII of 1955.

[74] See *Devarajiah v. Padmanna*, A.I.R. 1958 Mys. 84.

to worshippers (*vis-à-vis* attendants) in sacred places;[75] nor does it include that 'untouchability' which follows upon expulsion or excommunication from caste.[76] It is confined to that untouchability ascribed by birth rather than attained in life. Further, it does not include every instance in which one is treated as untouchable in certain respects because of a difference in religion or membership in a different or lower caste. It does not include the use of varna distinctions to demarcate Shudras from the twice-born.[77] It includes, in the words of the first court to pass on the issue explicitly, only those practices directed at 'those regarded as 'untouchables' in the course of historic development'; that is, persons relegated 'beyond the pale of the caste system on grounds of birth in a particular class'.[78] Thus untouchability would not include practices based on avoidance due to a difference of religion or caste, except insofar as the caste was traditionally considered 'untouchable' and 'outside the pale of the caste system'. Thus disabilities imposed, for example, by one group of Brahmans on other Brahmans, by Brahmans on non-Brahmans, by 'right-hand' on 'left-hand' castes, would all fall outside the prohibition of Article 17.

The meaning of untouchability then is to be determined by reference to those who have traditionally been considered 'Untouchables'. But it is no easier to define Untouchables than it is to define 'untouchability'. 'Beyond the pale of the caste system' is a misleading and unworkable formulation. Even the lowest castes are within the system of reciprocal rights and duties; their disabilities and prerogatives are articulated to those of other castes. Presumably the Mysore court means by this phrase, outside the four varnas of the classical lawbooks. In reference to their customary rights, Untouchables have sometimes, particularly in southern India, been referred to as a fifth varna, below the Shudras.[79] But in other places they were regarded as Shudras.[80] For purposes of personal law, the courts have never attempted to distinguish Untouchables from Shudras.[81] Even where Untouchables are popularly regarded as Shudras, they

[75] *Parameswaran Moothathu v. Vasudev Kurup*, I.L.R. 1960 Ker. 73.

[76] *Hadibandhu v. Banamali*, A.I.R. 1960 Or. 33; *Saifuddin Saheb v. State of Bombay*. A.I.R. 1962 S.C. 853.

[77] *Sangannagonda v. Kallangonda*, A.I.R. 1960 Mys. 147.

[78] *Devarajiah v. Padmanna*, A.I.R. 1958 Mys. 84 at 85. See *Kandra Sethi v. Metra Sahu*, XXIV Cuttack L. T. 364 at 366 (1963).

[79] See, for example, *Sankaralinga Nadan v. Raja Rajeshwari Dorai*, 35 I.A.C. 176 (1908).

[80] See, for example, *Atmaram v. King-Emperor*, A.I.R. 1924 Nag. 121.

[81] See *Muthusami v. Masilimani*, I.L.R. 33 Mad. 342 (1901); *Maharajah of Kolhapur v. Sundaram Awar*, A.I.R. 1925 Mad. 497 at 521; *Manickam v. Poogavanammal*, A.I.R. 1934 Mad. 323; *Bhola Nath v. Emperor*, A.I.R. 1924 Cal. 616; *Sohan Singh v. Kabla Singh*, A.I.R. 1927 Lah. 706.

cannot be equated with them, since there are non-Untouchable groups which belong to this category. Thus, the tests used for distinguishing Shudras from the twice-born cannot be used as a satisfactory measure of untouchability.

Although the abolition of untouchability amounts to a kind of negative recognition of the sacral order of Hinduism, it is not likely that the jurisprudence recognizing that order will find new employment for the purpose of identifying 'Untouchables'. In attempting to identify Untouchable groups for the purpose of giving them benefits and preferences, the government has not tried to apply general criteria but has adopted the device of compiling lists of castes in each locality. These were drawn up mainly with an eye to low ritual standing, but there is an uneven admixture of other social and economic factors.[82]

Thus the 'untouchability' forbidden by law is confined to discrimination against certain not readily defined classes of persons. It includes not every discrimination against them but only those imposed because of their position in the caste system. The provisions making untouchability an offence attempt to distinguish between those disabilities and exclusions imposed on grounds of caste position, and those which derive from religious and sectarian difference. Crucial sections of the Untouchability Offences Act are qualified to make it an offence to exclude Untouchables only from places 'open to other persons professing the same religion or belonging to the same religious denomination or section thereof'.[83] Thus the scope the rights conferred on Untouchables by the UOA depends on the meaning of the phrases 'the same religion' and 'the same religious denomination or section thereof'. To the extent that caste distinctions are conceived of as religious or denominational differences, the rights of Untouchables are limited. Thus exclusion of Untouchables by Jains is not forbidden insofar as it is on the ground that they are non-Jains rather than because of their caste.[84] In spite of some

---

[82] Such lists derive from earlier attempts (in the 1930s) to find a single set of criteria to measure 'untouchability.' (These included such tests as whether the caste in question was 'polluting' or 'debarred' from public facilities — which may admit of no unequivocal answer — and whether they were served by 'clean' Brahmans — which has only a local and comparative reference.) All attempts to set up tests based on the assumption that 'Untouchables' are set off by some uniform and disitinctive pattern of practices proved inadequate to isolate the groups which local administrators felt deserving of inclusion. Additional criteria of poverty and illiteracy had to be added. The government lists then give little guide to the meaning of untouchability. There is no adequate inclusive list of all groups considered Untouchable or any single set of criteria for identifying them. For a discussion of the problem of identifying the 'Untouchables,' see Dushkin (1957, 1961).

[83] Sec. 3(1).

[84] *Devarajiah v. Padmanna*, A.I.R. 1958 Mys. 84; *State v. Puranchand*, A.I.R. 1958 M.P. 352.

attempt by the lawmakers to minimize such distinctions,[85] courts have (on solid textual grounds) been reluctant to read the act as obviating these distinctions. In *State of Kerala v. Venkiteswara Prabhu*,[86] Untouchables were prevented from entering the *nālambalam* of a temple belonging to the Gowda Saraswat Brahman community. Since only members of this community ordinarily entered this part of the temple, the court held that exclusion of Untouchables was not an offense, since they did not belong to the same 'denomination or section thereof'. The acceptance by the court of denominational lines within Hinduism as limiting the operation of the temple-entry provisions may produce some unanticipated results. For the 'religion' and 'denomination' qualifiers also appear in other provisions of the Untouchability Offences Act.[87] Thus judicial solicitude for the sectarian prerogatives of groups within Hinduism may severely limit the rights granted by some of the central provisions of the UOA.[88]

Since untouchability has been interpreted to include only discriminations against Untouchables, the legislation against it has not touched discriminations against other classes of Hindus. Troubled by the anomalous situation in which, while it is an offence to exclude Untouchables from temples, classes of touchable Hindus may be excluded with impunity, several states have responded by enacting supplementary legislation. A Bombay act, for example, makes it an offence to prevent 'Hindus of any class or sect from entering or worshipping at a temple to the same extent and in the same manner as any other class or section of Hindus.'[89] These laws extend protection to non-Untouchables, and they also overcome the sectarian and denominational limitations which the courts have found in the UOA. Although the states are limited in their power to

---

[85] See the 'Explanation' attached to Sec. 3 of the UOA.

[86] A.I.R. 1961 Ker. 55.

[87] The qualification appears in the provisions relating to: use of utensils and other articles kept in restaurants, hotels, etc.; use of wells, water sources, bathing ghats, cremation grounds; the use of 'places used for a public or charitable purpose'; the enjoyment of benefits of a charitable trust; and the use of *dharmshalas, sarais*, and *musāfirkhānās*. Sections 4(ii), 4(iv), 4(v), 4(ix). Strangely enough, it does not appear in Section 4(x) regarding 'the observance of any . . . religious custom, usage or ceremony or taking part in any religious procession.' Thus Untouchables seem to have access to the religious processions of Hindu denominations and sects, but not to their wells, etc.

[88] Courts have also read severe private property restrictions into the Act. Thus, a privately owned well which the owner allows to be used by villagers is not covered. *Benudhar Sahu v. State*, I.L.R. 1962 Cuttack 256. Nor are religious performances 'in an open space' to which the public is invited. *State of M.P. v. Tikaram*, 1965 M.P.L.J. (Notes of Cases) 7. See *Kandra Sethi v. Metra Sahu*, XXIX Cuttack L.T. 364 (1963).

[89] Bombay Hindu Places of Public Worship (Entry Authorization) Act, 1956 (Bombay Act No. XXXI of 1956). Unied Provinces Temple Entry (Declaration of Rights) Act, 1956 (U.P. Act. No. XXXIII of 1956).

legislate directly on the subject of untouchability, this legislation will substantially broaden the rights of Untouchables as well, for the rights of the latter under the Untouchability Offences Act are automatically elevated to a parity with the new rights which the state legislation confers on caste-Hindus.[90] It remains to be seen whether the denominational prerogatives preserved by the Untouchability Offences Act will be found to have a constitutional foundation.[91]

'Protective Discrimination'

The attack on discrimination is only one side of the attempt to remove the disabilities of the lower castes. For the purpose of securing equality, the government is authorized to depart from indifference to caste in order to favour the Untouchables, tribals, and backward classes. These provisions for 'protective discrimination' are the principal exceptions to the constitutional ban on the use of communal criteria by government. The Constitution authorizes government to provide special benefits and preferences to previously disadvantaged sections of the population. Reserved posts in government service, reserved seats in legislatures,[92] reserved places in educational institutions, and an array of preferences and welfare measures have been made available to the Scheduled Castes and, to a lesser extent, to the 'backward classes.'

As mentioned earlier, the selection of Scheduled Castes has been on the basis of ritual standing, supplemented by social and economic criteria. No uniform tests have been used. Pollution and impurity in the local scale are important tests, but considerations of varna as such played a minor role in the process of selecting these castes.[93]

The provisions for 'protective discrimination' extend not only to Untouchables but to 'other socially and educationally backward classes'. Although the Constitution refers to backward *classes*, caste groups have commonly been the units selected as backward. Low standing in the ritual order was clearly one of the bases upon which caste groups have been deemed backward. But varna as such was

[90] *State of Bombay v. Yagna Sastri Purushadasji*, 61 Bom. L. Reporter 700 (1958). (Under the Bombay Act, Satsangis [members of Swaminaraya Sampradaya] could not exclude non-Satsangi Harijans from their temples.)

[91] See discussion on pages 172–3 below. Generally, on the relation of denominational prerogatives to temple entry and the abolition of untouchability, see Galanter (1964b).

[92] As originally enacted, the Constitution provided reserved seats in Parliament and the state legislatures for the Scheduled Castes (that is, Untouchables) and the state legislatures for the Scheduled Castes (that is, Untouchabilies) and the Scheduled Tribes for a ten year period. This has been extended for another ten year period. Constitution (Eighth Amendment) Act, 1959.

[93] See note 80 above.

never used for this purpose.[94] Increasing criticism within and without the government, and the increasing willingness of the courts to subject preferences for backward classes to close scrutiny, have caused a trend away from caste in favour of noncommunal economic and educational criteria. The Supreme Court in *Balaji v. State of Mysore* struck down a scheme for reservations in colleges for backward classes on the ground that they were selected primarily on the basis of caste; that is, the groups were chosen on the basis of their ritual and social standing.[95] It appeared to be constitutionally permissible for the state to use castes or communities as the units it designated as backward,[96] so long as it selected these units by other criteria; that is, by social, economic, and educational indices. So far, the Supreme Court seemed to reject the sacral model in favour of something along the lines of the organic model. Caste groups might be used, if selected as backward by nonreligious criteria on a relative scale in which they were measured against other caste groups. However, more recently, the Court has indicated that the caste group may be used to select beneficiaries of preferences on a much narrower scale than the *Balaji* case seemed to imply. In *Chitralekha v. State of Mysore*, the Court suggested that, while it is permissible to use caste (presumably in the sense of standing or rank) 'in ascertaining the backwardness of a class of citizens', it is by no means necessary to take caste into account 'if [government] can ascertain the backwardness of a group of persons on the basis of other relevant criteria'.[97] This confirms the *Balaji* view that caste standing is neither a necessary nor a sufficient criterion of backwardness. The Court then goes on to say that, while caste may be relevant in determining the backwardness of individuals, caste groups are themselves not the classes whose backwardness is to be ascertained. It suggests that if the whole caste is backward it should be included among the Scheduled Castes.

Apparently, caste groups are not to be used as the units or classes which are deemed backward. However, the first High Court to which this case was cited had no difficulty in approving the use of caste units. It upheld the classification of Ezhuvas as a backward class, indicating that this was not based on an assessment of their impurity but rather of the continuance of disabilities to which they

---

[94] A 'varna' test was advocated by Shri S.D.S. Chaurasia, a member of the Backward Classes Commission, but his contention that the Backward Classes should be equated with Shudras was rejected by the commission (Report 1956: I, 44; III. 22ff.).

[95] *Balaji v. State of Mysore*, A.I.R. 1963 S.C. 649.

[96] Id; *Ramakrishna Singh v. State of Mysore*, A.I.R. 1960 Mys. 338.

[97] A.I.R. 1964 S.C. 1823 at 1833. One of the remarkable and confusing aspects of these cases is that the courts manage to use the word caste to mean successively (1) 'communal unit' and (2) 'caste rank or standing' without feeling any need for clarification or distinction. I have tried here to sort out these meanings.

were subjected. 'Habits of thought die hard and slow, and occupations like toddy-tapping carry their social stigma from one generation to another and through decades of conduct'.[98] It is not their location in the varna scheme, but their location in the estimate of others and the impact of this on their opportunities which the state may take into account.

In the designating of the beneficiaries of protective discrimination, the Scheduled Castes are selected by a mixture of ritual and social-economic characteristics. In this respect the government has used the shadow or negative of the sacral view of caste. However, in selecting backward classes, the courts have narrowly confined the use of caste standing: it can be used only an index of social and economic backwardness and then only in conjunction with other tests. The permissibility of using caste groups as units is open to doubt. Insofar as caste may be used in this area, it is caste conceived in the organic view — as a group with a relative share of social, educational, economic resources, rather than as a group with a given ritual standing.

With membership in caste groups a qualification for preferment of various kinds, it is not surprising that disputes have arisen concerning such membership. In order to qualify for preferences, one must be a member of the listed caste. In *Chatturbhuj Vithaldas Jasani v. Moreshwar Pareshram*, the Supreme Court decided that a Mahar who had joined the Mahanubhava Panth,[99] a Hindu sect which repudiated the multiplicity of gods and the caste system, remained a Mahar (thus eligible to stand for a reserved seat in the legislature). The Court arrived at this conclusion on the ground that he had continued to identify himself as a Mahar and had retained full acceptance by the Mahar community. The Court concluded that conversion to this sect imports little beyond an intellectual acceptance of certain ideological tenets and does not alter the convert's caste status'.[100] Thus the Court saw no distinctive religious content in membership in the caste; its bonds are 'social and political ties.' 'If the individual . . . desires and intends to retain his old social and political ties', and if the old order is tolerant of the new faith and does not expel the convert, the conversion does not affect his caste right [to stand for a reserved seat] is a right of the old body, the right membership.[101] However, the Court recognizes that there is a religious level to caste affiliation as well. It is not only the convert's own choice that must be taken into account, 'but also the views of

[98] *State of Kerala v. Jacob Mathew*, I.L.R. 1964(2) Ker. 53 at 60.
[99] On the role of this sect in Mahar tradition, see Miller (1966).
[100] [1954] S.C.R. 817 at 840.
[101] Id. at 839.

the body whose religious tenets he has renounced because here the conferred upon it as a special privilege to send a member of its own fold to Parliament'.[102]

The same question came before the Mysore High Court in the case of *Shyamsunder v. Shankar Deo*,[103] where the question was whether the candidate had lost his membership in the Samgar caste by joining the Arya Samaj, a Hindu sect which rejects idolatry and ascription of caste by birth. The Court said there would be no deprivation of caste unless there was either expulsion by the old caste or international abandonment or renunciation by the convert. Since there was no evidence of expulsion or ostracism by the old caste, the question was whether there had been a break from the old order 'so complete and final that . . . he no longer regarded himself as a member of the Samgar caste'.[104] Here the Court felt this was refuted not only by his activities but by his testimony that he believed in idols and in texts repudiated by the Samajists. Again while religious criteria played a secondary role in defining membership in caste, the Court, like the *Jasani* Court, conceived of the caste as having some body of religious tenets. One might be a member while repudiating them, but adhering to them was evidence that one regarded oneself as a member. In these cases, the view of caste fits what we have called the associational model. It is a group characterized by a constellation of social and political ties; it has 'religious tenets', though adherence to them is not a requisite for membership so long as the other ties are not severed.

In *V. V. Giri v. D. Suri Dora* the question before the Supreme Court was whether a candidate had lost his membership in the Moka Dora tribe by becoming a Kshatriya. The candidate was born a Moka Dora, and his family had described itself as such in all documents from 1885 to 1923. Since that time they had described themselves as Kshatriyas. There was evidence that the family had adopted Kshatriya customs, celebrated marriages in Kshatriya style, was connected by marriage to Kshatriya families, employed Brahman priests, and wore the sacred thread in the manner of Kshatriyas.[105]

---

[102] Id. at 839. Perhaps the 'religious tenets' language is here only because the Court used as authority the case of *Abraham v. Abraham*, 9 M.I.A. 199 (1863), which involved conversion from one religion to another with retention of personal law.

[103] A.I.R. 1960 Mys. 27.

[104] Id. at 32. The caste referred to as 'Samgar' here is evidently the same as that listed as 'Samagara' in Census of India, Paper no. 2 (1960): Scheduled Castes and Scheduled Tribes Arranged in Alphabetical Order, p. 84. The Sonar caste, mentioned below, are traditionally goldsmiths and of a higher social standing, usually associated with Vaishya status.

[105] A.I.R. 1959 S.C. 1318. Apparently the candidate's family was one of a number of families of Mokasadars or large landholders who, according to the Election Tribunal, 'would not like to be called Moka Doras but considered themselves Kshatriyas'.

The candidate's election was challenged on the ground that he was no longer a Moka Dora and was therefore ineligible to stand for a seat reserved for Scheduled Tribes. The Supreme Court solved the question by deciding that he had not in fact become a Kshatriya because 'the caste status of a person in this context would necessarily have to be determined in the light of the recognition received by him from the members of the caste in which he seeks entry'. Finding no evidence of such recognition, the court said that 'unilateral acts cannot be easily taken to prove that the claim for the higher status . . . is established'.[106] This recognition test is essentially a variant on the reputation test for the varna standing of caste groups. It is notable that it completely excludes any religious test of Kshatriyahood. One judge (J. L. Kapur), dissenting, vigorously rejected the majority notion that caste is determined in the first instance by birth and can be varied (at least upward) only by recognition of his claims by members of the group to which he aspires. He put forward a theory that caste rank varies as a consequence of the *gunas*, karma, and *subhāvanā* and is dependent on actions; he found that the candidate had 'by his actions raised himself to the position of a Kshatriya. . . .'[107] The majority did not accept this but did regard the varna order as hierarchical. It was a hierarchy determined by mutual social acceptance, rather than by possession of traits indicative of religious capacity or attainments.

So long as they are dealing with caste within Hinduism, whether it is the precedence or rights of a caste or membership in it, the courts have been unwilling to describe and rationalize these differences in terms of the sacral model of caste. They assign only a minor role to the religious content of caste and avoid invoking the notion of an overarching sacral order in which all castes are hierarchically arranged. The use of their 'untouchability' as the criterion for selecting the Scheduled Castes implies a kind of reverse recognition of the Hindu ritual order. However, it is clear that such recogniton cannot be extended to the selection of the 'backward classes'. The only instance so far in which we have seen implicit reference to a hierarchical ordering is in the case of the tribals. In the Moka Dora case, the Kshatriya status was denied on grounds that implied such a hierarchy, even though it had no specially religious content. However, when we move to questions which concern persons and groups

---

XV E.L.R. 1 at 38 (1957). The tribunal found that the candidate had 'totally given up feeling himself to be a member of the Moka Dora tribe and considers himself a a Kshatriya.' Id. For a comparison of the divergent approaches of the Election Tribunal, the High Court and the Supreme Court in this case, see chap. 6.

106 A.I.R. 1959 S.C. at 1327.

107 Id. at 1331. 'Gunas', etc. means roughly material nature, deeds and temperament.

outside 'Hinduism' we find that the religious content of caste re-emerges.

The 'Hindu' Component of Caste

The Constitution forbids religious discrimination on the part of the state[108] and guarantees freedom of religion.[109] The courts have been vigilant in invalidating governmental measures framed along religious lines.[110] Nevertheless, in some instances religion has been made a qualification for preferential treatment. The President's order specifying Scheduled Castes provided that 'no person professing a religion different from Hinduism shall be deemed a member of a Scheduled Caste'.[111] Who is a Hindu? What is the role of caste in deciding who is a Hindu? What is the role of Hinduism in determining membership in a caste group?

The legal definition of Hinduism, developed for the purpose of applying appropriate personal law, was not a measure of religious belief, nor was it a description of social behaviour so much as a civil status describing everyone subjected to the application of 'Hindu law' in the areas reserved for personal law.[112] Heterodox practice, lack of belief, active support of non-Hindu religious groups,[113]

[108] Arts. 15, 16.

[109] Arts. 25, 26, 30. On Indian secularism generally, see Smith (1963); Galanter (1965); Seminar 67 (1965).

[110] *State of Rajasthan v. Pratap Singh*, A.I.R. 1960 S.C. 1208; *Nain Sukh Das v. State of U.P.*, A.I.R. 1953 S.C. 384; *State of Jammu and Kashmir v. Jagar Nath*, A.I.R. 1958 J and K 14.

[111] Constitution (Scheduled Castes) Order, 1950, para. 3. An exception was included for Sikh members of 4 of the 34 Scheduled Castes listed for the Punjab. In 1956, the main provision was expanded to include all Sikhs, so that it now reads 'no person who professes a religion different from the Hindu or Sikh religion shall be deemed a member of a Scheduled Caste'. The Scheduled Castes and Scheduled Tribes Orders (Amendment) Act, 1956 (Act 63 of 1956), para. 3. Cf. the Government of India (Scheduled Caste) Order, 1936, para. 3, which provided: 'No Indian Christian shall be deemed a member of a Scheduled Caste'. The Constitution (Scheduled Tribes) Order, 1950, contains no religious qualifications.

[112] Or, more accurately, all who would be subject to Hindu law in the absence of proved special custom or of a contingency such as marriage under the Special Marriage Act (III of 1872). On the other hand, it does not include all persons subject to Hindu law. Hindu personal law has sometimes been applied to Christians (see *Abraham v. Abraham, supra,* note 100) and to Muslims (until the passage of the Muslim Personal law (Shariat) Application Act [XXXVI of 1937]). See also *Chinnaswamy v. Anthonyswamy* A.I.R. 1961 Ker. 161 (Tamil Vaniya Christian Community of Chittur Taluk governed by Hindu law in matters of inheritance and succession); cf. *Mira Devi v. Aman Kumari,* A.I.R. 1962 M.P. 212 (tribals may be subject to Hindu law although they are not Hindus).

See the broad definition in the Hindu Marriage Act, 1955, Sec. 2, and discussion in Derrett (1963b: Sec. 17ff.).

[113] *Bhagwan Koer v. Bose,* 30 I.A. 249 (1903). One remains a Hindu even when he joins a sect which has non-Hindu members. *State of Bombay v. Yagna Sastri Purushadasji,* 61 Bom. L. Reporter 700 (1960).

expulsion by a group within Hinduism[114]—none of these removed one from the Hindu category, which included all who did not openly renounce it or explicitly accept a hostile religion.[115] The individual could venture as far as he wished over any doctrinal or behavioural borders; the gates would not shut behind him if he did not explicitly adhere to another communion.[116] In *Chandrasekhara Mudaliar v. Kulandaivelu Mudaliar*[117] the Supreme Court had to decide on the validity of a consent to adoption by a sapinda who disavowed belief in the religious efficacy of adoption, in Hindu rituals and scriptures, in atman, and salvation. But the Court found that 'the fact that he does not believe in such things does not make him any the less a Hindu. . . . He was born a Hindu and continues to be one until he takes to another religion . . . whatever may be his personal predilections or views on Hindu religion and its rituals. . . .'[118]

In the post-constitutional cases involving preferences, the same broad conception of Hinduism has been carried over from the area of personal law. To 'profess' Hinduism merely means to be a Hindu by birth or conversion.[119] Unorthodoxy or lack of personal belief in its tenets does not mean lack of profession for this purpose. In effect the test seems to amount to a willingness to refrain from calling oneself

[114] *Ratansi D. Morarji v. Admr. General of Madras*, A.I.R. 1928 Mad. 1279 at 1283.

[115] The concept of 'Muslim' is treated somewhat differently. There are expressions of the same negative (no conversion) test. See *Bhagwan Baksh Singh v. Dribijai Singh*, I.L.R. 6 Luck. 487 (1931). Generally it is held that adherence to some minimum of beliefs (the unity of God, the mission of Mohammed, the authority of the Koran) is necessary and sufficient to make one a Muslim. *Narantakath v. Parakkal*, I.L.R. 45 Mad. 986 (1922); *Jiwan Khan v. Habib*, I.L.R. 14 Lah. 518 (1933). Repudiation of these beliefs, even without conversion to another religion, makes one not a Muslim. *Resham Bibi v. Khuda Bakhsh*, I.L.R. 19 Lah. 277 (1938).

[116] No proof of formal abandonment of his new religion is necessary for the convert to effect a successful reconversion to Hinduism. While a mere declaration is not sufficient to restore him to Hinduism, acceptance by a Hindu community with whatever formalities it deems proper — even none at all — is sufficient. *Durgaprasada Rao v. Sundarsanaswami*, A.I.R. 1940 Mad. 513; *Gurusami Nadar v. Irulappa Konar*, A.I.R. 1934 Mad. 630. However, cf. *Marthamma v. Munuswami*, A.I.R. 1951 Mad. 888 at 890, where the primary test is the 'intention' of the reconvert; the court says 'the religious persuasion of a man now-a-days depends on his 'subjective preference' for any religion.'

For purposes of at least certain preferences, re-converts to Hinduism who were born in Scheduled Castes are deemed members of the Scheduled Castes. But those born in another religion (for example, whose fathers were converts) are not treated as members born in another religion (for example, whose fathers were converts) are not treated as members of Scheduled Castes 'whatever may be their original family connections.' *Report of the Commissioner for Scheduled Castes and Scheduled Tribes* (1953, p. 132). In the personal law cases, acceptance by the community was a measure of one's success in re-entering Hinduism; here, Hindu birth is the condition of gaining membership in the community.

[117] A.I.R. 1963 S.C. 185.

[118] Id. at 200.

[119] *Michael v. Venkataswaran*, A.I.R. 1952 Mad. 474.

something else. Thus, where the election to a reserved seat of an active supporter of Dr Ambedkar's Buddhist movement was challenged on the ground that he was not a Hindu, the Court found that 'it has to be established that the person concerned has publicly entered a religion different from the Hindu . . . religion'. Mere declarations falling short of this would not be sufficient.[120] The candidate had supported the movement for mass conversion by serving on the reception committee, editing a newspaper supporting the movement, and attending a rally where an oath, 'I abandon the Hindu religion and accept the Buddha religion' was administered by Dr Ambedkar. When those who wished to convert were asked to stand, the candidate stood. But there was no evidence that he did in fact take the oath; the Court held that in the absence of evidence of such a declaration, he remained a Hindu.[121] The same test of public declaration was recently upheld by the Supreme Court in *Punjabrao v. D. P. Meshram*, where the Court found that a conversion had in fact occurred.[122]

Once it is established that the candidate has 'professed' Buddhism, the question arises whether this is a religion 'different from the Hindu religion' within the meaning of the Scheduled Castes Order. Converts to Christianity and Islam are, of course, non-Hindus.[123] But Buddhists, Jains, and Sikhs are treated as Hindus for many purposes.[124] Hindu is an equivocal term, sometimes used to refer to adherents of more or less 'orthodox' vedic and brahmanical communions and at other times used to embrace the full array of 'heterodox' groups, including Sikhs, Jains, and Buddhists. Yet Jains and Buddhists are considered non-Hindus for purposes of preferences; Sikhs were

---

[120] *Karwadi v. Shambharkar*, A.I.R. 1958 Bom. 296 at 297. On the Buddhist movement, see Zelliot (1966).

[121] *Karwadi v. Shambharkar*, Id. at 299. The vagaries of the declaration test are illustrated in *Rattan Singh v. Devinder Singh*, VII E.L.R. 234 (1953), XI E.L.R. 67 (1955), where the candidate had at various times described himself as a Mazhabi Sikh, a Harijan Hindu, a Balmiki, and a Balmiki Hindu.

[122] A.I.R. 1965 S.C. 1179.

[123] *Michael v. Venkataswaran*, A.I.R. 1952 Mad. 474.

[124] These groups are Hindu for purposes of personal law, but their separateness has been recognized in other contexts, for example, Jains are not 'the same religion' as Hindus for purposes of temple-entry legislation. *State v. Puranchand*, A.I.R. 1958 M.P. 352: *Devarajiah v. Padmanna*, A.I.R. 1958 Mys. 84. Yet they are 'Hindus' within the definitions contained in most temple-entry acts. Cf. Constitution of India, Art. 25(2) (b), Explanation I. Recently the Supreme Court has indicated that the boundaries of Hinduism for this purpose enfold even sects with historical Hindu connections who currently assert that they constitute a distinct religion. *Sastri Yagnapurushadasji v. Muldas Bhundardas Vaishya*, A.I.R. 1966 S.C. 1119. Differences in regard to the organization of religious trusts may be recognized by having a Hindu religious trusts law which covers Jains and Buddhists but not Sikhs. *Moti Das v. S.P. Sahi*, A.I.R. 1959 S.C. 942.

originally excluded and are now mentioned separately.[125] The Supreme Court concluded that 'Hindu' in the Scheduled Castes Order 'is used in the narrower sense of the orthodox Hindu religion which recognizes castes and contains injunctions based on caste distinctions'.[126] This comes close to suggesting a positive definition of Hinduism which goes beyond merely excluding Sikhs, Jains, and Buddhists, to throw doubt on the inclusion of those sects which in one fashion or another repudiate caste distinctions in doctrine or practice. This association of Hinduism with caste disctinctions is particularly puzzling in view of the Supreme Court's subsequent indication that invidious caste distinctions are 'founded upon superstition, ignorance and complete misunderstanding of the true teachings of Hindu religion. . . .'[127]

The *Meshram* Court, while inquiring what it means to 'profess' a religion 'different from the Hindu' reads the order as requiring that to be treated as a member of a Scheduled Caste 'a person . . . must be one who profess either Hindu or Sikh religion'.[128] And it notes that 'the word 'profess' in the presidential order appears to have been used in the sense of open declaration or practice by a person of the Hindu (or the Sikh) religion'.[129] In these dicta the Court comes perilously close to suggesting a positive religious test for membership in Scheduled Castes. In spite of the Court's remarks, it is clear that the Scheduled Caste Order itself does not establish or sanction such a positive religious test. It does not require anyone to profess Hinduism, much less practice it. It merely requires that he not profess a different religion. In determining this, the profession and practice of Hinduism need be considered, as they are in *Meshram*, only for their evidentiary value.

It is clear, then, that Buddhists are excluded by the presidential order. The Supreme Court merely notes that when a person 'has ceased to be a Hindu he cannot derive any benefit from that Order.'[130] Persistent efforts by Buddhists to be treated as members of Scheduled Castes have proved unavailing.[131] Their exclusion is usually justified on the ground that conversion to Buddhism operates as loss of caste.

[125] *Gurmukh Singh v. Union of India*, A.I.R. 1952 Pun. 143; *Rattan Singh v. Devinder Singh*, VII E.L.R. 234 (1953), XI E.L.R.67 (1955). See note 109 above.

[126] A.I.R. 1965 S.C. at 1184.

[127] *Sastri Yagna Purushadasji v. Muldas Bhundaras Vaishya*, *supra* note 122, at 1135. In this case the court propounds a normative view of Hinduism which seems to withdraw the mantle of religion from such practices.

[128] A.I.R. 1965 S.C. at 1184.

[129] Id.

[130] Id.

[131] A bill to this effect was defeated in the Lok Sabha. *New York Times*, Aug. 30, 1961, p. 2, col. 6. On the problems that the 'Hinduism test' presents for the neo-Buddhists, see Isaacs (1965, p. 117 ff.).

'As Buddhism is different from the Hindu religion, any person belonging to a Scheduled Caste ceases to be so if he changes his religion. He is not, therefore, entitled to the facilities provided under the Constitution specifically for the Scheduled Castes.'[132] The central government, recognizing that conversion itself is unlikely to improve the condition of the converts, has recommended that the state governments accord the Buddhists the concessions available to backward classes. Such preferences, less in scope and quantity than those for Scheduled Castes, have been granted in some cases.[133]

In *Punjabrao v. Meshram*, the Supreme Court never reaches the question of whether the 'Hinduism' test for recipients of preferences infringes the constitutional ban on religious discrimination by the state.[134] The constitutional challenge has been raised in several earlier cases in the High Courts. The judicial response to this challenge presents the problem of characterizing the relation of the caste group to Hindism.

In *S. Gurumukh Singh v. Union of India*,[135] a Bawaria Sikh protested his exclusion from the Scheduled Castes, in which the President had included Hindu Bawarias. The Court found that Article 341 empowered the President to select those 'parts of castes' which he felt should be included and that he could select these parts on the basis of religion. He did not violate Article 15(1), the ban against religious discrimination, since Article 15(4), which authorizes preferential treatment of the backward classes, operated as an exception to the prohibitions of Article 15(1). The Court conceded that Scheduled Castes were to be designated on the basis of their backwardness. But, finding that the Constitution vested in the President

---

[132] *Report of the Commissioner for Scheduled Castes and Scheduled Tribes* (1957–8, I, 25; This ruling is based squarely on the 'Hinduism' requirement of the President's Order. See the statement of Pandit Pant, *Times of India*, Aug. 21, 1957, p. 12, col. 3.

[133] *Report of the Commissioner for Scheduled Castes and Scheduled Tribes* (1957–8, I, 25); II, 60). While some states have included Buddhists within the Backward Classes, others have continued to treat them like Scheduled Castes for some purposes, and still others have withdrawn all preferential treatment.

In Maharashtra, where over three-quarters of the Buddhists live, they enjoy all concessions and facilities extended to the Scheduled Castes by the state government. They do not enjoy the constitutional provisions for reserved seats or the benefits provided by the central government, especially post-matriculation scholarships and job reservations.

[134] The question was never raised. In the Election Tribunal, the candidate contested the fact of the conversion. Apparently, counsel for the challenged candidate wer: eager to argue that Buddhists were indeed Hindus. The High Court excluded this issue as a factual question which should have been pleaded and proved by evidence, but decided in favour of the candidate, finding no conversion had occurred. The Supreme Court reversed, finding there was a conversion. At no stage was the constitutional issue raised.

[135] A.I.R. 1952 Punj. 143.

the entire power to make such determinations, the Court refused to review his order by considering whether the Sikh Bawarias were in fact sufficiently backward to be included. In this situation, it is conceded that these non-Hindus either constitute or are members of a caste group; what is decided is that the President's exclusion of that group (or part of the group) is unreviewable.[136]

In *Michael v. Venkataswaran*[137] the religious requirement was upheld against a Paraiyan convert to Christianity who wished to stand for a reserved seat. Even if there are cases in which both the convert and his caste fellows consider him as still being a member of the caste, the Court found, 'the general rule, is [that] conversion operates as an expulsion from the caste . . . a convert ceases to have any caste'.[138] The presidential order, according to the Court, proceeds on this assumption and takes note of a few exceptions. The Court declined to sit in judgment on the President's determination that similar exceptional conditions do not prevail in other instances. Thus the presidential order was upheld not because of an absence of judicial power to review it but because of its accuracy in the general run of cases.

In *In re Thomas*[139] another bench of the Madras court considered a convert case which did not involve the presidential order. The Madras government had extended school-fee concessions to converts from Scheduled Castes 'provided . . . that the conversion was of the . . . student or of his parent. . . .' A Christian student whose grandfather had converted could not, it was held, complain of discrimination, since converts did not belong to the Harijan community. By conversion they had 'ceased to belong to any caste because the Christian religion does not recognize a system of castes'.[140] The concessions to recent converts were merely an indulgence, and the state could determine the extent of this indulgence.

136 The unreviewability of the presidential order would seem open to question in light of subsequent cases which have firmly established judicial power to review the standards used by government to designate the recipients of preferential treatment. *Balaji v. State of Mysore*, A.I.R. 1963 Mys. 649. There is no indication in the Constitution that executive action, even in pursuance of expressly granted and exclusive constitutional powers, is immune from judicial review for conformity with constitutional guarantees of fundamental rights. See Article 12. The position in the *Gurmukh Singh* case must be seen as one of judicial restraint rather than judicial powerlessness. The restraint there expressed seems out of line with later judicial assertiveness in this area.

137 *Michael v. Venkataswaran*, A.I.R. 1952 Mad. 474.

138 Id. at 478.

139 A.I.R. 1953 Mad. 21.

140 Id. at 88. The exclusion of neo-Buddhists from the preferences for Scheduled Castes has been similarly justified by the notion that 'Buddhism [does] not recognize castes'. Statement of Mr. B.N. Datar in Rajya Sabha, Aug. 26, 1957. Reported in *Times of India*, Aug. 27, 1957, p. 10, col. 1.

The theory that acceptance of a non-Hindu religion operates as loss of caste reflects the continued force of the sacral view of caste. The question arises in two kinds of factual situations: first, those involving a caste group or a section of a caste made up of members who are non-Hindus; second, those involving an individual convert. In the first type, there is little dispute that such persons as the Sikh Bawarias in the *Gurmukh Singh* case are, in fact, members of a caste in one of the segmental senses of caste encountered in the law. The existence of such caste groups among non-Hindus in India is well known and has long been recognized by the judiciary.141 To refuse to recognize caste membership among such non-Hindu groups implies that the 'caste' of which the court is speaking is not a caste in the sense of a body of persons bound by social ties, but a caste in the sense of a body which occupies a place in the ritual order of Hinduism.142

The second type of case is that of the individual convert. In such cases, the question facing the court would seem to be whether the individual convert's acceptance of Christianity, Islam, or Buddhism evidences a loss of membership in the caste group to which he belonged at the time of the conversion. This can be treated as a question of fact, to be answered by evidence about his observable interactions with other members of the group. This was the approach taken in the cases dealing with conversions to sects within Hinduism. It is the approach taken in dealing with converts among the Scheduled Tribes.143 It is presumably the approach that would be taken with Scheduled Caste converts to Sikhism. But it is not taken with Scheduled Caste converts to other religions. Yet in at least some cases of conversion outside Hinduism, there is evidence that the convert continues to regard himself and to be regarded by others as member of the old caste.144 In dealing with these conversions to

141 Cf. *Report of the Backward Classes Commission* (1956, I 28ff.).

142 The inclusion of Sikhs along with Hindus in 1956 (see note 109 above) renders this view even less tenable than it might have been earlier. As it stands, caste groups are to be recognized outside of the Hindu fold, but only among Sikhs and not among Christians, Muslims, and Buddhists.

143 *Kartik Oraon v. David Munzni*, A.I.R. 1964 Pat. 201; *Gadipalli Paroyya v. Goyina Rajaryya*, XII E.L.R. 83 (1956).

144 The reports are replete with cases in which converts have lived so indistinguishably with their caste fellows that the courts retrospectively infer a tacit reconversion without either formal abjuration of the new religion or formal expiation and readmit-to Hinduism. *Durgaprasada Rao v. Sundarsanaswaram*, A.I.R. 1940 Mad. 513; *Gurusami Nadar v. Kurulappa Konar*, A.I.R. 1934 Mad. 630; *Venkataramayya v. Seshayya*, A.I.R. 1942 Mad. 193. The 'indulgence' extended by the state in the *Thomas* case, *supra* note 137, seems to reflect an awareness that recent converts, if not effective members of their old castes, are at least subject to similar disabilities. And cf. *Muthasami Mudaliar v. Masilami*, I.L.R. 33 Mad. 342 (1909), where lifelong Christians were accepted as members of a Hindu caste.

religions outside Hinduism, the courts have forsaken this empirical approach and have treated the conversion as depriving him of his membership as a matter of law. This conclusion derives not from the facts of the individual case but from a view of castes as the components in the sacral order of Hinduism. When that overarching scheme is abandoned, so is caste membership.

Caste Autonomy

Notwithstanding the common rhetoric about the 'casteless' society, the Constitution is quite unclear about the position of the caste group in Indian life. While there are guarantees to preserve the integrity of religious and linguistic groups,[145] there are none for the caste group. It would not seem to enjoy any constitutional protection as such. The silence may represent an anticipation that caste will wither away and have no important place in the new India, or it may represent an implicit ratification of the old policy of noninterference.

There is a desire to minimize the impact of caste groupings in public life. The government has discouraged the use of caste as a legal identification.[146] Appeals to caste loyalty in electoral campaigning are forbidden.[147] Promotion of enmity between castes is a serious criminal offence.[148]

---

[145] Articles 25–30, 350A, 350B.

[146] The Registration Act has been amended to exclude the requirement of caste as personal identification. Indian Registration (Amendment) Act, 1956 (Act XVII of 1956) Sec. 2.

[147] Representation of the People Act (Act XLIII of 1951), Sec. 123. It is not entirely clear just what sort of appeals are barred by this provision. Appealing to Chambhars to elect a Chambhar brother has been held to be a corrupt election practice. *Lachhiram v. Jamuna Prasad Mukhariya*, 9 E.L.R. 149 (Elec. Tribunal, Gwalior, 1953). So also with appeals to Gadarias to vote for Gadarias and not for Brahmans. *Shiv Dutt v. Bansidas Dhangar [No. 2]*, 9 E.L.R. 325 (Elec. Tribunal, Faizabad, 1954). But an appeal to 'all Yadav brethren to do their duty to the country by voting Congress' is not an appeal on grounds of caste. Nor is an appeal to Vishwakarmas and Kalakars to vote for a candidate because of his service to their professions. *Rustom Satin v. Dr Sampoornanand*, 20 E.L.R. 221 (Allahabad High Court, 1958). Nor is an appeal to voters to support a candidate because their caste organizations support his party (Congress). *Sant Prasad Singh v. Dasu Sinha*, A.I.R. 1964 Pat. 26. Nor is an appeal to a community not to vote for a party on the ground of alleged misdeeds to the community. *Raja Vijai Kumar Tripathi v. Ram Saran Yadav*, 18 E.L.R. 289 (Allahabad High Court, 1958). Nor are appeals to the loyalty of ascriptive groupings smaller than caste (*gotra*). *Pratap Singh v. Nihal Singh*, 3 E.L.R. 31 (Elec. Tribunal, Patiala, 1953). Nor are appeals to classes of persons made up of a number of castes (Backward Classes). *Raja Vijai Kumar Tripathi v. Ram Saran Yadav, supra*. It appears, then, that the forbidden appeals are only appeals to members of a caste to vote for a member of that caste on grounds of his membership in that caste (or against a member of another caste on grounds of his membership in that other caste). Presumably all but the most unsophisticated politicians can manage to avoid this class of appeals.

[148] Section 153–A of the Penal Code (first enacted as Act IV of 1898), which provided up to two years' imprisonment for promoting feelings of enmity between

Apart from explicit restrictions on caste discrimination, there is a tendency to discourage any arrangements which promote the coherence and integrity of the caste group as such. Thus, for example, the Supreme Court struck down as unreasonable restrictions on property rights laws providing for pre-emption on the basis of vicinage. The Court held that the real purpose of these laws was to promote communal neighbourhoods, a purpose which could have no force as public policy, since the desire to promote such exclusiveness could no longer be considered reasonable.[149]

What is left of caste autonomy? What remains of the prerogatives previously enjoyed by the caste group? The caste retains the right to own and manage property and to sue in court. Section 9 of the Civil Procedure Code, with its bar on judicial cognizance of 'caste questions', is still in force. Courts still refuse to entertain suits involving caste questions (for example, fitness of an officer to manage property),[150] and castes retain their disciplinary powers over their members (for example, a court refused to declare invalid the assessment of a fine for an alleged breach of caste rules).[151] The caste retains its power of excommunication.[152] It is still a good defence to a criminal action for defamation word to assert the privilege of communicating

---

different classes of Indian citizens, was replced by a new enactment in 1961. (Act XLI of 1961). The new 153-A provides:

Whoever—

(a) by words, either spoken or written, or by signs or by visible representations or otherwise, promotes, or attempts to promote, on grounds of religion, race, language, caste or community or any other ground whatsoever, feelings of enmity or hatred between different religious, racial or language groups or castes or communities, or

(b) commits any act which is prejudicial to the maintenance of harmony between different religious, racial or language groups or castes or communities and which disturbs or is likely to disturb the public tranquility, shall be punished with imprisonment which may extend to three years, or with fine or with both.

For an interesting American analogy, see *Beauharnais v. Illinois*, 343 U.S. 250 (1951).

[149] *Bhau Ram v. Baij Nath*, A.I.R. 1962 S.C. 1476. Compare *Ram Swarup v. Munshi*, A.I.R. 1963 S.C. 553, where the Supreme Court took a more lenient view of pre-emption which aimed at preserving the integrity of the village and familial expectations. In this regard the Court's view runs parallel to the government's preference for the village unit over the communal one, for example, in implementation of Panchayati Raj.

[150] *Kanji Gagji v. Ghikha Ganda*, A.I.R. 1955 N.U.C. 986.

[151] *Bharwad Kama v. Bai Mina*, A.I.R. 1953 Saur. 133.

[152] *Varadiah v. Parthasarathy*, 1964 (2) Mad. 417 (in spite of the changing social order 'where an individual has done something wrong or prejudicial to the interests of his community, the members of his community which, by virtue of custom or usage, is competent to deal with such matters [can] take a decision by common consent; and so long as such a decision does not offend law, it can be enforced by the will of the community'. Id. at 420). But composite groups such as 'non-Brahmans' and 'caste Hindus' do not enjoy the privilege when they undertake to outcaste for caste offenses. *Ellappa v. Ellappa*, A.I.R. 1950 Mad. 409.

news of an excommunication to one's caste fellows.[153] Yet these powers are subject to some restriction. The UOA makes inroads on caste autonomy by making it an offence to expel caste fellows or boycott outsiders for their failure to enforce disabilities.[154] And the Representation of the People Act forbids the use of caste disciplinary machinery for political purposes.[155]

In one sense the autonomy of the caste group is enhanced by the constitutional provisions. One of the basic themes of the Constitution is to eliminate caste as a differentia in the relationship of government to the individual—as subject, voter, or employee. The Constitution enshrines as fundamental law that government must regulate individuals directly and not through the medium of the communal group. The individual is responsible for his own conduct and cannot, by virtue of his membership in a caste, be held accountable for the conduct of others. Thus the imposition of severe police restrictions on specified castes in certain villages on grounds of their proclivity to crime was struck down as unconstitutional, since the regulation depended on caste membership rather than on individual propensity.[156] Similarly, the Supreme Court held unconstitutional a punitive levy on a communal basis, since there were some law-abiding citizens in the penalized communities.[157] Thus it would appear that regulative or penal measures directed at certain castes are beyond the power of government; a caste, then, enjoys a new

---

The powers of the caste closely parallel those of the club. Cf. *T. P. Daver v. Victoria Lodge*, A.I.R. 1963 S.C. 1144 (expulsion for 'Masonic offenses'). The grounds for judicial intervention in such private groupings are analogous to those employed in judicial review of special governmental tribunals. See *Jugal Kishore v. Sahibganj Municipality*, (reviewable only for lack of jurisdiction, procedural irregularities, or violations of natural justice). See, generally, Chakraverti (1965).

153 *Panduram v. Biswambar*, A.I.R. 1958 Or. 256. For a description of the mechanics of outcasting in the caste here involved (Orissa Telis), see Patnaik (1960). *Manna v. Ram Ghulam*, A.I.R. 1950 All. 619 (a caste member is bound to publish the resolution 'for saving himself and the caste from the defilement which would take place by acting against the verdict of excommunication.' Id. at 620).

154 Untouchability (Offences) Act, 1955, Sec. 3(2). But cf. *Sarat Chanda Das v. State*, A.I.R. 1952 Or. 351, where it was held that imprecations against a priest for associating with Untouchables and taking food from them was admonition and did not amount to criminal defamation.

155 The Representation of the People Act, 1951 Sec. 123(2), makes it 'undue influence' and a corrupt election practice if one:

... (i) threatens any candidate, or any elector, or any person in whom a candidate or an elector is interested, with injury of any kind, including social ostracism or excommunication or expulsion from any caste or community; or

(ii) induces or attempts to induce a candidate or an elector to believe that he, or any person in whom he is interested, will become or will be rendered an object of divine displeasure or spiritual censure. ...'

156 *Sanghar Umar Ranmal v. State*, A.I.R. 1952 Saur. 124.

157 *State of Rajasthan v. Pradap Singh*, A.I.R. 1960 S.C. 1208.

protection from regulation directed at it as a corporate whole.

The autonomy of the caste group is also affected by the provisions of the Constitution which guarantee the prerogatives of religious groups. Article 26 guarantees to every 'religious denomination or section thereof' the right to establish and maintain religious and charitable institutions, to own and administer property and to 'manage its own affairs in matters of religion'. It is in the application of these denominational rights that we can see the courts envisaging castes in our sectarian model.

In *Sri Venkataramana Devaru v. State of Mysore* the government sought to apply the Madras Temple Entry Act to a temple which the trustees claimed was exempt as a denominational temple belonging to the Gowda Saraswat Brahman community. The government contended that the temple was 'only a communal and not a denominational temple' unless it could be established that there were 'religious tenets and beliefs special to the community. . . .'[158] Finding that members of the community brought their own idols to the temple, that they recognized the authority of the head of a particular *math*, and that others were excluded from certain ceremonies, the Supreme Court concluded that they were indeed a 'religious denomination'. A denomination's right to manage its own affairs in matters of religion included not only matters of doctrine and belief but also practices regarded by the community as part of its religion, including the restriction of participation in religious services.[159] However, the Court found that the temple-entry rights granted by Article 25 included such denominational temples and overrode the denomination's rights to exclude Untouchables completely. The denomination's rights were not entirely without effect, nevertheless. The Court held that the denomination's rights may be recognized where 'what is left to the public is something substantial and not merely the husk of it'. Since the other occasions of worship were sufficiently numerous that the public's rights were substantial, the Court was willing to recognize the right of the denomination to exclude all non-Gowda Saraswat Brahmans during special ceremonnies and on special occasions.

Thus we find that the caste's assertion of its denominational character enables it to enjoy certain prerogatives. But this view of the caste is of a sect or denomination; its claim rests not on its position

[158] A.I.R. 1958 S.C. 255 at 263.

[159] More recently, the Supreme Court has indicated that maintenance of caste distinctions in places of worship may not in its view be a genuine tenet of Hinduism entitled to the protection of freedom of religion. See *Sastri Yagnapurushadasji v. Muldas Bhundardas Vaishya*, A.I.R. 1966 S.C. 1119, where the Court indicated that such considerations were superstitious and inconsistent with 'the true teaching of Hindu religion.'

in the overarching Hindu order but on its religious distinctiveness.

In *Saifuddin Saheb v. State of Bombay*, the Supreme Court held that the power to excommunicate for infractions of religious discipline is part of the constitutional right of a religious denomination to manage its own affairs in matters of religion.[160] The case, involving excommunication from a Muslim sect, held unconstitutional a Bombay act making excommunication a criminal offence. This does not imply a similar protection for caste groups as such; it would presumably protect only those that can qualify as religious denominations. It probably would not protect excommunication that was merely social and was not 'to preserve the essentials of religion'. Even if the excommunication were a matter of religious discipline, it would presumably not be constitutionally protected if the breach of discipline involved failure to obeserve untouchability[161] or if its purpose were political.[162]

Once a caste is recognized as a religious denomination, then as a religious group it is presumably a 'minority . . . based on religion' and as such enjoys a constitutional right under Article 30(1) 'to establish and administer educational institutions of [its] choice'. Article 30(2) provides that, in granting aid to educational institutions, the state shall not 'discriminate against any educational institution on the ground that it is under the management of a minority, whether based on religion or language'. (On the other hand, once it receives state aid it cannot discriminate on caste lines in admissions.)[163]

To the extent that its religious (or other) distinctiveness can be construed as giving it a 'distinct . . . culture of its own', the caste group may merit the protection afforded by Article 29(1), which provides that 'Any section of . . . citizens . . . having a distinct language, script, or culture of its own shall have a right to conserve the same'. Article 29 has rarely been considered by the courts inde-

160 A.I.R. 1962 S.C. 853.

161 Excommunication on this ground is forbidden by the Untouchability (Offences) Act, Sec. 7(2). I have seen no post-constitutional case involving this section. (Cf. *Sri Sukratendar Thirtha Swami of Kashi Mutt v. Prabhu*, A.I.R. 1923 Mad. 582, for the old law). The conclusion here is by analogy with the *Devaru* case, *supra* note 156, where the temple-entry rights were held to override the denominational rights. The instant situation might of course be distinguished on the ground that here the excommunicated party is asserting no constitutional right of his own, but is only vicariously asserting the Article 17 rights of the Untouchable. However, Article 17 might be read as conferring directly on every person a right to immunity from caste action against him for the purpose of enforcing untouchability. Cf. *Barrows v. Jackson*, 346 U.S. 249 (1953).

162 Representation of the People Act, Sec. 123 (see note 153 above). Cf. *Ram Dial v. Sant Lal*, A.I.R. 1959 S.C. 855, where Sat Guru's threats of expulsion from the Namdhari sect of Sikhs were found to constitute undue influence, a corrupt practice sufficiently serious to void the election of this candidate.

163 Art. 29(2).

pendently; usually it has been mentioned only in the context of the assertion of rights under Article 30(1). Apparently, every religious denomination or section thereof might qualify as a cultural group. Their right to 'conserve' their culture clearly includes the right to transmit this culture. In the *Bombay Education Society* case, the Supreme Court referred to 'the right to impart instruction in their own institutions to children of their own community in their own language' as the 'greater part' of the contents of Article 29.[164] However, more recently the Supreme Court has held that this right extends to political action to preserve the distinctive characteristics of the group, even where this involved resistance to, and disparagement of the cultural claims of others.[165]

The potential protections of Articles 29 and 30 have been radically enhanced by several recent judgments of the Supreme Court, which refer to these rights as 'absolute', in contrast to most fundamental rights, which are subject to 'reasonable restrictions' with a view to various public interests. In *Rev. Sidhrajbhai Sabbaji v. State of Gujarat*, the Court invalidated a government order threatening withdrawal of aid and recognition from a teacher-training institution which refused to reserve 80 per cent of its places for government appointees. It held that the right of a Presbyterian Society under Article 30(1) to run its schools was 'absolute' and 'not to be whittled down by so-called regulative measures conceived in the interest not of the minority educational institution, but of the public or the nation as a whole'.[166] Thus governmental regulation of minority educational institutions must be not only reasonable, but 'conducive to making the institution an effective vehicle of education for the minority community. . . .' In *Jagdev Singh v. Pratap Singh*, the Supreme Court held that election appeals to vote for a candidate on the ground that he would conserve Hindi (by opposing introduction of Punjabi as a second lanuage in the schools) was not a corrupt election practice, but was instead protected by Article 29(1). 'The right conferred upon the section of the citizens . . . to conserve their language script or culture is made by the Constitution absolute. . . .'[167]

Presumably, then, any group which can characterize itself as either 'a minority based upon religion' or a 'section of citizens with

[164] *State of Bombay v. Bombay Education Society*, 17 S.C.J. 678 (1954).

[165] *Jagdev Singh v. Pratap Singh*, A.I.R. 1965 S.C. 183.

[166] A.I.R. 1963 S.C. 540. Compare the less stringent views in *In re Kerala Education Bill*, A.I.R. 1958 S.C. 956; *Dipendra Nath v. State of Bihar*, A.I.R. 1962 Pat. 101 at 108. (Articles 29 and 30 concern 'the sphere of intellect and culture' and do 'not involve dispensation from obedience to general regulations made by the state for promoting the common good of the community.'); *Arya Pratinidhi Sabha v. State of Bihar*, A.I.R. 1958 Pat. 359.

[167] A.I.R. 1965 S.C. 183 at 188. In *Kultar Singh v. Mukhtiar Singh*, A.I.R. 1965 S.C. 141 (decided after *Jagdev Singh*), the Supreme Court, without mentioning Artic¹ᵉ

a distinct . . . culture' may qualify for a wide range of protections. The characterization of the caste group by the sectarian model puts it in the constitutionally privileged status of a religious denomination. Once so characterized, the group enjoys, to some extent at least, constitutional protection not only in its right to control its religious premises but also in its rights to excommunicate dissidents, to maintain educational institutions free from government regulation which is not in *its* interest, and to 'conserve' its distinctive culture by political means. Of course this applies only to those castes which could qualify as 'religious denominations or section[s] thereof'. It seems unlikely, however, that any government could allow these privileges to some castes and not to others; and in any event it seems probable that all castes could produce enough distinctive ritual or doctrine to qualify as denominations. This view of caste would seem to present difficulties to those proponents of the casteless society who advocate prohibition of communal charities and educational institutions.[168]

New Models for Old

Before suggesting some of the implications of this new dispensation, let me summarize briefly the recent changes in the legal characterization of caste. Since independence, the sacral view has been drastically impaired. In the personal law, varna distinctions (and with them the necessity of determining the varna standing of caste groups) have been eliminated, at least for the future, although these matters persist for a time. In the area of precedence and disabilities, there has been a withdrawal of all support for precedence based on ritual standing. This withdrawal is embodied in provisions against caste discrimination, in the abolition of untouchability, and in temple-entry laws. The government has reversed its previous policy by now intervening to prevent the imposition of disabilities and to give preferential treatment to those at the bottom of the socio-religious order. In administering these preferences, the courts have avoided giving recognition to the sacral view, at least when dealing with transactions within Hinduism, although the shadow or mirror of it appears in the definition of Untouchables and it appears in an attenuated form in dealing with non-Hindus.

Where the sacral order implicitly remains, its religious content is relatively diffuse and indefinite. But in other post-independence

29(2) excluded certain religious appeals from the coverage of corrupt practices: 'Political issues which form the subject-matter of controversies at election meetings may indirectly and incidentally introduce considerations of language or religion.'

168 See, for example, Narayan (1955, pp. 72, 75); Karve (1961, p. 154). ('Contributions to funds intended to benefit castes or communal groups should be stopped by law.').

developments we see caste given a more positive religious treatment. Alongside the remnant of the older one, a different image of the caste group is found, seeing the caste group as a religious unit, denomination, or sect distinguished by its own idiosyncratic cult, doctrine, and ritual. This we found in the cases involving temple entry and in the protection of denominational rights.

The associational model which sees the caste as an association characterized by a complex of characteristics (including but not limited to religious ones) has been strengthened since independence. The area of caste autonomy, where it previously prevailed, is largely unimpaired and in some respects enhanced. It retains a minor but important role in the personal law area. It has been accepted in the cases involving group membership, at least within Hinduism.

The organic model, which stresses the relative economic, educational, and occupational attainments of the caste group, is used (along with a mirror image of the sacral view) in the selection of Scheduled Castes for preferential treatment. For a time, this organic view led a rather vigorous life in the area of preferences for the backward classes. Its use for this purpose is now greatly restricted and very possibly fated to extinction.

In short, there has been a decline in the use of the sacral model and increasing reliance on the other models. We may think of the courts during the British period as conceiving of castes primarily as graded components in the sacral order of Hinduism and secondarily as autonomous associations. In administering the law, they were sensitive to vertical differences amng castes (expressed in varna distinctions and pollution) as well to horizontal differences (expressed in sectarian distinctiveness and in caste autonomy). The Constitution now forbids the courts to give recognition and support to the vertical, hierarchical distinctions; but other constitutional provisions (guarantees to religious denominations and of the integrity of groups) enjoin the courts to recognize and support the horizontal distinctions. The Constitution can be read as the 'disestablishment' of the sacral view of caste—the courts can give no recognition to the integrative hierarchical principle, yet the Constitution recognizes the claims of the component parts of the system. Claims based on the sacral order are foreclosed (in personal law reform, temple entry, abolition of untouchability, non-recognition of exclusionary rights), but claims based on sectarian distinctiveness or group autonomy are not.

While the integrative and hierarchic elements of the caste system may elicit no (or little) legal recognition or protection, the new dispensation does not prevent caste groups from using their resources (including legal protections and powers) to advance their claims in both 'old' (socioritual) and 'new' (economic, political) hierarchies. Indeed, the new view of caste is one that seems well suited to a situation in which castes have overlapping claims, func-

tions, and positions in each of several hierarchies; in which 'horizontal' solidarity and organization within caste groups grows at the
expense of 'vertical' integration among the castes of a region; in
which relations between castes are increasingly independent and
competitive and less interdependent and co-operative.[169] In particular, the new legal view of caste furnishes recognition and protection for the new social forms through which caste concerns
may be expressed: caste associations, educational societies, political
parties, religious societies, unions, etc. It offers scope not only to the
endogamous *jāti* but to 'the caste-like units which are so active in
politics and administration in modern India'.[170] We may anticipate
that the new legal view of caste will not only sanction but stimulate
and encourage new forms of organization, new self-images, and new
values within caste groups. And the disestablishment of the predominant organizing model of cultural unity may give new vitality
to lesser traditions and new scope for innovation.

We can visualize the judiciary as mediating between the Constitution's commitment to a great social transformation and the
actualities of Indian society. The court must combine and rationalize
the various components of the constitutional commitment — voluntarism and respect for group integrity on the one hand, and equality
and nonrecognition of rank ordering among groups on the other.
They must do this in the process of applying these constitutional
principles to claims and conflicts which arise within the existing
structure of Indian society. In working out the application of these
principles, the judiciary have produced a picture of caste which no
one proposed and no one anticipated.

We may also visualize the judiciary as mediating between the
aspirations of the 'non-caste' (or anti-caste) people, those politicians,
publicists, and intellectuals who envision a radical transformation to
a 'casteless' society, and the attitudes and concerns of the 'caste
people'. The courts provide one forum in which conflicting policies,
claims, and ambitions are reconciled. In this process, the implementation of strictures against caste does not lead to the results
anticipated by the 'no-caste' people. And operating through new
channels and with new concepts, caste will be transformed in ways
not envisioned by the 'caste people'. It is just because traditional
proponents and modern opponents of caste are linked together by
the courts (and other mediating institutions, such as political parties)
that it can confidently be predicted that what will emerge will be
neither the caste society of the imagined past nor the 'casteless' society
of the imagined future.

---

[169](Srinivas, 1962, Introduction, chaps. 1,4,6; Gould, 1963; Rudolf and Rudolf,
1960).
[170] (Srinivas, 1962, 6).

IV

PURSUING EQUALITY
IN THE LAND OF
HIERARCHY

8

Pursuing Equality in the Land of Hierarchy: An Assessment of India's Policies of Compensatory Discrimination for Historically Disadvantaged Groups*

Independent India embraced equality as a cardinal value against a background of elaborate, valued and clearly perceived inequalities. Her constitutional policies to offset these proceeded from an awareness of the entrenched and cumulative nature of group inequalities. The result has been an array of programmes that I call, collectively, a policy of compensatory discrimination. If one reflects on the propensity of nations to neglect the claims of those at the bottom, I think it is fair to say that this policy of compensatory discrimination has been pursued with remarkable persistence and generosity (if not always with vigour and effectiveness) for the past thirty years.

These compensatory discrimination policies entail systematic departure from norms of equality (such as merit, evenhandedness, and indifference to ascriptive characteristics). These departures are justified in several ways: first, preferential treatment may be viewed as needed assurance of personal fairness, a guarantee against the persistence of discrimination in subtle and indirect forms. Second, such policies are justified in terms of beneficial results that they will presumably promote: integration, use of neglected talent, more equitable distribution, etc. With these two — the anti-discrimination theme and the general welfare theme — is entwined a notion of historical restitution or reparation to offset the systematic and cumulative deprivations suffered by lower castes in the past. These multiple justifications point to the complexities of pursuing such a policy and of assessing its performance.

India's policy of compensatory discrimination is composed of an array of preferential schemes. These programmes are authorized by Constitutional provisions that permit departure from formal equality for the purpose of favouring specified groups.

The benefits of 'compensatory discrimination' are extended to a wide array of groups. There are three major classes. First, there are those castes designated as Scheduled Castes on the basis of their 'untouchability'. They number nearly 80 million (14.6 per cent of population) according to the 1971 Census. Second, there are the

* Reprinted from *Social and Economic Development in India: a Reassessment*, Dilip K. Basu and Richard Sisson, eds., New Delhi, Sage Publications India Pvt. Ltd. (1986).

Scheduled Tribes who are distinguished by their tribal culture and physical isolation and many of whom are residents of specially-protected Scheduled Areas. They number more than 38 million (6.9 per cent of the population in 1971). Third, there are the 'Backward Classes' (or, as they are sometimes called, 'Other Backward Classes',) a heterogeneous category, varying greatly from state to state, comprised for the most part of castes (and some non-Hindu communities) low in the traditional social hierarchy, but not as low as the Scheduled Castes. Also included among the Other Backward Classes are a few tribal and nomadic groups, as well as converts to non-Hindu religions from the Scheduled Castes and in some areas the Denotified Tribes. It has been estimated[1] that there were approximately 60 million persons under the Other Backward Classes heading in 1961 — roughly the magnitude of the Scheduled Caste population at that time (64 millions). (Today the portion of the population designated under this heading is probably larger).

For the most part, preferences have been extended on a communal basis. Members of specified communities are the beneficiaries of a given scheme and all members of the community, however, prosperous, are entitled to the benefits. However, some schemes use a means test to supplement the communal one — only members of the listed communities with incomes below the specified ceiling are eligible. In a few instances, the communal test has been replaced by an economic one — income or occupation or a combination of the two — and a few schemes use tests neither communal nor economic.

Preferences are of three basic types: first, there are reservations, which allot or facilitate access to valued positions or resources. The most important instances of this type are reserved seats in legislatures, reservation of posts in government service, and reservation of places in academic institutions (especially the coveted higher technical and professional colleges). To a lesser extent, the reservation device is also used in the distribution of land allotments, housing and other scarce resources. Second, there are programmes involving expenditure or provision of services — e.g., scholarships, grants, loans, land allotments, health care, legal aid — to a beneficiary group beyond comparable expenditure for others. Third, there are special protections. These distributive schemes are accompanied by efforts to protect the backward classes from being exploited and victimized. Forced labour is prohibited by the Constitution (Art. 23 (2) and in recent years there have been strenuous efforts to release the victims of debt bondage, who are mostly from Scheduled Castes and Tribes. Legislation regulating money lending, providing debt relief, and restricting land transfers attempt to protect Scheduled Castes

[1] Dushkin (1961:1665).

and Tribes from economic oppression by their more sophisticated neighbours. Anti-untouchability propaganda and the Protection of Civil Rights Act attempt to relieve Untouchables from the social disabilities under which they have suffered. This legislation is not 'compensatory discrimination' in the formal sense of departing from equal treatment to favour these groups; it enjoins equal treatment rather than confers preferential treatment. But in substance it is a special undertaking to remedy the disadvantaged position of the Untouchables.

The Array of (Alleged) Costs and Benefits

Few in independent India have voiced disagreement with the proposition that the disadvantaged sections of the population deserve and need 'special help'. But there has been considerable disagreement about exactly who is deserving of such help, about the form this help ought to take, and about the efficacy and propriety of what the government has done under this head.

There is no open public defence of the *ancien regime*. Everyone is against untouchability and against caste. Public debate takes the form of argument among competing views of what is really good for the lowest castes and for the country. These views involve a host of assertions about the effects — beneficial and deleterious — of compensatory discrimination policies.

Here, I would like to sketch in the most general terms the full range of claims that are made as to benefits and costs — the various ways in which the policy of compensatory discrimination allegedly helps or hurts the protected groups, others and India as a whole.

Rough and redundant as it is, this anthology of claims will provide us with a checklist that will help in devising appropriate standards for evaluating specific schemes. For convenience each claimed benefit is paired with the opposite claim of cost. It would simplify matters if each of these pairs represented points on a single dimension that could be unambiguously measured. Unfortunately, the most that can be claimed for them is that each is a composite of sometimes reinforcing but occasionally conflicting qualities that may conveniently be grouped together. Since the lines between the claimed effects are not always distinct, some overlap and redundancy is unavoidable. It was necessary, for convenience in labelling, to devise some rubrics that are not found in ordinary talk of these matters, imparting a somewhat stilted quality to the list. These claims are listed here in an order that proceeds roughly from the most focussed and immediate to the most capacious and remote, from those which speak of impacts primarily on the benefiiciary groups to those which concern the shape and career of the whole society.

Alleged Benefits and Costs of
Policy of Compensatory Discrimination

1. RE-DISTRIBUTION	*vs.*	DIVERSION
Preferences provide a direct flow of valuable resources to the beneficiaries in larger measure than they would otherwise enjoy.		These resources are enjoyed by a small segment of the intended beneficiaries and do not benefit the group as a whole.
2. REPRESENTATION	*vs.*	MISREPRESENTATION
Preferences provide for participation in decision-making by those who effectively represent the interests of the beneficiaries, interests that would otherwise be underrepresented or neglected.		By creating new interests which diverge from those of the beneficiaries, preferences obstruct accurate representation of their interests.
3. INTEGRATION	*vs.*	ALIENATION
By affording opportunities for participation and well being, preferences promote feelings of belonging and loyalty among the beneficiaries, thereby promoting the social and political integration of these groups into Indian society.		By emphasizing the separateness of these groups, preferences reduce their opportunities for (and feelings of) common participation.
4. ACCEPTANCE	*vs.*	REJECTION
Preferences induce in others an awareness that the beneficiary groups are participants in Indian life whose interests and views have to be taken into account and adjusted to.		Preferences frustrate others by what they consider unfair favoritism and educate them to regard the beneficiaries as separate elements who enjoy their own facilities and have no claim on general public facilities.

5. INTEGRITY	*vs.*	MANIPULATION
Preferences permit forms of action that promote pride, self-respect, sense of achievement and personal efficacy that enable the beneficiaries to contribute to national development as willing partners.		Preferences subject these groups to manipulation by others, aggravate their dependency, and undermine their sense of dignity, pride, self-sufficiency and personal efficacy.
6. INCUBATION	*vs.*	OVER-PROTECTION
By broadening opportunities, preferences stimulate the acquisition of skills and resources needed to compete successfully in open competition.		Preferences provide artificial protection which blunts the development of the skills and resources needed to succeed without them.
7. MOBILIZATION	*vs.*	ENERVATION
By cultivating talents, providing opportunities and incentives and promoting their awareness and self-consciousness, preferences enhance the capacity of the beneficiary groups to undertake organized collective action.		By making them dependent, blunting the development of talent, undermining self-respect, preferences lessen the capacity for organized effort on their own behalf.
8. STIMULATION	*vs.*	SEDATION
By increasing the visibility of the beneficiary groups, promoting their placement in strategic locations, and emphasizing the national commitment to remedy their condition, preferences serve as a stimulus and catalyst of enlarged efforts for their uplift and inclusion.		By projecting an image of comprehensive governmental protection and preferment, preferences stir the resentment of others, allaying their concern and undermining intiatives for measures on behalf of the beneficiary groups.

SELF-LIQUIDATION	*vs.*	SELF-PERPETUATION
The benefits of preferential treatment are mutually reinforcing and will eventually render unecessary any special treatment		These arrangements created vested interest in their continuation, while discouraging the development of skills, resources and attitudes that would enable the beneficiaries to prosper without special treatmet.
10. FAIRNESS	*vs.*	UNFAIRNESS
Preferences compensate for and help to offset the accumulated disablements resulting from past deprivation of advantages and opportunities		Preferences place an unfair handicap on individuals who are deprived of opportunities they deserve on merit.
11. SECULARISM	*vs.*	COMMUNALISM
By reducing tangible disparities among groups and directing attention to mundane rather than ritual standing, preferences promote the development of a secular society		By recognizing and stimulating group identity, preferences perpetuate invidious distinctions, thereby undermining secularism.
12. DEVELOPMENT	*vs.*	STAGNATION
Preferences contribute to national development by providing incentives opportunities and resources to utilize neglected talent.		Preferences impede development by misallocation of resources, lowering of morale and incentive, and waste of talent.

This catalogue is clearly not a set of explanations of why these policies were adopted, although such goals undoubtedly played a part. It is a set of standards for judging these policies. But whose standards? Obviously they are the author's, but I would claim that they are more than the author's — a claim supported by theeir provenance, for they are refinements and generalizations of arguments found in current Indian discourse about these policies. I would claim further that the list encompasses most of the standards that would occur to a disinterested policy-maker. (By this I mean a policy-maker concerned with these policies *per se* rather than with their implications for his political fortunes.) In this accounting of costs and benefits I have put to one side those benefits (and costs) that accrue to individual actors from supporting or opposing a particular programme (apart from some that enter incidentally under the heading of diversion). This is not because I suppose that policy makers omit consideration of the personal and political gains and losses that such positions entail. The course that policy takes is very much shaped by this second level of costs and benefits. But although any given actor may have his own schedule of priorities and his own admixture of second-level objectives, almost all would share at least some part of the goals implied by these standards. And virtually all would appeal for support in terms of these standards. That actors differ in their priorities and goals as well as in their estimates of fact does not reduce the usefulness of exposing and articulating standards for judging these policies.

The evaluation of these compensatory programmes involves a two stage inquiry. First there is what we might call the problem of performance: does the programme actually deliver the goods (more jobs or housing or better performance in schools or whatever)? In making such judgments we must be wary of all the pitfalls of measuring programme effects. Having satisfied ourselves that the programme has the projected effect, we then face what we might call the problem of achievement. Has the programme produced the results that it is supposed to achieve — do more jobs for Scheduled Castes produce considerate treatment by officials, or stimulate educational accomplishment, or produce social integration? To what extent does delivering the jobs entail the costs alleged by critics of preferential treatment — stigmatizing the beneficiaries, fomenting group resentments, lowering self-esteem, etc?

This is not a list of the direct effects of compensatory programmes — e.g. more jobs, higher literacy or better nutrition. It is a list of the good and bad effects that are alleged to flow from the performance promised by the programme including the effects attributable to the specifically preferential aspects of the programme. Even if all of these dimensions enter into estimation of the overall working of the compensatory discrimination policy, it is not implied that all of them are involved in every specific scheme. Nor is it implied that they are to be accorded equal weight in making such evaluations. Presumably specific schemes in different fields (education, housing, etc.) or for different groups (Scheduled Castes, Scheduled Tribes, Other Backward classes) have a different mix of intended effects. And, the relative weights to be assigned to those effects (and to unintended by-products) will differ among various participants and observers.

A Costly Success

Have these policies 'worked'? What results have they produced? And at what costs? Our tabulation of alleged costs and benefits suggests the complexity hidden in these apparently simple questions. Performance is difficult to measure: effects ramify in complex inter-action with other factors. Compensatory policies are designed to pursue a multiplicity of incommensurable goals in unspecified mixtures that vary from programme to programme, from time to time, and from proponent to proponent. Evaluation of a specific scheme for a specific group during a specific period is itself a daunting undertaking. In other places, I have attempted to use this checklist of claims in evaluating specific schemes.[2]

What I want to do here is draw a crude sketch of the effects of the compensatory discrimination policy in their largest outline. What has the commitment to compensatory discrimination done to the shape of Indian society and of lives lived within it?

The limited clarity of such a sketch is dimmed by the necessity of distinguishing between compensatory discrimination for the Scheduled Castes and Tribes on the one hand for the Other Backward Classes on the other. The following summary focuses on programmes for Scheduled Castes and Tribes and adds some qualifications in the light of experience with schemes for the Other Backward Classes.

Undeniably compensatory discrimination policies have produced substantial redistributive effects. Reserved seats provide a substantial legislative presence and swell the flow of patronage, attention and favourable policy to Scheduled Castes and Scheduled Tribes. The reservation of jobs has given to a sizable portion of the beneficiary groups earnings, and the security, information, patronage and prestige that goes with government employment. At the cost of enormous wastage, there has been a major redistribution of educa-tional opportunities to these groups. (Of course not all of this redis-tribution can be credited to preferential policies, for some fraction would presumably have occured without them.)

[2] Galanter (1979); Galanter (1984).

Such redistribution is not spread evenly throughout the beneficiary group. There is evidence for substantial clustering in the utilization of these opportunities. The clustering appears to reflect structural factors (e.g., the greater urbanization of some groups) more than deliberate group aggrandizement, as often charged.[3] The better situated among the beneficiaries enjoy a disproportionate share of programme benefits.[4] This tendency, inherent in all government programmes — quite independently of compensatory discrimination — is aggravated here by passive administration and by the concentration on higher echelon benefits. Where the list of beneficiaries spans groups of very disparate condition — as with the most expansive lists of other Backward Classes — the 'creaming' effect is probably even more pronounced.

The vast majority are not directly benefitted, but reserved jobs bring a manyfold increase in the number of families liberated from circumscribing subservient roles, able to utilize expanding opportunities and support high educational attainments. Although such families comprise only a tiny fraction — an optimistic guess might be 6 per cent[5] — of all Scheduled Caste families, they provide the crucial leaven from which effective leadership might emerge.

Reserved seats afford a measure of representation in legislative settings, though the use of joint electorates deliberately muffles the assertiveness and single-mindedness of that representation. The presence of SC and ST in legislative settings locks in place the other programmes for their benefit and assures that their concerns are not dismissed or ignored. Job reservations promote their presence in other influential roles and educational preferences provide the basis for such participation. Of course these positions are used to promote narrower interests — although we should not assume automatically that those they displace would bestow the benefits of their influence more broadly. If, for example, reserved seat legislators are dispro-

---

[3] Shah and Patel (1977:149ff).

[4] Malik (1979: 158).

[5] Isaacs (1965:111) estimates that perhaps as many as ten per cent of the Scheduled Caste population is 'coming up,' largely through reserved posts. Our figures suggest that his estimate is probably on the high side. By 1975, just about 180,000 Scheduled Castes were in Class III or higher service with the Central government. Let us assume (generously) about an equal number of Class III or higher in public sector enterprises and the same number in state services. That gives us about 500,000. If we make the very optimistic assumption that for every such person, there is another person similarly situated in local government, public sector, or private sector employment, we come to a total of 1,000,000. If we assume that each of these persons is the head of a family of five, we come to a total of five million persons in such mobile families — just over 6 per cent of the entire Scheduled Caste population. If we were willing to include Class IV employment as having similar potential for mobility, we would more than double our figures and come out with something close to Isaacs' 10 per cent.

portionately attentive to the concerns of those of their fellows who
already have something, it is not clear that this is more the case with
them than with legislators in general seats.

Legislative seats are occupied by members of national political
parties. They must aggregate broad multi-group support in order
to get elected and, once elected, must participate in multi-group
coalitions in order to be effective. In the office setting, too, there
are relations of reciprocity and interdependence. The broad participation afforded by reserved seats and reserved jobs is for many
others a source of pride and warrant of security.

If the separate and special treatment entailed by preferential
programmes wounds and alienates the members of beneficiary
groups, this is amplified by the hostility experienced on being
identified as a recipient. As sources of alienation, these experiences
must be placed against the background of more devastating manifestations of hostility, such as the much publicized assaults and
atrocities perpetrated on Scheduled Castes.

At the policy-making level, reserved seats have secured the
acceptance of Scheduled Castes and Tribes as groups whose interests
and views must be taken into account. In every legislative setting
they are present in sufficient numbers so that issues affecting these
groups remain on the agenda. Anything less than respectful attention
to their problems, even if only lip service, is virtually unknown.
Overt hostility to these groups is taboo in legislative and many other
public forums. But there is evidence that Scheduled Castes and
Tribes are not accepted politically. Very few members of these
groups are nominated for non-reserved seats and only a tiny number
are elected. There is massive withdrawal by voters from participation in election for reserved seats in the legislative assemblies.
Apparently large numbers of people do not feel represented by these
legislators and do not care to participate in choosing them.[6]

In the long term, education and jobs help weaken the stigmatizing
association of Scheduled Castes and Tribes with ignorance and incompetence, but in the short run they experience rejection in the
offices, hostels and other settings into which they are introduced
by preferential treatment.[7] Resentment of preferences may magnify
hostility to these groups, but rejection of them obviously exists
independently of compensatory programmes.

Compensatory programmes provide the basis for personal achievement and enlarge the beneficiaries' capacity to shape their own
lives. But in other ways the programmes curtail their autonomy.

---

[6] The evidence is presented in Galanter (1979).

[7] Cf. Malik's (1979:50) finding that middle-class Scheduled Castes experience
more exclusion than do their less educated fellows.

The design of the legislative reservations, the dependence on outside parties for funds and organizations and the need to appeal to consti- tuencies made up overwhelmingly of others — tends to produce compliant and accomodating leaders rather than forceful articula- tors of the interests of these groups. The promise of good positions offers a powerful incentive for individual effort. But reservations in government service — and educational programmes designed to provide the requisite qualifications — deflect the most able to paths of individual mobility that remove them from leadership roles in the community. Constraints are intruded into central issues of personal identity by eligibility requirements that penalize those who would solve the problem of degraded identity by conversion to a non-Hindu religion.

Although preferential treatment has kept the beneficiary groups and their problems visible to the educated public, it has not stim- ulated widespread concern to provide for their inclusion apart from what is mandated by government policy. (This lack of concern is manifest in the record of private sector employment — as it was in public undertaking employment before the introduction of reser- vations.) Against a long history of such lack of concern, it is diffi- cult to attribute its current absence to compensatory discrimination policy. But this policy has encouraged a tendency to absolve others of any responsibility for the betterment of Scheduled Caste and Scheduled Tribe on the ground that it is a responsibility of the government.[8] The pervasive overestimation of the amount and effectiveness of preferential treatment reinforces the notion that enough (or too much) is already being done and nothing more is called for.

Compensatory preference involves a delicate combination of self-liquidating and self-perpetuating features. Reservations of upper- echelon positions should become redundant as preferential treatment at earlier stages enables more beneficiaries to compete successfully, thus decreasing the net effect of the reservations. A similar reduction of net effect is produced by the extension to others of benefits pre- viously enjoyed on a preferential basis (e.g., free schooling). Judicial requirements of more refined and relevant selection of beneficiaries (and of periodic reassessment) and growing use of income cut-offs provide opportunities to restrict the number of beneficiaries.

Reserved seats in legislatures are self-perpetuating in the literal sense that their holders can help to produce their extension, but

[8] Cf. Dushkin's observation (1979:666) that 'In the course of my visits to India over two decades I have noticed an erosion and virtual disappearance of a liberal-minded public opinion supporting private efforts to improve opportunities for the Scheduled Castes.'

extension requires support from others. The periodic necessity of renewal provides an occasion for assessment and curtailment. Programmes for Scheduled Castes and Scheduled Tribes are for a delimited minority and pose no danger that the compensatory principle will expand into a comprehensive and self-perpetuating system of communal quotas. Although restrained by the courts, the provisions for the Other Backward Classes are open-ended: a majority may be the beneficiaries and the dangers of self-perpetuation cannot be dismissed.

The diversion of resources by compensatory discrimination programmes entails costs in the failure to develop and utilize other talents. The exact extent of this is unclear. It seems mistaken, for example, to consider compensatory discrimination a major factor in the lowering of standards that has accompanied the vast expansion of educational facilities since independence. The pattern in education has been less one of excluding others than of diluting educational services while extending them nominally to all. Similarly the effect of Scheduled Castes and Tribes on the effectiveness of a much enlarged government bureaucracy is overshadowed by a general lowering of standards combined with the assumption of a wide array of new and more complex tasks.

The most disturbing costs of preferential programmes may flow not from their exclusion of others but from their impact on the beneficiaries. What do the programmes do to the morale and initiative of those they purport to help? The numbers who fall by the educational wayside are legion. How rewarding is the educational experience of those who survive? Compensatory discrimination policies are not the source of the deficiencies of Indian education which impinge with special force on the beneficiaries.

As a forced draft programme of inclusion of Scheduled Castes and Scheduled Tribes within national life, compensatory discrimination has been a partial and costly success. Although few direct benefits have reached the vast mass of landless labourers in the villages, it has undeniably succeeded in accelerating the growth of a middle class within these groups—urban, educated, largely in government service. Members of these groups have been brought into central roles in the society to an extent unimaginable a few decades ago. There has been a significant redistribution of educational and employment opportunities to them; there is a sizable section of these groups who can utilize these opportunities and confer advantages on their children; their concerns are firmly placed on the political agenda and cannot readily be dislodged. But if compensatory discrimination can be credited with producing this self-sustaining dynamic of inclusion, there is at the same time a lesser counter-dynamic of resentment, rejection, manipulation and low self-esteem.

And these gains are an island of hope in a vast sea of neglect and oppression. This mixed pattern of inclusion and rejection, characteristic of urban India and of the 'organized' sector, is echoed in the villages by a pattern of increasing assertion and increasing repression.

Since independence India has undergone what might crudely be summarized as development at the upper end and stagnation at the bottom. With the boost given by compensatory discrimination a section of the Scheduled Castes and Scheduled Tribes has secured entry into the 'modern' class manning the organized sector. What does this portend for the bulk of untouchables and tribals who remain excluded and oppressed? Are they better or worse off by virtue of the fact that some members of their descent groups have a share in the benefits of modern India? The meaning of these achievements ultimately depends on how one visualizes the emergent Indian society and the role of descent groups in it.

Even this kind of crude characterization of the overall impact of policies is not possible in dealing with measures for Other Backward Classes. Policies diverge from state to state, and very different groups of people are involved. In some states the Other Backward Classes category is used to address the problems of a stratum of lowly groups who are roughly comparable in circumstance to the Scheduled Castes and Tribes. In other places this category has been used to tilt the distribution of government benefits in favour of a major section of the politically dominant middle castes. The latter doubtless produce substantial redistributive effects, if less in the way of including the most deprived. But these expensive preferences for Other Backward Classes are of immense consequence for the Scheduled Castes and Tribes. They borrow legitimacy from the national commitment to ameliorate the condition of the lowest. At the same time they undermine that commitment by broadcasting a picture of unrestrained preference for those who are not distinctly worse off than non-beneficiaries, which attaches indiscriminately to all preferential treatment. And because the Other Backward Class categories are less bounded and are determined at the state rather than at the Centre, they carry the threat of expanding into a general regime of communal allotments.

Fairness and History

Arguments about the utility of compensatory preference entwine with arguments about its fairness. Apart from the other costs, is compensatory preference so tainted by unfairness that it is illegitimate to promote the general welfare by this means? We have encountered several arguments about the fairness and unfairness of these programmes. We shall examine four here: (1) is it unfair to depart from judgments on 'individual merit' to favour the beneficiaries over contenders for valued resources? (2) Is it unfair to compensate members of some groups for injustice perpetrated on their ancestors? (3) Is it unfair to compensate some victims and not others? (4) Is it unfair that some should bear more of the burden of compensation than others? Before taking up these fairness arguments it may be helpful to recall the justifications that may be advanced for the compensatory discrimination policy. Although they are often entwined in practice, we can separate out three sorts of justifications for these measures, which I label the non-discrimination, the general welfare and the reparations themes.

The Non-discriminatory Theme

Compensatory discrimination may be viewed as an extension of the norms of equal treatment, an extension invited by our awareness that even when invidious discriminatory standards are abandoned there remain subtle and tenacious forms of discrimination and structural factors which limit the application of new norms of equality. Aspiring members of previously victimized groups encounter biased expectations, misperceptions of their performance, and cultural bias in selection devices; they suffer from the absence of informal networks to guide them to opportunities; entrenched systems of seniority crystallize and perpetuate the results of earlier discriminatory selections. Thus norms of non-discrimination in present distributions are insufficient to erase or dislodge the cumulative effects of past discrimination. Compensatory preference operates to counter the residues of discrimination and to overcome structural arrangements which perpetuate the effects of past selections in which invidious discrimination was a major determinant. In this view compensatory preference serves to assure personal fairness to each individual applicant. Group membership is taken into account to identify those individuals who require special protection in order to vindicate their claim for selection on 'merit' grounds. (The justification for much American affirmative action is often cast in these terms — as an extension of classical individualistic non-discrimination principles.)

The General Welfare Theme

On the other hand, compensatory discrimination may be advocated not as a device to assure fairness to individuals, but as a means to produce desired social outcomes — e.g. to reduce group disparities, afford representation, encourage the development of talent and so forth. Arrangements for reservations in British India were justified on such 'functional' grounds as are the various preferences for

'Oriental' Jews in Israel today.[9] Americans are familiar with the 'balanced ticket' and other arrangements by which shares are apportioned among various constituencies in the exception that abrasive disparities are kept in bounds, participation is spread out, representation is secured and responsiveness assured. The units in such functional 'welfare' calculations are groups rather than individuals. The chances of individuals are affected by the rearrangement of the chances of groups. But the purpose is not to rectify discriminatory selection among individuals, but to introduce a standard quite apart from personal desert.

The contrast of non-discrimination and welfare themes is displayed in imaginery alternative proposals for admitting more members of Group X to medical colleges. The non-discrimination proposal might argue that selection procedures be revised to eliminate subtle bias which impinged on individual Xs—e.g., culturally biased tests or differences in networks for acquiring recommendations. The general welfare proposal might argue that more Xs should be admitted in order to equalize distribution of medical services or for the representation of X views in making health policy or to afford non-X doctors the experience of fellowship with Xs. The goal is not a non-discriminatory selection of Xs among individual applicants, but a selection that optimises social goals. Such a selection might diverge from that which would be dictated merely by the prospects of individual performance on the job, because it defines the job to include the symbolic, representational and educational aspects that may not be included in the job description.

The Reparations Theme

In some cases, compensatory policies have another root—that a history of invidious treatment has resulted in accumulated disabilities which are carried by certain groups. No matter how fair and unbiased the measures presently employed for distributing benefits, the victims of past injustice will not fare well in terms of current performance. To distribute benefits by neutral standards will perpetuate and amplify unjust exactions and exclusions in the past. Fairness then demands that present distributions be arranged to undo and offset old biases, not to perpetuate them.[10]

9 On these programmes, see Smooha (1978); Adler, *et. al* (1975).

10 The notion of restitution for collective misdeeds is familiar in the indemnity and reparation payments exacted from defeated nations. See Angell (1934). Something closer to the compensatory discrimination situation is found in the reparations payments paid by Germany to Israel as the representative of Jews victimized by the Nazis. See Grossman (1954). In these instances the lapse of time between misdeed and reparation is relatively short. But cf. the occasional claims for 'three hundred years

Like the non-discrimination theme, this is a fairness argument rather than a welfare argument. But it emphasizes groups as the carriers of historic rights rather than as indicators of individual victimization. And it looks to a very different time frame. Welfare arguments are prospective; non-discrimination looks at the present situation and seeks to refine out lingering inequalities. The reparations theme sees the present as an occasion to reckon accounts for past injustice.

Do preferential programmes unfairly confer benefits on grounds that depart from evenhandedness, merit, etc. that should govern the distribution of opportunities and resources? Along the lines referred to above as the 'non-discrimination theme, proponents might respond that some of the preference accorded is not departure from evenhandedness but its extension in substance rather than form to individual members from the beneficiary groups. In this view, compensatory discrimination arrangements counter subtle discriminations and overcome the structural arrangements which entrench the results of past selections from which the beneficiaries were excluded.

In the Indian setting, few would argue that compensatory discrimination seeks only to protect merit against subtle or structural bias. Preferential treatment is accepted as a departure from merit selection in order to promote such goals as redistribution, integration and representation. Is it unfair to combine these with merit as a basis for distributing benefits?

Let us take merit to mean performance on tests (examinations, interviews, character references, or whatever) thought to be related to performance relevant to the position (or other opportunity) in question and commonly used as a measure of qualification for that position. (In every case it is an empirical question whether the test performance is actually a very good predictor of performance in the position, much less of subsequent positions for which it is a preparation.) Performance on these tests is presumably a composite of native ability, situational advantages (stimulation in the family setting, good schools, sufficient wealth to avoid malnutrition or exhausting work, etc.) and individual effort. The latter may be regarded as evidence of moral desert, but neither native ability nor situational advantages would seem to be. The common forms of selection by merit do not purport to measure the moral desert dimension of performance. Unless one is willing to assure that such virtue

---

of back pay' raised in the American setting. Lecky and Wright (1969); Schuchter (1970); Bittker (1973). But a scheme of preferences may lack this theme of justification entirely, as in the Israeli case (note 9 above) where the privileged and deprived groups had little previous contact.

is directly proportionate to the total performance, the argument for merit selection cannot rest on the moral deservingness of individual candidates. Instead it rests upon the supposed consequences: those with more merit will be more efficient or productive; awarding them society's scarce resources will produce more indirect benefits for their fellows. A regime of rewarding merit will maximize incentive to cultivate talents; the demoralizing effects of departing from merit outweigh supposed advantages, based on calculations of imponderables. The argument for merit is an argument for production of more social well-being.

Many sorts of effects flow from any allocation of resources: benefits are multiple, they include not only tangible production, but symbolic affirmations and the creation of competences. The allocation of education, government jobs or medical careers arguably has consequences for the distribution of incentives, levels of participation, disparities in the delivery of services. Which dimensions of benefit are to be taken into account in designing a given selection? In settings where there has been a broad concensus that a single-minded test of performance is appropriate, what is the argument for a shifting to a broader, promotional basis of selection? Compensatory discrimination schemes involve enlargement of the basis of selection to include other criteria along with the productivity presumably measured by 'merit-' — representation, integration, stimulation, and so forth. This enlargement is justified on the ground that without it society would be deprived of the various benefits thought to flow from the enhanced participation of specified groups in key sectors of social life. The argument is that the combination maximizes the production of good results. Of course there is always the empirical question of whether the promised results are indeed produced, but the supplementation of merit with other instrumental bases of selection hardly seems unfair in principle. Pursuit of other worthy results can be balanced against merit, as one result-oriented justification for unequal allocations against another.

Compensatory discrimination is both more and less than a reformulation of selection criteria. It is less because typically merit (in narrow performance terms) is left intact for the main part of the selection. The criteria are modified to require the inclusion of certain groups, an inclusion thought to produce a wide spectrum of beneficial results. But the new mixed standards are not applied across the board to the whole selection. So compensatory discrimination involves something more: the demarcation of those groups in whose behalf the broader promotional standards should be employed.

To prefer one individual over another on grounds of caste, religion (or other ascriptive criteria) is specifically branded as unfair by the anti-discrimination provisions of the Indian Constitution. The ban on the use of these criteria is, as we have seen, qualified to allow preferential treatment of a certain range of groups, whose history and condition seemed distinctive. There was agreement that some groups were burdened by a heritage of invidious discrimination and exclusion (and/or isolation) that made their condition distinct from that of their fellow citizens; the deprivations of their past and present members were thought to justify a special effort for their improvement and inclusion.

Spokesmen for backward classes sometimes call for measures specifically to remedy the wrongs of the past. If one thinks of the blighted lives, the thwarted hopes, the dwarfing of the human spirit inflicted on generations of untouchables, or of the oppression and exploitation of tribal peoples, the argument for measured vindication of these historic wrongs has an initial appeal. But there are many kinds and grades of victimization; deprivations are incommensurable. Perpetrators and victims sometimes stand out in stark clarity, but infirm and incomplete data often leave unclear precisely who were brutally exploitative, who willing or reluctant collaborators, who inadvertant beneficiaries of what we now see as systems of oppression. These arrangements interact with many other factors — climate, invasions, technology — in their influence on the present distribution of advantages and disadvantages. The web of responsibility is tangled and as we try to trace it across generations, only the boldest outlines are visible. Without minimizing its horrors, the past provides a shaky and indistinct guide for policy. It is beyond the capacity of present policy to remedy these wrongs: in the literal sense these injustices remain irremediable.

But if our perception of past injustice does not provide a usable map for distributing reparative entitlements, it can inform our vision of the present, sensitizing us to the traces and ramifications of historic wrongs. The current scene includes groups which are closely linked to past victims and which seem to suffer today from the accumulated results of that victimization. In a world in which only some needs can be met, the inevitable assignment of priorities may take some guidance from our sense of past injustice — thus providing the basis for a metaphoric restitution.

All remedies involve new distinctions and thus bring in their train new (and it is hoped lesser) forms of unfairness. Singling out these historically deprived groups for remedial attention introduces a distinction among all of the undeserved inflictions and unfairnesses of the world. One batch of troubles, but not others, are picked out for comprehensive remedy using extraordinary means. Those afflicted by other handicaps and misfortunes are left to the succor and aid that future policy-makers find feasible and appropriate within the framework of competing commitments (including commitments to equal treatment). But drastic and otherwise outlawed remedies were authorized for victims of what was seen as a fundamental flaw in the social structure. The special quality of the commitment to correct this flaw is dramatized by the Constitution's simultaneous rejection of group criteria for any other purpose.

The line of distinct history and condition that justifies compensatory discrimination is of course less sharp in practice than in theory. There are borderlines, grey areas, gradual transitions. There is disagreement about just where the line should be drawn. And once it is drawn, the categories established are rough and imperfect — summations of need and desert; there are inevitable 'errors' of under-inclusion and over-inclusion.

We arrive then at an ironic tension that lies at the heart of the compensatory discrimination policy. Since the conditions that invite compensatory treatment are matters of degree, special treatment generates plausible claims to extend coverage to more groups. The range of variation among beneficiaries invites gradation to make benefits proportionate to need. Those preferential policies create new discontinuities and it is inviting to smooth them out by a continuous modulated system of preferences articulated to the entire range of need and/or desert. But to do so is to establish a general system of group allotments.

Compensatory discrimination replaces the arbitrariness of formal equality with the arbitrariness of a line between formal equality and compensatory treatment. The principles that justify the preference policy counsel flexibility and modulation, We may shave away the arbitrary features of the policy in many ways. But we may dissolve the arbitrary line separating formal equality and preferential treatment only at the risk of abandoning the preference policy for something very different.

If there is to be preferential treatment for a distinct set of historically victimized groups, who is to bear the cost? Whose resources and life-chances should be diminished to increase those of the beneficiaries of this policy? In some cases, the costs are spread widely among the taxpayers, for example, or among consumers of a 'diluted' public service. But in some cases major costs impinge on specific individuals like the applicant who is bumped to fill a reservation. Differences in public acceptance may reflect this distinction. Indians have been broadly supportive of preferential programmes — e.g. the granting of educational facilities and sharing of political power — where the 'cost' of inclusion is diffused broadly. Resentment has been focussed on settings where the life chances of specific others are diminished in a palpable way, as in reservations of jobs and medical college places.

There is no reason to suppose that those who are bumped from valued opportunities are more responsible for past invidious deprivations than are those whose well-being is undisturbed. Nor that they were disproportionately benefitted by invidious discrimination the past. Reserved seats or posts may thus be seen as the conscription of an arbitrarily selected group of citizens to discharge an obligation from which equally culpable debtors are excused. The incidence of reservations and the effectiveness with which they are implemented tends to vary from one setting to another. Reservations impinge heavily on some careers and leave others virtually untouched. The administration of compensatory discrimination measures seems to involve considerable unfairness of this kind. If some concentration of benefits is required by the aims of the perference policy,[11] it seems clear that more could be done to distribute the burden among non-beneficiaries more widely and more evenly.

Secularism and Continuity

Fairness apart, to many Indian intellectuals compensatory discrimination policies seem to undermine progress toward the crucial national goal of a secular society. Secularism in this setting implies more than the separation of religion and state (religious freedom, the autonomy of religious groups, withdrawal of state sanction for religious norms, and so forth). It refers to the elimination (or minimization) of caste and religious groups as categories of public policy and as actors in public life.[12] In this 1950s and 1960s this was frequently expressed as pursuit of a 'casteless' society. Proponents of such a transformation were not always clear whether they meant the disestablishment of social hierarchy or the actual dissolution of caste units. But at the minimum what was referred to was a severe reduction in the salience of caste in all spheres of life.

The Constitution envisages a new order as to the place of caste in Indian life. There is a clear commitment to eliminate inequality of status and invidious treatment and to have a society in which government takes minimal account of ascriptive ties. But beyond this the posture of the legal system toward caste is not as single-minded as the notion of a casteless society might imply. If the law discourages some assertions of caste precedence and caste solidarity, in other respects the prerogatives previously enjoyed by the caste

[11] Cost spreading may itself entail costs in terms of the benefits delivered to the beneficiaries. For example, rotation of reserved legislative seats among constituencies might cure unfair clustering of costs, but it would vitiate the value of the reserved seats, rendering impossible the accumulation of experience, seniority and political strength.

[12] On the aspiration to secularism, see the literature cited in Galanter (1984: chap. 9, note 85).

group remain unimpaired. The law befriends castes by giving recognition and protection to the new social forms through which caste concerns can be expressed (caste associations, educational societies, political parties, religious sects).[13]

If the legal order's posture toward caste is ambivalent, public denunciation of caste has universal appeal. For lower castes it provides an opportunity to attack claims of superiority by those above them; for the highest castes it is a way to deplore the increasing influence of previously subordinate groups, either the populous middle castes that have risen to power with adult suffrage or the lowest castes whose inclusion is mandated by compensatory discrimination programmes. Looking up, the call for castelessness is an attack on the advantages retained by those who rank high in traditional terms; looking down, it denies legitimacy to the distributive claims of inferiors and insists on evenhanded application of individual merit standards.

The use of caste groups to identify the beneficiaries of compensatory discrimination has been blamed for perpetuation the caste system, accentuating caste consciousness, injecting caste into politics and generally impeding the development of a secular society in which communal affiliation is ignored in public life.[14] This indictment should be regarded with some scepticism. Caste ties and caste-based political mobilization are not exclusive to the backward classes. The political life within these groups is not necessarily more intensely communal in orientation:[15] nor are the caste politics of greatest political impact found among these groups. Communal considerations are not confined to settings which are subject to compensatory discrimination policies but flourish even where they are eschewed. Although it has to some extent legitimated and encouraged caste politics, it is not clear that the use of caste to designate beneficiaries has played a preponderant role in the marriage of caste and politics. Surely it is greatly overshadowed by the franchise itself, with its invitation to mobilize support by appeal to existing loyalties. But the avowed and official recognition of caste in compensatory discrimination policy combines with the overestimation of its effects to provide a convenient target for those offended and dismayed by the continuing salience of caste in Indian life.

---

[13] See chap. 7.

[14] See Galanter (1984: chap. 4, D).

[15] Consider, for example, the fascinating finding of Eldersveld and Ahmed (1975: 205), studying political participation, that 'the overwhelming majority of . . . activists are politically conscious of caste. They know how the caste leader voted. Among upper castes the ratio is 6 to 1 that activists are aware of caste, in the middle castes it is better than 8 to 1, and for the lower castes and Harijans it is 3 to 1 that activitists are caste conscious.'

The amount of preference afforded to the Scheduled Castes and Tribes is widely overestimated. The widespread perception of uniquitous and unrestrained preferment for these groups derives from several sources. First, there is the chronic overstatement of the effects of reservation: large portions of reservations (especially for cherished higher positions) are not filled; of those that are filled, some would have been gained on merit; diversion of benefits to a few may be perceived as a deprivation by a much larger number. The net effect is often considerably less than popularly perceived. Second, ambiguous nomenclature and public inattention combine to blur the distinction between measures for Scheduled Castes and Tribes and those for Other Backward Classes. The resentment and dismay engendered by use of the Other Backward Classes category to stake out massive claims on behalf of peasant-middle groups (particularly in some southern states) are readily transferred to discredit the more modest measures for Scheduled Castes and Tribes.

If caste has displayed unforeseen durability it has not remained unchanged. Relations between castes are increasingly independent and competitive, less interdependent and co-operative. 'Horizontal' solidarity and organization within caste groups have grown at the expense of 'vertical' integration among the castes of a region. The concerns of the local endogamous units are transformed as they are linked in wider networks and expressed through other forms of organization — caste associations, educational societies, unions, political parties, religious societies.

If secularism is defined in terms of the elimination of India's compartmental group structure in favour of a compact and unitary society, then the compensatory discrimination policy may indeed have impeded secularism. But one may instead visualize not the disappearance of communal groups but their transformation into components of a pluralistic society in which invidious hierarchy is discarded while diversity is accomodated. In this view compensatory discrimination policy contributes to secularism by reducing group disparities and blunting hierarchic distinctions.

The development of a secular society in which the hierarchic ordering of groups is not recognized and confirmed in the public realm is a departure from older Indian patterns. The compensatory discrimination policy is a major component in the dissestablishment of a central part of the traditional way of ordering the society. But this break with the past, itself is conducted in a familiar cultural and institutional style. The administration of preference programmes reflects older patterns in the fecund proliferation of overlapping schemes, the fragmentation of responsibility and the broad decentralization of authority under the aegis of unifying symbols. When these policies encounter the judiciary, what purports to be a pyramidal hierarchy establishing fixed doctrine turns out to be a loose collegium presiding over an open textured body of learning within which conflicting tendencies can be accomodated and elaborated.

The compensatory principle of substantive equality is added to the constitutional scheme of formal equality but it does not displace it. This juxtaposition of conflicting principles is an instance of what Glanville Austin admiringly describes as one of

India's original contributions to constitution-making [that is] accomodation . . . the ability to reconcile, to harmonize, and to make work without changing their content, apparently incompatible concepts — at least concepts that appear conflicting to the non-Indian, and especially to the European or American observer. Indians can accomodate such apparently conflicting principles by seeing them at different levels of value, or, if you will, in compartments not watertight, but sufficiently separate so that a concept can operate freely within its own sphere and not con-conflict with another operating in a separate sphere. . . .

With accommodation, concepts and viewpoints, although seemingly incompatible, stand intact. They are not whittled away by compromise but are worked simultaneously.[16]

The expectation that these principles could co-exist has been fulfilled. The compensatory principle has been implemented but it has not been allowed to overshadow or swallow up opposing commitments to merit and to formal equality.

The compensatory discrimination policy is not to be judged only for its instrumental qualities. It is also expressive: through it Indians tell themselves what kind of people they are and what kind of nation. These policies express a sense of connection and shared destiny. The groups that occupy the stage today are the repositories and transmitters of older patterns. Advantaged and disadvantaged are indissolubly bound to one another. There is a continuity between past and future that allows past injustices to be rectified. Independence and nationhood are an epochal event in Indian civilization which make possible a controlled transformation of central social and cultural arrangements. Compensatory discrimination embodies the brave hopes of India reborn that animated the freedom movement and was crystallized in the Constitution. If the reality has disappointed many fond hopes, the turn away from the older hierarchic model to a pluralistic participatory society has proved vigorous and enduring.

[16] Austin (1966: 317–18).

9

Missed Opportunities: The Use and Non-use of Law Favourable to Untouchables and Other Specially Vulnerable Groups

It is a common place observation that the law in action does not correspond to the law in the books. This is especially so with laws favourable to 'have not' groups. India has an abundance of laws which purport to favour labourers, untouchables, tenants, and other disadvantaged groups. Such laws are often unenforced. We speak of the failure of implementation: the law has flaws or the administration has the wrong priorities or the courts are too lenient. Indeed, in most instances it is not too difficult to discover such deficiencies. One important aspect of this 'failure of implementaton' is, sometimes, the failure of putative beneficiaries to extract possible benefits from such laws. Two case studies will illustrate this. The first of these is the enforcement of anti-disabilities legislation; the second is implementation of measures of compensatory discrimination in favour of backward classes.

The Paucity of Civil Rights Complaints

State Civil Disabilities Acts covered most of India by 1949.[1] In the

---

[1] In 1938 the Madras legislature passed the first comprehensive penal act to remove social disabilities, making it an offence to discriminate against Untouchables not only in regard to publicly supported facilities such as roads, wells, and transportation, but also in regard to 'any other secular institution' to which the general public was admitted, including restaurants, hotels, shops, etc. The Act also barred judicial enforcement of any customary right or disability based on membership in such a group. Violation was made a cognizable offence with a small fine for the first offence, larger fines and up to six months' imprisonment for subsequent offences.

Between the end of the Second World War and the enactment of the Constitution, with power passing entirely into Indian hands, acts removing civil disabilities of Untouchables were passed in most of the provinces and in many of the larger princely states. With some variation in detail, these statutes followed the general lines of the Madras Removal of Civil Disabilities Act. Enforcement of disabilities against Untouchables, variously described as Harijans, Scheduled Castes, excluded classes, Backward Classes or depressed classes, was outlawed in regard to public facilities like wells and roads and places of public accommodation like shops, restaurants, and hotels. Violations were made criminal offences, in most cases cognizable. Judicial enforcement of customs upholding such disabilities was barred.

208

early 1950s there were probably more than four hundred cases registered each year under these State Acts.[2] The Untouchability (Offences) Act (UOA), 1955[3] made more things punishable; it made all offences cognizable; it extended throughout India. It provided heavier penalties and contained a presumption designed to make proof easier. It was accompanied by a great deal of publicity. One would then expect that there would be more enforcement activity on the basis of this stronger and more extensive statute. During the first years that the UOA was in force, there was a slight increase in the number of prosecutions registered, but this tapered off in the early 1960s. By a rough projection from the data available, we arrive at an average of approximately 520 cases a year during 1956–64. Most of the increase in the reported cases over the period 1952–5 is accounted for by the inclusion of Rajasthan, which had no earlier State Act. Population increases more than offset the remainder of the increase. Contrary to expectations, the UOA did not have a significantly higher level of use than did the predecessor State Acts. In those areas covered by the State Acts, there were more anti-disabilities prosecutions in the early 1950s than in the early 1960s.

Not only was this new, stronger and more comprehensive statute used less frequently, but the percentage of cases ending in conviction dropped dramatically. If you look at the first three rows of Table 1 you will see that convictions decreased, compoundings and acquittals

[2] Material on the amount and character of anti-disabilities litigation before 1965 is drawn from Galanter, (1972). An earlier version of that paper appeared as Galanter (1969) in the excerpt presented at pp. 207–14.
Annual Number of January, 1969 (Vol. IV, Nos. 1 & 2).

[3] Article 17 not only forbids the practice of 'untouchability' but declares that 'enforcement of disabilities arising out of 'Untouchability' shall be an offence punishable in accordance with law'. Article 35 provides that 'Parliament shall have, and the Legislature of a State shall not have, power to makes laws . . . prescribing punishment for those acts which are declared to be offences under 'this Part' [i.e. the Fundamental Rights section of the Constitution]. Parliament is specifically directed to 'make laws prescribing punishment' for acts declared offences in Part III. 'as soon as may be after the commencement of this Constitution'. Until Parliament discharged this duty, existing laws prescribing punishment for such acts were continued in force 'until altered or repealed or amended by Parliament'. Thus existing state (i.e. provincial) law was continued in force, but was frozen in its then-existing form, beyond the power of the state legislatures to modify or repeal.

In 1955 Parliament exercized this exclusive power and passed the Untouchability (Offences) Act (UOA), which remains the culmination of anti-disabilities legislation to the early 1970s. The UOA outlaws the enforcement of disabilities 'on the ground of untouchability' in regard to, inter alia, entrace and worship at temples, access to shops and restaurants, the practice of occupations and trades, use of water sources, places of public resort and accommodation, public conveyances, hospitals, educational institutions, construction and occupation of residential premises, holding of religious ceremonies and processions, use of jewellery and finery. The imposition of disabilities is made a crime punishable by fine of up to Rs 500, imprisonment for up to six months, cancellation or suspension of licenses and of public grants.

grew larger. In the early 1950s the conviction rate in anti-disabilities cases was close to the average rate of convictions for all criminal trials in India. But by the early 1960s the convictions in anti-disabilities cases fell to less than half of the former rate.

Table 1

Disposition of Cases
under
State Anti-Disabilities Acts, Untouchability (Offences) Act,
Protections of Civil Rights Act.

Legislation	Time	Total Disposals	Disposition		
			Convicted	Compounded	Acquitted
State Acts (41 mos.)	Jan. 1952– May 1955	508	360 (70.9%)	58 (11.4%)	90 (17.7%)
U.O.A. (55 mos.)	June 1955– Dec. 1959	1,281	548 (42.8%)	427 (33.3%)	306 (23.9%)
U.O.A. (48 mos.)	1960– 1963	1,124	350 (31.1%)	421 (37.5%)	353 (31.4%)
U.O.A. (24 mos.)	1969– 1970	360	93 (27.2%)	178 (49.4%)	84 (23.3%)
PCRA (24 mos.)	1977– 1978	1,571	269 (17.1%)	——	1,302 (82,8%)

Sources: The first three rows are taken from Galanter (1972). The 1969 and 1970 figures are from the Report of the Commission for Scheduled Castes and Scheduled Tribes for 1970 Vol. 1, p. 59. The 1977 and 1978 figures are from the Report for 1978–9, Vol. II, p. 137.

I was puzzled by the fact that this more comprehensive and apparently stronger act was less used and led to inferior results. At the time I tried to sketch an explanation in terms of the situation in which potential users found themselves. Let me presume to give you a slightly abridged version of that 1968 explanation here and then proceed to ask how, if at all, that situation has changed in the interval.*

* The following account is excerpted from Galanter (1972).

*   *   *

The state anti-disabilities acts and later the UOA were passed with a great deal oi attendant publicity and public excitement. In the initial years there was a backlog of ready complaints and vulnerable targets. News of the Act might be expected to hearten potential complainants and impress them with a sense of the possibility of change. Police would be responsive to the exhortations of their superiors who urged them to enforce the Act vigorously. In the initial impetus view, public excitement faded, the backlog got used up, potential complainants were uninspired by the results of prosecutions, and new contenders would have stronger call on police efforts. This explanation leads to further questions. For an upward spiral of increasing use can be imagined as readily as a downward spiral decreasing use. Why should the level of police activity fall off rather than become routinized as police gain experience in successful prosecution under the Act? Why has the experience of early complainants not induced others to follow in their course by emboldening new groups of Untouchables and orienting them toward change? What evidence (besides the decline in the number of cases) makes the downward spiral more plausible?

. . . .

Although the UOA is a central law, responsibility for its enforcement, like that of virtually all central laws, lies with the states. There is no separate central enforcement apparatus for this or (typically) for other national policies. Nor is there any central agency which actively co-ordinates or directs enforcement activity in the various states. Within each state, the same decentralized pattern is replicated. There is no special agency or staff for the enforcement of these laws or co-ordination of the enforcement activities of local officials. There is not at any level any agency which systematically gathers information about the problems and policies of enforcement. Thus, initiative is extremely decentralized.

The total expenditure of law enforcement resources on anti-disabilities enforcement is miniscule compared, e.g., with prohibition. For example, in Bombay from 1946–54 there were over 100,000 prohibition cases registered (and over 44,000 convictions obtained). It will be recalled from Table 10.6 that there were far less than 1,000 anti-disabilities prosecutions during the comparable period. While dry states typically have special squads, task forces, co-ordinating officers and intelligence bureaus for prohibition enforcement, there are no special squads or staffs for enforcing laws against untouchability.

There are no non-governmental organizations which systematically undertake a purposeful role in the development of anti-disabilities legislation or in its enforcement. There is no group (like, for example, the American Civil Liberties Union or the legal staff of the National Association for the Advancement of Coloured People) which concerns itself with the strategy or tactics of anti-disabilities litigation. Social uplift organizations, especially the Harijan Sevak Sangh, are involved in a large portion of the cases brought under the UOA. But their involvement is typically initiated by their local workers in response to a particular situation, not as part of a co-ordinated programme of litigation. The Sangh has no specialized legal staff and ordinarily does not engage lawyers to handle these cases. There are no lawyers with specialized expertise in such matters. Untouchable lawyers, particularly those in politics, are involved in some UOA cases. But again, there is no co-ordinated programme of action nor are there any channels for sharing of experience. It should of course be added that opposition to anti-disabilities laws appears if anything more unorganized. Except in a few disputes involving major temples, there has been no organiz-ational support to defendants in such cases.

Among the educated strata, untouchability is generally a dead issue. Aside from occasional flashes of interest when it temporarily assumes some political import, it evokes little attention from politi-cians, administrators and intellectuals—with a few notable excep-tions. Except for intermittent atrocity stories the sparse public discussion and disconnected press coverage are mostly confined to political issues (or to misgivings about the government policies of preferential treatment). While the political scene is subject to waves of concern about regional, religious and caste cleavages, there has been little sustained concern about untouchability. There is no interest in the intellectual community in the mechanics of pro-grammes for attacking untouchability, no debate about alternatives, no assessment of prospects. The problems of anti-disabilities policies, their tactics and strategy, receive little systematic consideration.

Anti-disabilities prosecutions depend on the initiative of the local police and the sympathy of local magistrates, both of whom have obvious reason to be disinclined to antagonize the dominant elements of the local community. If they are not themselves members of the latter, they are heavily dependent upon their co-operation in order to do their jobs and gain promotions. (As an Untouchable Ph.D. succinctly put it, 'law means police and police means higher caste people'.) Police are often uninformed of the provisions of the UOA—or even of its existence. The vast majority of potential cases that come to the attention of the police are ignored or at best 'com-promised' without being registered. Even if the police were sym-pathetic—and they often are not—limited resources and career pressures would keep them from expending much effort on this unrewarding line of activity.

Initiative then must be supplied by the Untouchable to press his claim for police time and resources. Complainants and the witnesses they need to prove their cases are extremely vulnerable to intimidation and reprisal. Very often they are economically dependent upon the higher castes. They may face social boycott or reprisal in the form of eviction or denial of grazing rights, when they do not meet with physical coercion in the form of beatings, house burnings or worse. Except in cases where outsiders are involved, witnesses other than their caste fellows are not ordinarily forthcoming. A High Court judge, rebuking a magistrate for discounting the evidence of the complainant and his friend as interested, observed that 'in such cases it would be idle to expect disinterested witnesses to support the complainant's case'. If the complainant does have witnesses and a good case, he is likely to be subjected to pressure to compound it.

Although boycotts and reprisals are themselves offences, they are by their very nature extremely difficult to prove. They involve the behaviour, often covert, of a large number of people and successful prosecution requires proof of motivation. Witnesses will be even harder to find. So these secondary offences are even more difficult to prove than the original offence. As a social worker active in anti-disabilities campaigns put it: In the case of a boycott we apply to the Collector and he will go out and settle the matter. We do not look to the judiciary in such cases because it is too difficult to prove the existence of a boycott in court'.

If the complainant manages to withstand these pressures, there is often considerable delay in the disposition of cases. The Elaya-perumal Committee analysed 70 completed cases and found that the average time elapsed was more than six months. Repeated court appearances are time-consuming and expensive. The accused almost always have greater resources and can hold out longer. With the passage of time, witnesses may be less persuasive and less inclined to co-operate with the prosecution.

. . . .

In at least some cases there is a disparity of legal resources. Prosecutions in magistrates' courts are ordinarily handled by the police prosecutor. Usually this official is a lawyer but with little prestige in the profession, and as a civil servant whose superior is a police official, he does not represent a strong and independent prosecutory initiative. (In a few cases, state prosecutors have been deputed or private lawyers engaged to assist in the prosecution of UOA cases.) The accused are often represented by counsel, a circumstance that is credited by knowledgeable observers with increasing their chances of obtaining an acquittal. Untouchables are often too poor to engage lawyers, and local lawyers, who are themselves drawn from the higher castes and dependent upon the landowning classes for patronage, may be reluctant to take such cases. Most of the states extend some legal aid to Untouchables, although usually not to help them prosecute criminal cases. However, legal aid is so inadequate and so poorly distributed that it is completely insignificant.

Once the case is presented to the court there are formidable obstacles due to the ambiguities and loopholes of the Act. These defects are, as we have seen, difficult to overcome in the High Courts where there is a higher level of skill, more time for preparation and greater insulation from local pressures. It may be surmised that in the magistrates' courts the tendency of the Act's ambiguities and of the difficulties of proof to reduce chances of conviction are amplified. It is, simply, very hard to win one of these cases. Some indication of the difficulty of winning one of these cases is presented in a recent report of the Harijan Sevak Sangh. The Sangh, an organization of mild reformist outlook, prefers uplift and persuasion to litigation. However, its workers are involved in a large proportion of cases registered under the UOA. The Sangh's policy is that only 'when all persuasive measures fail and workers feel themselves at bay and there is no way out they very hesitatingly seek the help of law.' One may assume that the cases with Harijan Sevak Sangh sponsorship should be fairly good candidates for securing convictions. There should not be many false cases or abuses of the UOA for harassment or complainants who misunderstand the law. There would typically be an outside witness, an experienced social worker with previous experience with the law, and there would be some organizational resources to insure tenacity against dilatory tactics. Yet with all of these advantages, of the 476 Harijan Seva Sangh cases that were disposed of in the five-year period 1961–6, only 90 (18.9 per cent) resulted in convictions.

Even if a conviction is secured, the penalties imposed are so light as to have little deterrent effect (and to generate little favourable publicity). 'Ludicrously low' fines of only Rs 3 or 4 are not uncommon. The Elayaperumal Committee gathered information on 22 cases in which convictions were obtained and found that the median penalty was a fine of Rs 10. Judicial tolerance of this level of penalties is suggested by a recent case in the Allahabad High Court, where the Judge upheld the conviction of a couple who had been fined Rs 20 each.

Taking into account the strong prejudices which are still continuing in that part of the State, coupled with the fact that [complainant] had come to the tap after responding to the call of nature and while attempting to take water had touched the utensils of the [defendants] which in the natural course of events must have given them the cause of provocation, I think the ends of justice will be met if the . . . fine imposed on them is reduced to Rs. 10 each.

In the event of an appeal, there is a high probability that a conviction will be reversed.

In view of the trouble and expense, often accompanied by economic hazard and physical danger, the uncertainty of securing a conviction and the tiny effect of such a conviction, it is not difficult to see why few Untouchables would feel that there is anything to be gained by instituting a case under the UOA. It is sometimes suggested that the failure to invoke the Act or to abide by its provisions is due to lack of awareness of its provisions. There would seem to be little merit to' this explanation. The Art has been widely publicized. A national survey, the reliability of which cannot be ascertained, found that 77 per cent of the non-Scheduled Castes and 66 per cent of the Scheduled Castes were aware of the existence of an act outlawing untouchability. In any event, there can be no doubt that vast numbers of Untouchables are aware of the UOA. It is not unawareness of its existence that inhibits its use, but awareness of its hazards and weaknesses. Untouchables are, quite sensibly, more deterred by the formidable difficulties of using the UOA than caste Hindus are deterred by the remote and mild penalties that it threatens them with. Any moderately astute Untouchable, finding himself with the strategic resources to invoke judicial power in a dispute with his caste Hindu neighbours, might, one supposes, be far less attracted to the UOA than to the ordinary criminal law.

In view of the difficulties of using the UOA, it is not surprising that Untouchables who seek extra-local support may not find the Act a promising means of redress and may instead resort directly to executive officers or political figures rather than to the police and the courts. A very crude index of the extent of such recourse is given by the Commissioner for Scheduled Castes and Scheduled Tribes who has noted in his annual report the number of 'complaints' made directly to his office in New Delhi by aggrieved Untouchables. Even as the number of prosecutions under the UOA has declined, the number of complaints to this remote and relatively powerless official has continued to increase.

The Commissioner could offer nothing more than an attempt to sting local and state officials into action, a process subject to delays of years before getting even a satisfactory reply. To many, even this cumbersome and remote procedure seemed a more attractive remedy than the UOA. It is safe to assume that persons sophisticated enough to communicate with a high administrative officer in New Delhi are not ignorant of the existence of legal remedies close at hand.

The question, then, is not why so few cases are brought, but rather why any cases are brought! Probably very few are brought without the intervention of some 'outsider' to the local situation — usually political leaders, social workers or religious reformers. For example, a single organization, the Harijan Sevak Sangh helped to bring 721 cases in the four-year period from 1961–2 to 1964–5. This is somewhere over 40 per cent of the cases registered during that period. This intervention tends to concentrate on certain kinds of facilities. Temple-entry has often, particularly in the early years of anti-disabilities legislation, involved political sponsorship. Social workers tend to concentrate on public accommodations like tea shops, hotels and barbers. Such conspicuous and fixed establishments appear to be most vulnerable to anti-disabilities litigation. The identifiable offender with a fixed establishment and a public licence cannot melt away like the crowd at the village well or return later to intimidate the complainant. Where, however, disabilities are supported by self-help of coherent social groups it is unlikely that they will be deterred by the UOA, even when the presence of transient intervenors increases the probability that it will be invoked. In these settings, the problem is not that the UOA depends on outside intervention for its invocation, but rather that it is such a poor vehicle for getting intervention of the requisite strength and tenacity. It leaves Untouchables vulnerable to itinerant reformers who often cannot deliver the goods that they promise, after Untouchables have risked their well-being.

The conjunction of use of the UOA with outside intervention calls for two cautions in interpreting its impact. First, in many instances the successful use of facilities has a purely ceremonial and symbolic character. Under the wing of a politician, social worker, or government officer, Untouchables enter a temple, draw water from a well or are served in a tea shop with much fanfare. After Harijan Week is over and the outsiders have departed, the situation reverts to normal, perhaps until the next special occasion when the ritual of acceptance is re-enacted. This performance cannot be dismissed as without significance, but it is of a different order than the free access contemplated by the law. Second, assertion of their rights by Untouchables, especially when reinforced by outside intervention, may result not in sharing of common facilities but in provision of separate facilities for Untouchables. This may happen when Untouchables are successful and their entry is met by withdrawal of others. It also may happen through diversion of governmental resources to provide a separate well or meeting hall, etc. Thus, while the Act takes equal access to common facilities as its aim, it often serves to provide leverage for increase of separate facilities for Untouchables.

*   *   *

The impact of anti-disabilities legislation is not to be measured merely by the few cases that are brought. The aim of such laws is not to prosecute offenders, but to promote new patterns of behaviour. Undoubtedly the total effect of the UOA as propaganda, as threat and as leverage for securing external intervention outweighs its direct effect as an instrument for the prosecution of offenders. To some extent, however, its effectiveness as an educational device and as a political weapon depends upon its efficacy as a penal provision. The power of the law to elicit widespread compliance depends in part on its ability to deal with obvious cases of non-compliance. A law that permits effective prosecution of offenders may, of course, not succeed in inducing widespread compliance, but surely a statute which fails to enable effective prosecution is less likely to secure general compliance.

This is especially so in the case of laws like the UOA. For successful legal regulation depends upon what might be called the 'halo-effect' — the general aura of legal efficacy that leads those who are not directly the targets of enforcement activities to obey the law. The 'halo' may be generated by self-interest, approval of the measure, generalized respect for the law-making authority, the momentum of existing social patterns, expectation of enforcement—or a combination of these. In the case of anti-disabilities legislation which runs counter to the sentiments and established behaviour patterns of wide sectors of the public, the 'halo' of efficacy depends very heavily on the expectation of enforcement. By this measure, the UOA appears as an unwieldy and ineffective instrument; its shortcomings as a penal provision vitiate its capacity to secure general compliance.

However, even where there are massive inputs of governmental enforcement, the other components of compliance (approval, self-interest, momentum of existing patterns) must be successfully mobilized to some extent. In the case of the UOA, there is little in the way of these other components, except for generalized respect for government. The law goes counter to perceived self-interest and valued sentiments and deeply ingrained behavioural patterns. Thus the difficulties of securing general compliance in the case of anti-disabilities legislation are so formidable that to measure the Act in these terms sets too high a standard. A more modest approach would be to ask how the Act stands an enabling measure. First, to what extent does it enable Untouchables to improve their position vis-à-vis disabilities? Does it provide useful leverage for those willing to expend their resources and take risks? Second, does the level of enablement sustain itself? Is the Act effective enough so that its use as an enabling measure is cumulative and self-reinforcing?

Perhaps the greatest 'enabling' feature of anti-disabilities legislation is its general symbolic output. This legislation has an effect on the morale and self-image of Untouchables, who perceive government action on their behalf as legitimating their claims to be free of invidious treatment. By providing an authoritative model of public behaviour that they have a right to demand, it educates their aspirations. More generally, such legislation promotes awareness of an era of change in caste relations. Specifically it provides an alternative model of behaviour; it puts the imprimatur of prestigious official authority upon a set of values which are alternative to prevailing practice. Thus it presents a challenge to social life based upon hierarchic caste values. As Henry Orenstein observed in a Maharashtrian village, 'The villagers knew that there was a prestigious alternative to the traditional values of caste.'

Although mere knowledge of the existence of such legislation does not in itself bring about changes, there are instances in which these laws have contributed to 'widespread change in daily behaviour'. For at least some groups the Act provided leverage for securing favourable changes. However, this successful use did not lead to widespread use of the legislation.

We have seen that the UOA followed a 'downward spiral' of ineffectiveness rather than an upward spiral of increasing use. Obviously this downward spiral has happened to a different degree in different places and perhaps there are some places where it has not happened at all. But the overall pattern seems to be that both inputs from the government side (official concern, police initiative, judicial sympathy, etc.) and those from potential complainants and their supporters have declined. As the press and higher officials have diverted their exhortations to the new problems of the day, the police have ceased to make whatever special efforts they made in such cases, which are costly to them in terms of their normal functioning. Offenders become more subtle, lawyers find the soft spots in the Act and courts become less vigilant in protecting the Act against restrictive interpretation. There are fewer convictions and smaller penalties. Potential complainants and their supporters are discouraged from using the Act. The few outsiders who might get involved are heavily overcommitted to many programmes and projects. As instances of the use of the Act become rarer, they seem less matter of course. And insofar as the law's halo of efficacy depends heavily on expectation of enforcement, the law loses its ability to secure compliance from those who are not themselves the direct target of enforcement activity.

*   *   *

Subsequently the deficiencies of the UOA became apparent; and, after years of intermittent discussion and debate, a new improved version of the UOA, renamed the Protection of Civil Rights Act (PCRA), became law in 1976. A number of loopholes were closed and the Act was given a symbolic face-lift. Given a new title to emphasize that the practice of untouchability would be treated as a violation of 'civil rights'[4], the PCRA broadened the definition of 'public place' where untouchability could not be practised[5] and narrowed the range of religious practises and denominations exempt from the operation of the law.[6] The penalties were enhanced to a maximum of Rs 500 and six months imprisonment and a minimum of Rs 100 and one month's imprisonment.[7] Subsequent second and further convictions attracted an increased threshold of punishment.[8] The Act was extended to cover both the 'preaching and practice of untouchability',[9] and insulting Untouchables,[10] including justifying untouchability on philosophical or religious grounds or any tradition of the *caste* system.[11] Victimizing Untouchables or forcing them to engage in dirty occupations traditionally associated with them was illegal.[12] The enforcement provisions were strengthened. All these offences were cognizable and to be tried summarily.[13] The state government could impose a collective fine where the inhabitants of an area were thought to have committed an offence under the Act.[14] Special courts could be set up to hear untouchability cases and special measures created to implement the Act.[15] An Annual Report to Parliament was to constitute a monitoring mechanism.[16] While public servants who willfully neglected to pursue an investigation were deemed to have abetted an offence,[17] there were no legal proceedings against the government for any action done by the latter in good faith.[18]

[4] Section 3, the Untouchability Offences (Amendment and Miscellaneous Provisions) Act 1976 (hereafter UO(A)A) incorporated in Section 1 PCRA.
[5] Section 4 UO(A) A incorporated in Section 2 PCRA.
[6] Section 5 UO(A) A incorporated in Section 3 PCRA.
[7] Sections 5,6,7,8 and 10 UO(A)A incorporated in Sections 3,4,5,6,7,7A PRCRA.
[8] Section 11 UO(A)A (incorporated in Section 11 of the PCRA) which raised the maximum punishment for second and third (or subsequent) convictions to imprisonment for one year and two years respectively; and a *minimum* of six months and one year respectively.
[9] Section 2 UO(A)A incorporated in the long title.
[10] Section 9(b) UO(A)A incorporated in Section 7(1) (d) PCRA.
[11] Section 9(d)ii UO(A)A incorporated in Section 7, Explanation II(ii) PCRA.
[12] Section 10 UO(A)A incorporated in Section 7A PCRA.
[13] Section 17 UO(A)A incorporated in Section 15 PCRA.
[14] Section 13 UO(A) incorporated in Section 10A PCRA.
[15] Section 17 UO(A)A incorporated in Section 15A PCRA.
[16] Section 17 UO(A)A incorporated in Section 15(4) PCRA.
[17] Section 12 UO(A)A incorporated in Seetion 10 Explanation PCRA.
[18] Section 16 UO(A)A incorporated in Section 14A PCRA.

The PCRA came into effect on November 19, 1976. What happened? There are more cases than there were in the period that I studied. As you can see from the last row of Table 1, the number of cases disposed annually in 1977 and 1978 was about double in figure for twenty years earlier. And this probably understates the difference, because apparently more cases under the PCRA are not challaned by the police than was the case under the UOA.

Did the provisions of the PCRA lead to an increased number of prosecutions? Two bits of evidence make us qualify such a conclusion. First, the big increase had occurred earlier in the 1970s, although the records available to me don't make it clear just when it came and how sustained it was. In 1973 there were at least 2456 cases registered under the UOA.[19] In 1976 there were 7047 and 3425 in 1977.[20] Second, each time new anti-disabilities legislation comes into effect, there is a surge of prosecutions which declines after a few years. Since our data in Table 1 is from the first few years of the PCRA, it is too early to conclude what the level of use is going to be.

One thing that is clear is that the PCRA did not lead to more success for the complainants. As you will note from the last row of Table I, convictions are even more difficult to obtain than was earlier the case. The rate of convictions during the two years 1978 and 1979 fell to a mere 18 per cent — very low indeed for any offence in India. The long term trend toward fewer convictions continues. Given a stronger law, how come convictions are harder to get? I don't have any data from which to address this, but let me advance a few speculations. First, I suspect there is resistance from many judges, who are reluctant to impose what strike them as disproportionately severe penalties. I have no direct evidence, but this phenomenon has been reported in several studies of American judges who resisted imposing mandatory sentences in drunk driving cases and drug possession cases. Second, faced with higher penalties and no chance to compound, accused will fight harder by marshalling witnesses, intimidating or suborning prosecution witnesses, and all those familiar tricks. They may be willing to invest more in lawyers services. The Commissioner for Scheduled Castes and Scheduled Tribes suggests that 'many cases are acquitted due to inadequate presentation of evidence by the public prosecutor'.[21] So, by raising the stakes the new act has created more determined and resourceful

---

[19] *Report of the Commissioner of Scheduled Castes and Scheduled Tribes*, 1973–4: 181.

[20] Sixth Lok Sabha Committee on the Welfare of Scheduled Castes and Scheduled Tribes 1978–9 (Thirty First Report) 42.

[21] *Report of the Commissioner of Scheduled Castes and Scheduled Tribes*, 1978–9 I, 20.

opponents, but it has done nothing to upgrade the performance of users of the law. There are provisions for various officers and committees to co-ordinate prosecutions on the government side, but in these early years at any rate, that all remained on paper[22] and I would wager that it has not proceeded much beyond that.

This pattern of use for the PCRA must be examined in the light of the increasing level of violence against the Untouchables. A legal response to that violence would involve prosecutions for offences more serious than PCRA violations. But one wonders why the PCRA should be less successful in those instances where it is invoked.

My tentative hypothesis would be that the new act has widened the disparity in legal services. The accused have the incentive and the means to invest enough to thwart particular applications of the PCRA. The potential beneficiaries of the Act, on the other hand, for the most part lack the incentive or means as individuals to invest in making it effective. This would require a pooling of resources, co-ordination, development of expertise and persistance in following a long-run strategy. Few individual complainants are able to support this kind of effort or to elicit it from the government. It may be that anyone with resources to invest in a legal battle would find a more promising vehicle than the PCRA. Or, it may be that some local experiments in the effective use of this act are hidden in the dismal aggregates.

The Absent Beneficiaries: Non-Litigation and Compensatory Discrimination

Let us turn now to the Constitutional commitment to provide preferential treatment to Scheduled Castes and Scheduled Tribes in respect to education, jobs, housing, loans, and other matters.

Almost all of the cases if the field of preferences have originated in the High Courts or in the Supreme Court under their special jurisdiction to issue writs, orders, and directions for the protection of Fundamental Rights. The popularity of the writ procedure is not difficult to understand: writ petitions are expeditious and they are relatively cheap. The filing fee is almost nominal, compared with the high *ad valorem* court fees which must be paid in advance in an ordinary civil suit or appeal. Lawyers' fees (the costliest item in such litigation) and other costs are lower because the expensive process of presenting evidence is eliminated — the facts being established by affidavit. Starting at the top, there is less likelihood of

[22] Id.

being embroiled in subsequent appeals. Perhaps the most attractive feature of the writ is the direct access that it affords to a high and respected authority.

The procedure for hearing writ petitions differs from court to court. Some High Courts hear all writs by a Division Bench; others hear them by a single judge, with a possibility of appeal to a larger bench. In all courts there is some kind of screening process — an admissions stage in which writ petitions are examined to see whether they should be heard on the merits. A large majority — one knowledgeable lawyer estimated 80 per cent — are eliminated at this stage. Those eliminated include cases involving disputed questions of fact. In writ proceedings no evidence is taken. In theory the facts are determined by the affidavits of the parties. Generally, writs are not admitted if there are serious questions of fact. However, in the compensatory discrimination litigation, I observed that courts vary considerably in their willingness to undertake extended factual inquiries.[23]

In virtually every instance, it is the government or a public body that is the opposite party, since the Fundamental Rights, with few exceptions, run only against the government (in the broad sense, including public corporations, nationalized banks, universities, and other components of the growing public sector) and do not afford protection against the restriction of these rights by private action. These exceptions are not minor ones from the point of view of India's specially vulnerable groups. It is specifically the Fundamental Rights to be free of untouchability (Art. 17) and forced labour (Art. 23) that run against private persons as well as against government. Yet for thirty years there was not, to my knowledge, any attempt to use the writ jurisdiction to protect against privately inflicted denials of these Fundamental Rights!

Courts in India are viewed with a curious ambivalence; they are simultaneously regarded as foundations of justice and cesspools of manipulation. Litigation is widely regarded as infested with dishonesty and corrupt manipulation. But courts, especially the High Courts where this compensatory discrimination litigation takes place, are among the most respected and trusted institutions. Compensatory discrimination litigation tends to display the benign face of the Indian legal process. It is relatively inexpensive and quick; it focuses on genuine disputes — there are few distracting

---

[23] Thus, in *Kajari Saha v. State*, A.I.R. 1976 Cal. 359 at 369, the judge emphasizes that 'this Court in its Constitution Writ Jurisdiction is not a Court of facts. This Court does not sit in appeal over the decision of an authority whose order is challenged in this proceeding.' But cf. *Sardool Singh v. Medical College*, A.I.R. 1970 J. &. K. 45, at 47.

false issues (with the partial exception of some election cases) and there is little occasion for the fabrication of evidence. The other face of Indian litigation peeks through in the controversies over group membership and is fully exposed in some of the election controversies.

Virtually all of the litigation about compensatory discrimination has involved reservations, even though preferences in the form of provisions of facilities, resources, and protections directly affect a much larger number of recipients. A rough indication of this concentration is that in 113 reported litigations about compensatory discrimination which I found, only 12 concerned matters other than reservations — chiefly land acquisition and debt relief. (See Table 2.)

Table 2

Subject-Matter of Reported Litigation on
Compensatory Discrimination
1950–77

Subject-Matter of Preference	Cases		Total
	Involving Reservation Device	Not Involving Reservation	
Education	41	1	42
Government Employment	26	1	27
Elective Office	31	1	32
Other	3	9	12
TOTAL	101	12	113

Source: Galanter (1984: 501).

This litigation has, by and large, been initiated, not by the beneficiaries of protective discrimination, but by those complaining of schemes which affect their interest. Thus of the 113 cases, if we put aside the 32 battles over reserved legislative seats, we find that only 27 of the remaining 81 cases involved claims brought to the courts by would-be beneficiaries and that 54 were initiated by those challenging the existence or operation of the preference.

Table 3

Initiating and Successful Parties in
Compensatory Discrimination Litigation
1950–77

Subject-Matter of Preferential Treatment	Initiating Party Seeks To			
	Overturn, Restrict Curtail the Government's Preference		Enlarge, Restore or Amplify the Coverage of the Government's Preference	
	Successful	Unsuccessful	Successful	Unsuccessful
Education	12	12	4	14
Government Employment	10	10	0	7
Other	5		0	2
TOTAL	27	27	4	23

Source: Galanter (1984: 502).

As tables 2 and 3 suggest, litigation about compensatory discrimination has not been spread evenly over all applications of that principle. It has concentrated heavily on the use of the reservation device. Litigation has focussed on a few kinds of opportunities. There are a few persistent scenarios.

The most common of all litigants in these cases is the prospective student seeking admission to a higher technical or professional college — in most cases a medical college — who is excluded because of what he regards as an unwarranted reservation. Of the 42 education cases, 41 concern reservations in medical and engineering colleges in 11 states. Although some are filed by single petitioners, these petitions often come in large batches, so that these cases involve many hundreds of students. The largest involved 473 petitioners. Although all members of the affected class might eventually benefit from a revision of the government's order, interim relief is available only to the individual litigant. Thus those interested in obtaining admission in the current year would be inclined to file a writ petition of their own, and typically many coattail-riding petitions are filed whenever a college admission is under attack.

All of these admissions cases are concerned with reservations by the state governments rather than the Centre. And virtually all of them are concerned with reservations for the Other Backward Classes, not for the Scheduled Castes and Scheduled Tribes.[24] As table 3 suggests, those student petitioners seeking to overturn or curtail schemes of reservations have enjoyed a high degree of success.[25] (Their 50 per cent rate of success score is higher than that for any class of parties in Gadbois' survey of Supreme Court litigation.)[26] As we have seen, they have succeeded in establishing a variety of limits on the designation of Backward Classes, the extent of reservations, and the standards for the operation of preferences.

Not all of the professional-college litigation is initiated by non-beneficiaries. In 16 of our cases, it was initiated by putative beneficiaries who claimed that government policy wrongly confined the scheme or excluded them from it. Not only were they less frequently successful, winning only 3 of 16 cases, but their successes were of a limited and correctional nature, holding the government to its announced policy rather than broadening that policy.

A less numerous class of litigants are competitors for government jobs who feel that they have been unfairly denied employment or promotion by unfounded or excessive reservations. Actually, there has been surprisingly little litigation in regard to initial recruitment into government service. In twenty-four years there has been only a single reported case concerning initial recruitment into central government service! Of the 27 government employment cases that I have found, 19 were concerned with reservations in promotions (and a promotion element seems to be present in at least 2 others). Of the remaining 8 cases, 5 involved the post of munsif (minor civil judge); that is, the applicants were lawyers, presumably more inclined and better equipped to litigate. (At least one of the promotion cases also involved a legally trained petitioner.) Thus the employment cases may be viewed less as communities fighting over

24 Electoral cases are omitted from this calculation because almost all of them are group membership problems in which rival claimants from the beneficiary group dispute as to the boundaries of that group. It is difficult to identify those claims in terms of the classifications used here.

25 Gadbois 1970: 48, calculated the rates of success for claimants under various Fundamental Rights provisions. Overall, the individual claimant won in 36 of 120 decisions involving the Constitution's equality provisions Arts. 14–18).

26 This was the same as the average for all Fundamental Rights claims. Gadbois also has rates of success for different classes of parties (1970: 46). Only employees and unions with a 55 per cent success rate did better in contests with the government than our medical applicants. For example, political candidates prevailed only 21 per cent of the time in contests with the government.

entrance into government service than as government servants (and lawyers) fighting over opportunities for advancement. The battle focusses not at the top, where reservations have little impact on the chances of others, nor at the bottom, where they are hardly needed (and the others might not easily afford litigation), but in the middle — particularly in the upper ranges of Class III and the jump from Class III to Class II. Naturally, such litigation clusters where reservations are present and most effective — e.g. within the central government in those offices which put into practice the provision for reservation in promotions. Eleven of our 13 central government cases concern the Railways, Posts and Telegraphs and the Central Secretariat. All of the litigation about central schemes concerned reservations for Scheduled Castes and Scheduled Tribes, but 9 of the 14 cases about state schemes concerned reservations for the Other Backward Classes.

Not all of these were attempts by government servants to fend off reservations. In 7 cases, putative beneficiaries took the initiative and attempted to get courts to extend (in 2 cases, reinstate) pre-ferences. None were successful. The government servants, on the other hand, were successful in limiting the extent of reservations and eliminating practices unfavourable to them.[27] In several states they succeeded in knocking out reservations for Other Backward Classes (at least temporarily), and in a number of states they brought about major revisions in state policy, including severe curtailment of the scope of preferential treatment.

A third and much smaller and scattered group of litigants are those licensees, creditors, or property owners who object to particular government schemes for backward classes (debt relief, land acquisi-tion) that interfere with their property or business. There are only 10 of these cases, all concerning various state schemes, of which 5 were successful. No pattern is evident here.

Finally, there are the litigants is electoral cases — usually losing candidates, but in a few cases would-be candidates whose nomination papers were rejected because of dispute about their membership in the preferred group. Three cases involved the question of reservations as compartments or guaranteed minimums; in all, the courts decided in favour of the minimum. All the other cases involved claims of group membership. Again, no pattern is evident in the gross out-comes.

---

[27] Comparison with Gadbois' analysis of Supreme Court litigation provides a rough indication of just how successful they have been. He found that the overall success ratio of government employees litigating with the government was less than one in three (1970: 53); claimants under the various equality provisions of the Constitution prevailed in only 30 per cent of cases (1970: 48). Claims of the type considered here are then a successful sub-group of both these classes.

Except in the election disputes and a few cases involving protections (debt relief, etc.), the government is the opposite party. In many cases, where appointment or admission of others would be affected by the decree, it is necessary for the petitioner to join them as respondents. This is part of the reason that litigation is more popular when the number of others affected is small (e.g. in promotions or competition for post of *munsif*, than for example an annual recruitment of clerks by the state government, which might involve hundreds of potential parties).

Those who invoke the courts are those who not only have something substantial at stake but also have financial or organizational resources to defend or pursue it — e.g., excluded medical students, unpromoted middle-level government employees, defeated candidates. What is striking to me, on reflection, is what is not here and who is not here. The student who fails to get his stipend on time or the disappointed aspirant for a reserved government post are not likely to litigate — much less the hutment dweller with no decent water supply. Plaintiffs do not include the frustrated beneficiaries of programmes that delivered too little too late or the might-have-been beneficiaries of programmes that were mandated by law but failed to get established at all. For example, for many years most states have carried on their books provision for legal aid to Scheduled Castes, programmes so restrictively arranged and so lackadaisically administered that only a tiny fraction of the allocated funds were ever spent. Yet there was never any use of the courts to challenge this or a hundred other failures to implement policies which favoured the have-nots.

Considering the varied and diverse government programmes of compensatory discrimination, how can we account for this clustering of litigation around these few problems? Both electoral and medical-admission situations involve high stakes and claimants with status and resources (including, in the electoral cases, organizational backing). The government service promotion cases involve high stakes for those concerned — government servants in the middle grades — and often have common stakes around which group support can be organized. In all of these cases there is at stake a substantial interest in some life-opportunity for the petitioners. Few, if any, cases are deliberately brought as test cases — that is, for the primary purpose of establishing a principle rather than advancing the interests of an individual litigant.

But these cases do sometimes become a focus of group or communal struggle. A lawyer may develop a reputation as a champion of one particular interest, providing an informal focus for organized effort. In several of these cases we find low-grade government servants supporting expensive litigation that appears to be beyond their individual means, and in fact such litigants are often assisted either by organized associations or by *ad hoc* informal groups of supporters. Occasionally, organized communal associations have taken the initiative in instituting litigation, but more typically they have intervened in cases which have already been filed by interested litigants. For example, in Kerala, in the *Jacob Matthew* case,[28] the Nair Service Society provided financial assistance to those challenging reservations, while the S.N.D.P.Y., an Ezhuva association, hired a leading lawyer to argue in favour of the reservations. A somewhat different pattern of intervention is found in *Balaji's* case,[29] where several communal associations formally intervened, attacking the reservation on the grounds that they have been wrongfully excluded from the Backward Classes. Although such formal intervention is relatively rare in Indian litigation, it seems to be slightly less rare in these compensatory discrimination cases.

None of this intervention has been by 'universalistic' associations remote from the immediate parties or by disinterested *amici curiae* who seek the establishment of an abstract principle of law. There are, as we saw above, instances of participation by pre-existing and multi-purpose communal organizations, built around traditional ties. There are also instances of participation by organizations of a more 'modern' type (e.g. the associations of government servants of a particular grade). An intermediate instance is provided by composite communal interest groups which are bound together specifically by an interest in the reservation or other schemes of preference—for example, the Central Government Scheduled Castes and Scheduled Tribes Employees Welfare Association, which formally intervened in the *C.A. Rajendran*[30] case; or the statewide association of Backward Classes, which took the initiative in attempting to restore the Andhra Pradesh reservations by litigation;or the Kerala Harijan Samskarika Kshema Samity, at whose urging the Kerala government issued the extensions that led to the *Thomas*[31] litigation.

Some associations of beneficiaries of preferences are little more than paper organizations or mere personal organs. But others, especially associations of Scheduled Castes and Tribes government employees, are enduring organizations with a wide membership base — although the central government has adamantly refused to recognize them. Such associations have been active in making representations in individual grievances and pressing collective de-

[28] *State of Kerala v. Jacob Mathew, AIR 1964 Kerala 316.*
[29] *M.R. Balaji v. State of Mysore, AIR 1963 S.C. 649.*
[30] *C.A. Rajendran v. Union of India, AIR 1968 S.C. 507.*
[31] *State of Kerala v. N.M. Thomas, AIR 1976 S.C. 490.*

mands by lobbying and demonstrations, but have only infrequently resorted to litigation. For example, in early 1978 the All-India Scheduled Castes and Scheduled Tribes Railway Emplyees Association was supporting a case to disqualify a non-Scheduled Caste person who had obtained a railway job with the help of a false caste certificate. Although they have a host of grievances about the situation of Scheduled Caste and Tribe employees — ranging from administration of reservations, interviews, transfers, quarters, to non-recognition of their organization to the excessive complexity and ambiguity of the orders — limitations of financial resources and legal imagination left them unable to visualize the courts as a forum for securing redress.

The propensity to litigate these questions depends on a number of factors. One is clearly the feeling of being deprived of a scarce opportunity. Another is the question of expense: the cost of obtaining that opportunity by litigation must appear to be less than getting it by some other means. It is estimated that in Mysore, where the alternative to a state medical or engineering college is a private college with a high capitation fee, litigation was regarded as a much less expensive means of obtaining admission than payment of the fees in a private college. The tendency to have very large number of writ petitions whenever a reservation is challenged can be explained in part by decreasing marginal cost. Once a writ petition is filed, additional petitioners can usually arrange to be parties at much lower cost. Lawyers typically reduce their fees when there are a number of writ petitioners and arrange to process their petitions on a mass basis. Also, junior lawyers will file writ petitions along the line of one already filed by a senior, for a very much lower fee. Again, the publicity attendant upon such litigation tends to draw more petitioners in, and the success of one petition encourages others to try this mode of obtaining resources.

For the propensity to litigate depends, too, on the expectation of success. Let us take the negative case. One of the striking things about the nationwide pattern of litigation is that after the 1951 amendment of the Constitution there was no reported litigation about reservations in Madras until 1967. Sizable reservations for the Backward Classes in both employment and professional-college admissions exist in Madras; the Backward Class lists were formulated on the 'caste basis' and thus were arguably open to attack on the basis of the 1962 *Balaji* case.[32] Yet no litigants came forward to challenge the Madras scheme. In part this may be due to the fact that the Madras reservations are more modest than those that formerly obtained in Mysore, and thus the 'advanced communities'

32 *Supra* n. 29.

do have substantial opportunities to enjoy the resources involved. Lack of contest in regard to government employment might be attributed in part to the shift in ambition among Madras Brahmins away from government service towards the private sector. But this would not explain the absence of contention about reservations in professional and technical colleges, which are the prime preparation for advancement in non-governmental employment. It appears, then, that the communities not on the backward list regarded the reservations as good as it was possible to get under circumstances. There was a feeling that litigation would not be effective to prevent some kind of preference for the 'Backward Classes'. As one lawyer put it, 'If they get rid of it in one way, the majority communities would spare no efforts to introduce it in another way' — a remark that must be placed in the context of the existence of selection communities as well as of reservations. That is, it was felt that the courts could not provide ultimately effective relief. This might explain the lack of organized backing, since no widespread and permanent benefit could be anticipated, but it does not explain the lack of individual litigants, since these might gain some tangible advantage even though the excluded communities did not secure any lasting benefit.

Litigation, it seems, requires not only parties who would benefit but an example or model of success and an incident to serve as a catalyst. That is, present discomforts, possible gains, and acceptable costs and risks of litigation must be made vivid and palpable.[33] In Madras the model was provided by *Balaji* and the spate of similar litigation in other southern states. The catalyst was provided in 1967 by the peculiar circumstance that the marks list in the medical school competition had been destroyed. This was widely publicized, and as one of the lawyers described it, a large number of disappointed applicants, hoping that the entire selection might be struck down, filed petitions. The paucity of 'affirmative' litigation by the beneficiaries of preferences may be explained in part by lack of resources. But it also reflects the absence of such 'models of success' — there has been virtually no successful affirmative litigation — and of 'catalytic events' to trigger the attempt.[34]

[33] In Mysore, the catalytic event was provided in 1958 by the issuance of a new Backward Classes Order after the State's 1956 reorganization. The new Order impinged on groups outside old Mysore who were not previously covered. In Andhra, the event was the conjunction of *Balaji's* case with the State Government's expansion of preferences. In Kerala and elsewhere it was the decision in *Balaji's* case with its promise of successful challenge of existing policy.

[34] The two affirmative cases with a broader thrust were triggered by 'catalytic events' — the withdrawal (in *Dasa Rayudu v. Andhra Pradesh Public Service Commission*, A.I.R. 1967 A.P. 353) or curtailment (in *C.A. Rajendran v. Union of India, supra* n. 30) of previously existing reserations.

Summing up, we may surmise that the gross effect of litigation on the compensatory discrimination policy has been to curtail and confine it. Those who have attacked compensatory discrimination schemes in court have compiled a remarkable record of success, while those seeking to extend compensatory discrimination have been less successful. One way of expressing this gross effect is to compare the level of preferential treatment before and after judicial intervention.

Thus, we see that after litigation the level of preference was about the same as that previously propounded or administered by the government in 49 of the 80 cases; it was less in 27 cases and more in only 4 cases. This scoring probably understates the curtailing effect of the judiciary. Courts, reluctant to strike down government schemes, will uphold arrangements while warning of the necessity of correcting unconstitutional practices, thus coaching the government to introduce changes favourable to the petitioners.

Looking back over the pattern of litigation, we see that the central focus of the judicial encounter with the compensatory discrimination policy has been not the general outline and performance of that policy, but rather the way in which government's attempts at re-distributing certain highly valued scarce opportunities have impinged on the chances of other competitors with the resources and incentive to raise such challenges. The initiative has lain largely with these competitors. Thus the beneficiaries of compensatory discrimination have been at a disadvantage in the development of the law; the other competitors have occupied the strategic heights in the litigation battle. The working out of principles is inevitably shaped by the contingencies of the factual situations which formed the context in which they are formulated. It is the most expensive schemes and the most flagrant abuses which have come to court. The first round of litigation (the 1951 cases) came up from the old Madras system of communal quotas rather than from a scheme specifically designed to advance the Backward Classes. Subsequently, it was the Mysore situation, which was little removed from communal quotas, that produced litigation and Supreme Court law. Debate on Backward Classes policy has been strongly affected by the fact that the crucial and determinative litigation has come out of Mysore, where reservations for Backward Classes were the most flagrant and expansive. Had some state with a more modest scheme of preferences been the source of the important litigation, the law might have taken a somewhat different shape.

More generally, it should be noted that litigation (and public debate) has tended to centre almost exclusively around schemes for Other Backward Classes rather than the more extensive schemes for Scheduled Castes and Scheduled Tribes. Thus the law of Articles

15(4) and 16(4) is being made largely in reference to situations involving the Backward Classes. The courts have erected restrictions on these articles to counter excessive and unwarranted preferences for the Backward Classes. But these redound to restrict what the Government may do for the Scheduled Castes and Scheduled Tribes. Similarly, in public debate and scholarly commentary, the well-publicized abuses of the Backward Class category easily attach to the resentment against the preferences for the Scheduled Castes — a confusion compounded by the widespread inability of the public to distinguish these groups (i.e. to distinguish the inclusive and narrower senses of 'backward classes').

Thus Scheduled Castes and Tribes may be indirectly 'misrepresented' by spokesmen for the Other Backward Classes. More generally, the configuration of parties in this litigation may fail to make it a forum in which all the relevant interests can be heard. *Chitralekha*[35] provides an example. In the High Court proceeding, Judge Hegde upheld the State scheme, while finding serious deficiencies in its failure to do enough for those who should have been beneficiaries. The appeal was taken by the petitioners, who wanted the State order struck down — i.e. their criticisms were precisely in the opposite direction from Judge Hegde's. The State on the other hand wanted to maintain its Order. So both parties were hostile to his observations about the deficiencies of the State's scheme: There was no one to press the interests of those he felt wrongly excluded. The absence of adversary control may be responsible for the conceptual muddle that emerged. Although this is an extreme case, it is not difficult to believe that there may be a similar tendency in cases where the contending parties are the state and non-backward objectors to the state's schemes for the backward.

While providing a forum responsive to the 'too much' and 'wrong thing' criticism of compensatory discrimination policy, the courts have not been responsive to the 'not enough' or 'not the right thing' criticisms. They have done little directly to offer remedies for the deficiencies of implementation of existing schemes — to see that reservations are filled, scholarships are distributed on time, programmes are well suited to circumstances and pursued with sufficient energy, resources, and attention. In part, this is due to the posture of the Constitution, which provides no explicit authorization for affirmative judicial action. Claims under Directive Principles are not justiciable — so there is no justiciable claim that preferences be awarded. But courts, abetted by lack of inventiveness at the bar, have extended this passivity to claims that those preferences which are awarded be implemented effectively.

[35] *Chitralekha v. State of Mysore, AIR* 1964 S.C. 1823.

Conclusion

What do we learn from these two accounts? In each, we have seen a set of beneficiaries (or potential beneficiaries) who failed to get as much benefit as the law seemed to promise. In each case, the laws involved have deficiencies and the courts may indeed harbour some attitudes unfavourable to their claims. But I submit that the limited use of legal opportunities also points to another factor — the absence of suitable legal services. By this I mean more than the availability of skilled lawyers. I mean the presence of lawyers (or others) who are capable of providing such clients with continuing 'strategic' legal services. This means a legal services operation that is sufficiently continuous and enduring to make possible the development of expertise in the affairs of the client group. It means the development of a proactive strategy for using legal opportunities, both in courts and in other arenas. It means realizing economies of scale by aggregating claims that would otherwise be unmanageably small. It means persistance and inventiveness in pursuing long range goals of client groups. But such legal services cannot be provided by lawyers alone. They imply client groups that are organized and capable of using more aggressive, more continuous and more inventive lawyering.

V
Judges, Lawyers and Social Reform

10

Hinduism, Secularism, and the Indian Judiciary *

One of the most striking developments in independent India is the successful emergence of an avowedly secular State encompassing the bulk of the world's Hindus.[1] There is disagreement about what this secular State implies — whether it implies a severe aloofness from religion, a benign impartiality toward religion, a corrective oversight of it, or a fond and equal indulgence of all religions. But there seems to be a general agreement that public life is not to be guided by religious doctrines or institutions. There is a widespread commitment to a larger secular order of public life within which religions enjoy freedom, respect, and perhaps support but do not command obedience or provide goals for policy. At the same time Hinduism is undergoing a vast reformulation and transformation. In fact, these processes are closely interlinked. The nature of the emerging secular order is dependent upon prevalent conceptions of religion, and the reformulation of religion is powerfully affected by secular institutions and ideas.

This paper discusses some aspects of India's legal system as a link or hinge between the secular public order and religion. The modern legal system has transformed the way in which the interests and concerns of the component groups within Indian society are accommodated and find expression.[2] In traditional India, many groups (castes, guilds, villages, sects) enjoyed a broad sphere of legal autonomy, and where disputes involving them came before public authorities, the latter were obliged to apply the rules of that group. That is, groups generated and carried their own law and enjoyed some assurance that it would be applied to them. In modern India, we find a new dispensation — the component groups within society have lost their former autonomy and isolation. Now groups find expression by influence on the making of general rules made at centres of power — by representation and influence in the political sphere, by putting forth claims in terms of general rules applicable to the whole society. The legal system, then, provides a forum in

* Reprinted from *Philosophy East & West*, Vol. 21 No.4 (October 1971).

[1] Among the literature on the legal and theoretical aspects of secularism in India are: D. Smith (1963); D. Smith, ed. (1966); Luthera (1964); Galanter (1965); Sharma, ed. (1966); Subrahmanian (1966); Seminar (1965); Derrett (1968); Sinha, ed. (1968).

[2] Galanter (1966b); Galanter (chap. 2).

which the aspirations of India's governing modernized western-educated elite confront the ambitions and concerns of the component groups in Indian society. In this forum the law as a living tradition of normative learning encounters and monitors other traditions of prescriptive learning and normative practice. We shall be concerned with India's secularism both as a programme for the relation of law to religion and as an instance of the general problem of the relation between law and other traditions of normative learning.

Temple Entry and the Boundaries of Hinduism

We shall approach these general matters by considering in some detail in the 1966 case of *Sastri Yagnapurushdasji v. Muldas Bhundardas Vaishya*,[3] in which the Supreme Court of India attempted to define the nature and boundaries of Hinduism. In it we find the interplay of these various themes — secularism and Hinduism, traditional groupings and Westernized elite, parochial concerns and national aspirations, legal doctrine and religious learning — presented dramatically and not without a measure of ironic reversal and comic byplay.

Whatever the tenor of its encounter with religion, the law cannot entirely avoid questions of religious identity. Even in a secular State, civil authorities, including the courts, find themselves faced with the necessity of ascertaining what is religious. For example, courts in the United States find themselves having to decide what is religious for the purpose of avoiding forbidden 'establishments', for the purpose of determining the scope of religious freedom, for the purpose of administering statutory dispensations like tax exemptions and conscientious objection to military service. A second kind of problem of religious identity may also arise — whether a particular person or organization in fact belongs to a particular religion. This arises more rarely in the United States, but it does appear in church-property disputes, in adoption cases, and in carrying out testamentary purposes.

In India, the courts face quite as many problems in ascertaining religion in general and a great deal more in the way of fixing particular religious identities. This is because the Indian Constitution and legal system embody a different relation of law to religion. Indian law permits application of different bodies of family law on religious lines; permits public laws, like those of religious trusts, to be dif-

---

[3] A.I.R. 1966 S.C. 1119. (Decided 14 Jan. 1966 by a bench of five, consisting of P. B. Gajendragadkar, C.J.; K. N. Wanchoo; M. Hidayatullah; V. Ramaswami; and P. Satyanarayana Raju. JJ.) The high court judgment from which appeal was taken appeared in I.L.R. 1960 Bom. 467 [=LXI Bom. L. Rep. 700 (1968)].

ferentiated according to religion; and permits protective or compensatory discrimination in favour of disadvantaged groups, and these may sometimes be determined in part by religion. The penal law in India is extraordinarily solicitous of religious sensibilities and undertakes to protect them from offence. The electoral law attempts to abolish religious appeals in campaigning. In all these areas courts must determine the nature and boundaries of a particular religion. But beyond this the State is empowered generally to use its broad regulative powers to bring about reforms in religious institutions and practices, and this power is wider in respect to Hinduism than it is in respect to other religions. The State is empowered to assure that Untouchables have full access to Hindu religious institutions.[4]

The issue of temple entry historically came to symbolize the question of the inclusion of Untouchables within the Hindu community.[5] The vexed question of whether Untouchables were within the Hindu fold was generally settled in the course of the independence movement and the whole solution is crystallized in the Constitution itself which permits the State to enforce public religious acceptance of Untouchables upon other Hindus. At one time it was widely assumed that religious acceptance was the key to dissolving all the disabilities of the Untouchables. However in recent years attention has shifted to provision of more tangible advancements; the matter is generally one of low priority for Untouchables and reformers. Nevertheless temple-entry measures are of continuing symbolic importance, and conceptually they have remained troublesome to commentators on Indian secularism. Some critics have found the assertion of public control over temples a flaw in the pattern of Indian secularism either because it violates the integrity of religious premises or interferes in the internal affairs of religious bodies.[6] Other commentators have found in temple entry reason to reject as inappropriate for India this 'separation-of-powers' model of secularism with its underlying distinction between the intrinsically religious and

---

4 See the qualifications in Articles 25 and 26 of the Constitution of India, especially Art. 25(2)(b).

5 Galanter (1964b). For a sampling of the broad opposition to temple entry see Sorabji (1933); Aiyar (1965: 107ff); Durkal (1941: 2, 99); Iyengar (1935); Krishnamacharya (1930). Restrictions on temple entry are critically analysed in Pillai (1933). Legislation against these restrictions is reviewed in Venkataraman (1946). The changing (and eventually pro temple-entry laws) views of Gandhi can be traced in Gandhi (1954).

6 D. Smith (1966: 241–3). See Luthera (1964: 108) where he argues that the eradication of untouchability 'by the state as far as religious institutions are concerned is not consistent with the concept of the secular state. . . . In a secular state the nature of the relations between the Church and its believers is to be settled between themselves.'

secular, sacred and profane, church and state.[7]

Our case involves a group called the Swaminarayana Sampradaya, that is, followers of Swami Narayanan (1780–1830).[8] I will refer to them simply as Satsangis. They are a puritanical Vaishnavite sect, with several hundred thousand followers, mainly in Gujarat, founded in the 1820s. The sect, it appears, has been conventional and conservative in its views and practices regarding caste distinctions. In Manilal C. Parekh's sympathetic account of the Swaminarayana movement we learn that: 'Even the Untouchables were not excluded from the Satsang. It is true that these were not admitted to the inner part of the temples, but they were made disciples and in one or two places they built temples of their own.'[9] Parekh concludes that in regard to caste distinctions: 'Swami Narayana acted as one who was in no sense a social revolutionary. In spite of the fact that he, in common with all the great teachers of the Bhagwat Dharma, opened the way of salvation to all people including the Untouchabless and the non-Hindus, he would do nothing which would even distantly suggest that he was out to subvert the Hindu social order.'[10]

In 1947, the province of Bombay passed a temple-entry act, which provided that 'every temple shall be open to Harijans [that is, Untouchables] for worship in the same manner and to the same extent as to any member of the Hindu community or any section thereof. . . .' Early in 1948 some followers of the sect, anticipating an attempt by local Harijans to enter their temples, filed a suit alleging that their temples were not covered by the 1947 Act. During the pendency of the suit, the Constitution came into force in 1950 and the plaint was amended to urge that the act was violative of the constitutional guarantee of freedom of religion, which gave every

---

[7] Blackshield (1966: 54) observes that 'the very fact that temple-entry authorization is inconsistent with the 'separation of powers' model of secularism ought to suggest that the 'separation of powers' model must be wrong." He concludes that secularism in India must include active state determination of the boundaries of religion. Derrett (1968: 510 ff.) suggests that such state regulation of religious expression and caste relations in the public sphere is a role fully supported by Hindu tradition. Unlike separationists who subsume Hindu religious premises under the private sphere of 'church' as opposed to state, he sees these premises impressed with a public quality and subject to public regulation. Thus he finds that temple-entry laws 'may interfere with the continuance of ancient customs of a religious character, but no human beings are deprived of their religion, for they can decamp and set up private institutions where they can worship in their own way. . . . Those whose religious scruples include a belief in Untouchability have been prohibited from practising this: their remedy, if they continue in this belief, is to stay at home and to use only private facilities from which they can still exclude untouchables.'

[8] On this movement, see Parekh (1936); Monier-Williams (1891).

[9] Parekh (1936: 126).

[10] Id., 282.

denomination the power to manage its own affairs in matters of religion.[11] Furthermore, the Satsangis contended, their temples were not within the ambit of the act since they were a distinct and separate religious sect unconnected with the Hindu religion. They asserted that although they were socially and culturally Hindus, religiously they were distinct.

There is no single accepted legal test of who is a Hindu.[12] 'Hinduism' as used officially is an equivocal term with shifting denotation. Sometimes it is used in an inclusive sense which embraces all the heterodox communions—Jainism, Buddhism, Sikhism, etc.[13] At other times it is used in varying narrower senses meaning followers of more or less Vedic and Brahmanical communions.[14] The draftsmen of temple-entry laws have aimed at inclusiveness.[15]

The trial court concluded that the Satsangis formed a section of the Hindu community. But the court found that it was not established that their temples were used by non-Satsangis and granted them an injunction.[16] This was in 1951. Because of appeals on interlocutory orders the State's appeal on the merits did not reach the High Court until 1957. In the meantime the Bombay Act had been

11 Article 26 (see Appendix).

12 The most prevalent legal test of a Hindu is that developed for the purpose of applying appropriate personal law, that is, deciding who is to have Hindu law applied to him. Historically, this definition was neither a measure of religious belief nor a description of social behaviour; rather, it was a civil status describing everyone subjected to Hindu law in the areas reserved for personal law. Heterodox practice, lack of belief, active support of non-Hindu religious groups, expulsion from a group within Hinduism—none of these removed one from the Hindu category. The individual could venture as far as he wished over any doctrinal or behavioural borders; the gates would not shut behind him so long as he did not explicitly adhere to another religion. This negative definition prevails today. A few years ago the Supreme Court had to decide on the validity of consent to an adoption by one who disavowed belief in the religious efficacy of adoption, in Hindu rituals and scriptures, in *ātman* and salvation. The court found that 'the fact that he does not believe in such things does not make him any less a Hindu. . . . He was born a Hindu and continues to be one until he takes to another religion . . . whatever may be his personal predilections or views on Hindu religion and its rituals.' *Chandrasekhara Mudaliar v. Kulandaivelu Mudaliar*, A.I.R. 1963 S.C. 185, 200. See chap. 6.

13 See, for example, the explanation to Section 3 of the Untouchability (Offences) Act (see Appendix); Hindu Marriage Act, Sec. 2.

14 See, for example, the use of Hindu in the narrower sense (excluding Jains, Buddhists, etc.) in the President's Order defining Scheduled Castes, *Punjabrao v. Meshram*, A.I.R. 1965 S.C. 1179. The ambiguity is analogous to that of the term 'Protestant' in contemporary American usage.

15 See, for example, the Central and Bombay laws in the Appendix. On the shortcomings of these attempts, see chap. 6.

16 That is, on the ground that the Bombay Temple-Entry Act only extended rights to Untouchables who were members of the same denomination, an interpretation vindicated by *Bhaichand Tarachand v. Bombay*, A.I.R. 1952. Bom. 233.

supplanted by the Central Untouchability (Offences) Act of 1955.[17]
The Untouchability (Offences) Act had proved to have an unex-
pected flaw. As the judges read it, it seemed to preserve certain
denominational prerogatives so that Untouchables only gained rights
of entry if they were members of the Hindu denomination or sect
which managed the temple in question.[18] To obviate this difficulty,
the state of Bombay in 1956 passed supplementary legislation, which
provided that every temple open to any class of Hindus should be
open to all Hindus.[19] The High Court, now reviewing the case on
the merits in 1958, found that the new act was constitutional and
that it applied to the Satsangi temples.

When the case reached the Indian Supreme Court in 1966, the
principal argument put forward by the Satsangis was that they are
'a religion distinct and separate from the Hindu religion' and
consequently outside the scope of the Bombay Act.[20] The immediate
question for decision then was whether the Satsangis were Hindus
for purposes of the application of the Bombay Act. Rather than
undertaking a narrow and technical inquiry into the scope of the
temple-entry power and its exercise, the Supreme Court opted to
address a much broader question: 'We must inevitably enquire
what are the distinctive features of the Hindu religion.'[21] The court
set off to ascertain the nature of Hinduism.

In the course of this inquiry, the court propounded three different
views of Hinduism; or, more accurately, it considered Hinduism
from three quite different standpoints. Without acknowledgment it
shifted from one to the other in the course of its opinion.

The first standpoint is a descriptive one, which sees Hinduism
as a complex, indefinable inclusive aggregation of ways of life.

> We find it difficult, if not impossible, to define Hindu religion or even
> adequately describe it. Unlike other religions in the world, the Hindu
> religion does not claim any one prophet; it does not worship any one God;
> it does not subscribe to any one dogma; it does not believe in any one
> philosophic concept; it does not follow any one set of religious rites or
> performances; in fact, it does not appear to satisfy the narrow traditional
> [for traditional, read Western] features of any religion or creed. It may
> broadly be described as a way of life and nothing more.[22]

[17] Act XXII of 1955. This act was passed in exercise of exclusive central competence in penal measures against untouchability. On the extent to which this field
was preempted from state action, see Galanter (1961a).

[18] *State v. Puranchand*, A.I.R. 1958 M.P. 352. See Galanter (1964b).

[19] Bombay Hindu Places of Public Worship (Entry Authorization) Act, 1956 (Act
31 of 1956). Similar legislation has been passed in several other states.

[20] A.I.R. 1966 S.C. at 1127.

[21] Id., at 1127.

[22] Id., at 1128.

There is much more of the same: ... under Hindu philosophy, there is no scope of excommunicating any notion or principle as heretical and rejecting it as such.'[23] 'Unlike other religions and religious creeds, Hindu religion is not tied to any definite set of philosophic concepts as such.'[24]

How then, if no boundaries can be established, can it be ascertained whether a group is within Hinduism or not? In order to do so the court takes a second standpoint, which we might call the analytic. Beneath the diversity of Hindu philosophy, it finds, 'lie certain broad concepts which can be treated as basic'.[25] These include 'the acceptance of the Veda as the highest authority in religious and philosophical matters', the 'great world rhythm', and 'rebirth and preexistence'.[26] Having discerned the glimmerings of unity, the court goes on to ask, 'what, according to this religion, is the ultimate goal of humanity?'[27] And it has little difficulty in answering: 'It is release and freedom from the unceasing cycle of births and rebirths ... which is the ultimate aim of Hindu religion and philosophy. . . .'[28] On the means to attain this end 'there is great divergence of views . . .'[29] but '. . . all are agreed about the ultimate goal. Therefore it would be inappropriate to apply the traditional [again, read Western] tests in determining the extent of the jurisdiction of the Hindu religion. It can safely be described as a way of life based on certain basic concepts to which we have already referred.'[30] The court adds yet another 'working formula' borrowed from B. G. Tilak which it says 'brings out succinctly the broad distinctive features of Hindu religion'[31] — 'the acceptance of the Vedas with reverence, recognition of the fact that the means or ways of salvation are diverse; realization of the truth that the number of gods to be worshipped is large.'[32]

Having thus defined the undefinable, the court then proceeds to see whether or not the Satsangis are within the 'Hindu brotherhood'.[33] It finds their claim to be 'a distinct and separate religion different from the Hindu religion is entirely misconceived'.[34] First,

[23] Id., at 1129.
[24] Id., at 1130.
[25] Id., at 1130.
[26] Id. The court adapts this formulation from Radhakrishnan (1923: 26–27).
[27] A.I.R. 1966 S.C. at 1130.
[28] Id.
[29] Id.
[30] Id., at 1131.
[31] Id.
[32] Id. This is attributed to Tilak's *Gitarahasya*, but I have not been able to locate
[33] Id., at 1131.
[34] Id., at 1134.

the sect's founder was a Hindu saint. 'Acceptance of the Vedas with reverence, recognition of the fact that the path of Bhakti or devotion leads to Moksha, and insistence on devotion to Lord Krishna unambiguously and unequivocally proclaim that Swaminarayan was a Hindu saint . . . who wanted to restore the Hindu religion to its original glory and purity.'[35] The Sastangis had put forth a catalogue of traits that purportedly distinguished them from Hindus —membership by initiation rather than birth, openness to members of all religions, and the worship of Swaminarayan himself. The court, unimpressed by this catalogue, finds the sect is like others which grew out of the activities of reforming saints who 'basically subscribed to the fundamental notions of the Hindu religion and the Hindu philosophy'.[36]

If the Satsangis are, then, Hindus and within the ambit of the temple-entry acts, a further question arises: namely, whether their constitutional freedom as a denomination to manage their own affairs in matters of religion (Article 26(b)is violated by the temple-entry laws. Is admission to temples 'a matter of religion'? A few years earlier, another bench of the Supreme Court was faced with a similar conflict between temple-entry powers of the State and a denomination's right to control admission to its premises.[37] On that occasion, control over participation in temple service was indeed a matter of religion and therefore deserving of some constitutional protection. 'Under the ceremonial law pertaining to temples, who are entitled to enter them for worship and where they are entitled to stand and worship and how the worship is to be conducted are all matters of religion. . .'.[38] While upholding the primacy of the State's temple-entry power, the earlier decision affirmed that what were matters of religion depended on the historically grounded self-estimate of the group in question.

The court in the present case does not treat this issue explicitly;

---

[35] Id.

[36] Id., at 1134. The thinness of these contentions is apparent. The Satsangis could muster little evidence of departure from the mainstream of Hindu beliefs, symbols, and practices. A perusal of the *Sikṣā Patri* [Epistle of Precepts], which Monier-Williams refers to as 'their code of instruction,' reveals no such departures. Rather it points to their acceptance of the *varṇa* order, the special spiritual functions of Brahmins, the Vedas and Purāṇas, the Hindu pantheon, sundry philosophic concepts and the Hindu legal tradition. Thus:

> 103. Dharma is the good practice which is enjoined both by the Shrutis (the Vedas) and by the Smritis (the body of Law as delievered originally by Manu and other inspired legislators). . . .

> 97. When a question in regard to usage, practice and penance is to be determined, my followers should refer to the Yagnavalkya Smriti with its commentary called the Mitakshara. (Reprinted in Parekh 1936: 336–7.)

[37] *Sri Venkataramana Devaru v. State of Mysore*, A.I.R. 1958 S.C. 255.

[38] Id., at 265.

rather, it extends its disquisition on Hinduism to get around it and to undermine the earlier view. The Satsangis' apprehension about the pollution of their temple, says the court, 'is founded on superstition, ignorance and complete misunderstanding of the true teachings of the Hindu religion and of the real significance of the tenets and philosophy taught by Swaminarayan himself.'[39] The court is no longer speaking analytically of 'basic tenets' but of 'true teachings' as opposed to superstition and of 'real significance' as opposed to misunderstanding. This is a far cry from the earlier notion that Hinduism is so all-embracing that no principle is heretical. Having moved from the descriptive to the analytical, the court has now shifted its stance once again, this time to a normative perspective from which the different strands within Hinduism can be evaluated and judged. The court does not attempt explicitly to reconcile its divergent standpoints. How can Hinduism be at one moment so diffuse and inclusive as to defy description and at the next readily analysable into fundamentals? And what do the fundamentals have to do with the court's evaluation? It is nowhere suggested that the caste views of the Satsangis violate the 'basic concepts' regarding the Vedas, rebirth, multiplicity of means to salvation, etc. Rather than apply its basic concepts, the court propounds a view of Hinduism, its unity and continuity, that enables it to discern what is authentic and essential and what is not. Underneath the divergent teachings of saints and reformers, 'there is a kind of subtle indescribable unity which keeps them within the sweep of the broad and progressive Hindu religion'.[40] These saints and reformers have 'by their teachings . . . contributed to make Hindu religion ever alive, youthful and vigorous'.[41] Indeed, 'as a result of the teachings of Ramakrishna and Vivekananda, Hindu religion flowered into its most attractive, progressive and dynamic form'.[42]

In the eyes of the court, the true teachings of Hinduism are those which make it attractive, progressive, dynamic, alive, youthful, and vigorous. But how is the living stream of progressive and dynamic Hinduism to be separated from the dross of superstition and ignorance? It is clear that not all Hindus agree on which is which. In describing Satsangi beliefs, the court notes that their scriptures forbid ceremonial intercourse with low caste people.[43] Apparently

---

39 A.I.R. 1966 at 1135.

40 Id., at 1130.

41 Id., at 1134.

42 Id., at 1130.

43 The *Siksa Patri* upholds caste distinctions at a number of places. Participation is graded by *varna* and special provision is made for 'Untouchables,' for example:
  19. Nor shall anyone eat or drink except in Jagannathpuri from a person of a caste lower than one's own. . . .

at least some present-day Satsangis place a different value on this part of their tradition than does the court. The court does not explain the source of its mandate to overrule them on this question and to substitute its own views of the true teachings of Hinduism. It may seem somewhat uncharitable to chastise people for their complete misunderstanding of the true teachings of Hinduism after rejecting their contention that they are not Hindus.

Exactly what is the court saying about Hinduism? It seems to imply that regulation of temple worship is not a 'matter of religion' within the constitutional protection, and furthermore that invidious caste distinctions are not a part of Hinduism deserving legal recognition. These implications are at variance with the general tenor of the Supreme Court's earlier encounters with caste and Hinduism. Although generally unfriendly to manifestations of caste in public life,[44] the court has been rather sanguine in its estimation of the caste system and its place within Hinduism.[45] As recently as fifteen months before the Satsangi decision, another bench of the court observed that the term 'Hindu' (in the narrower sense) referred to 'the orthodox Hindu religion which recognizes castes and contains injunctions based on caste distinctions.'[46] Is the court now saying that invidious caste distinctions are not a part of Hinduism? Presumably these distinctions, at least in relation to temple attendance, are based upon notions of differential purity and ritual pollution. The court indicates in its opinion that it is willing to countenance these notions in at least some contexts.[47] How can it determine where distinctions based on ritual purity (for example, that between worshipper and officiating priest) are legitimately religious — and

---

24. None shall give up the performance of the duties that are imposed upon the class or the religious order to which he belongs, nor shall he adopt the duties that are enjoined on others. . . .

41. Those twice-born persons that have received initiation into the worship of Krishna from a proper spiritual preceptor shall always wear on their neck a double-necklace . . . and they shall make an upright mark on their forehead. . . .

44. Those pure Shudras, who are devotees of Krishna, while practicing their own peculiar duties, shall, like the twice-born wear the double neklace and make the vertical mark on the forehead.

45. The Shudras who are lower still have always wear a double necklace like the others and shall make the round mark on their forehead while eschewing the upright mark.

91. The twice-born should perform at the proper seasons and according to their means, the twelve sacraments, the six daily duties, and the Shradha offerings to the spirits of departed ancestors.

96. All my twice-born disciples who wish good to themselves should read and hear these noble Scriptures. (Reprinted in Parekh (1936: 327 ff.))

44 See, for example, *Bhau Ram v. Baij Nath*, A.I.R. 1962 S.C. 1476, and chap. 7.

45 For example, *D. Sura Dora v. V. V. Giri*, A.I.R. 1959 S.C. 1318, and chap. 6.

46 *Punjabrao v. Meshram*, A.I.R. 1965 S.C. 1179 at 1184.

47 A.I.R. 1966 at 1127.

thus entitled to governmental protection — and where they are merely superstition and subject to governmental overruling?[48] The court does not address this as a question of fact; it does not appeal to usage, to popular understanding, or to Hindu learning. Instead it acts as if it enjoyed a mandate to make such determinations on its own authority.

Was this the only way that the court could reach a desirable result? It is submitted that the court might have found a more direct and more craftsmanlike route to the same result.[49] The result itself is by no means a strained or contrived one. In crucial repects the Constitution *is* a charter for the reform of Hinduism. The wording of Articles 25 and 26 establishes the primacy of public interests over religious claims and provides a wide scope for governmentally sponsored reforms. Articles 15 and 17 forbid a whole cluster of usages that are intimately connected with popular Hinduism and have some sanction in Hindu learning. While denominational differences within other religions lie outside State power, the Constitution embodies the notion that divisions within Hinduism need not be accorded the same respect. Article 25(2)(b) establishes that the State may act to overcome caste and denominational barriers within Hinduism. The law may be used to create an integrated Hindu community by conferring common rights of entry in religious premises. To authorize State use of this power in the case of Satsangi temples, it was necessary only to construe the scope of this power. It was not necessary to consider the nature of Hinduism *per se*. On the verbal level, it would have been sufficient to confirm that this power extends to all who are 'socially' and 'culturally' Hindus, as the Satsangis conceded themselves to be.[50] Or, going deeper, the ambit of the temple-entry power could have been determined in

[48] The court here is cutting into a whole complex of learning about the dynamics of temples, idols, purity, etc. For many Hindus, temples are 'divine powerhouses, no mere prayer halls' (Krishnamacharya (1936: 8)) — nucleations of divine energy from which radiate spiritual energies which would be dissipated by contact with unfavourable emanations. In order to conserve the efficacy of these repositories of spiritual energy, there are detailed rules regarding the location and construction of temples, installation and consecration of idols, maintenance of requisite states of purity, and avoidance of contamination of various sorts through specific rites and observances. While not all Hindus take concern for contamination so literally or unbendingly, the court does not indicate the basis in Hindu doctrine for its selectivity. (Of course its basis in public policy is evident.)

[49] Any other disposition of this claim would have permitted appellants' lawyers to elicit a determination of the status of the entire group which violated previous understandings, without any evidence that this new characterization was widely shared by members of the group or that the plaintiffs had a mandate to represent all Satsangis in this matter. Some Satsangi witnesses had contested this contention below. LXI Bom. L.R. at 705. The High Court noted that there had been no previous claim of this kind and there was evidence of long-standing acquiescence in census enumation as Hindus.

[50] LXI Bom L.R. at 703–4.

the light of its purpose. Was this the evil that the Constitution makers were providing against? Was this a case of invidious caste distinctions operating to restrict religious participation? It was established doctrine that rights conferred under the temple-entry power prevail over denominational claims to exclude outsiders as part of their freedom of religion.[51] Thus it would not have been necessary to determine whether the exclusionary rules of the Satsangis were properly part of their religion or not.

Why, then, did the court choose the more circuitous and throny path? In part, there is a personal explanation. The judgment is the work of the then chief justice, P. B. Gajendragadkar, a man of strong reformist views within Hinduism and with an intensely activist judicial posture on the bench.[52] The chief justice was faced by compulsory retirement two months later. As a militant advocate of a reformist brand of secularism, he was displeased by an earlier series of decisions in which the Supreme Court had declared that what is religion is a matter to be determined by the doctrines of each religious community itself.[53] He felt strongly that this 'auto-determination' of religious rights was a pernicious doctrine which would give great scope to obscurantist religionists and would place beyond State power practices that were inimical to progress.[54] In several earlier cases, he had taken the opportunity to supply *dicta* which tended to undermine this view by asserting that matters which were 'obviously secular' or 'based on superstition' would not be deemed religious even if their adherents regarded them as such.[55] In other words, he strongly asserted the right of the courts to an independent power of decision as to what was the 'religion' protected by Articles 25 and 26. He may have hoped that before his retirement, there would be an opportunity to overrule the 'auto-determination' view. The Satsangi case was as close to such an opportunity as he could hope to find. The temptation to make the most of it may have been

[51] *Sri Venkataramana Devaru v. State of Mysore*, A.I.R. 1958 S.C. 255.

[52] Gajendragadkar (1951); Gajendragadkar: (1965). For an assessment of his judicial work, see *Journal of the Indian Law Institute* (1966). For a premonition of the views on Hinduism set forth in the Satsangi case, see the chief justice's inaugural address to the seminar on secularism just a few months before the judgment was delivered (in Sharma (1966: 1).

[53] For example, *Saiffudin Saheb v. State of Bombay*. A.I.R. 1962 S.C. 853.

[54] Cf. the criticism of Tripathi in Sharma (1966). Blackshield (1966: 61) finds 'the Indian Supreme Court's experiments with 'auto-determination' by religious institutions of what matters are 'religious' *do* seem inconsistent with secularism.' Smith, (1963: 105) had earlier observed that such an approach could not be accepted in India 'unless one were prepared to abandon all plans of social progress and modernization'. For an opposing view, see Ghose (1965).

[55] *Durgah Committee v. Hussain Ali*, A.I.R. 1961 S.C. 1402, 1415; *Shri Govindlalji v. State of Rajasthan*, A.I.R. 1963 S.C. 1638, 1660–1.

particularly strong because the case concerned untouchability, a subject on which the chief justice was an outspoken and militant reformer. This was one subject on which he might have felt little hesitation in confidently asserting the nonreligious character of the claimed practice, because he was confident of his grasp of the 'true teachings' of Hinduism.[56]

But the personal predilections of the judgment writer are hardly a sufficient explanation. The other four judges sitting concurred without being moved to dissociate themselves from the opinion in any way. We must then take this as more than an idiosyncratic view. It represents in a pronounced form one tendency which is to be found among the judiciary and more generally among the Westernized, educated ruling elite — a tendency to active reformulation of Hinduism under government auspices in the name of secularism and progress.

Alternative Modes of Secularism

The question brought into focus by the Satsangi judgment is the mode in which the secularism embodied in Indian law is to contribute to the transformation of religion in India. In assessing the thrust of India's secularism, it is important that we avoid equating secularism with a formal standard of religious neutrality or impartiality on the part of the State. No secular State is or can be merely neutral or impartial among religions, for the State defines the boundaries within which neutrality must operate. For example, the First Amendment of the United States Constitution is often said to enjoin state neutrality in religious matters. But it is clear that in the larger sense, American law is not neutral among religions except in a purely formal sense. In defining the boundaries within which neutrality must operate, the First Amendment is a charter for religion as well as for government. It is the basis of a regime which is congenial to those religions which favour private and voluntary observance rather than to those which favour official support of observance. It favours those religions which prefer social and spiritual sanctions over those which would employ official force to support their social and ritual perscriptions. It favours groups which are not exclusive in their claims and which are willing to tolerate the presence of alternative and hostile views. It is not so congenial to religions which purport to supply obligatory principles for the governing of society or which believe that it is necessary to extirpate error.

A secular State, then, propounds a charter for its religions; it

[56] The point is one he has made in innumerable public addresses. In asserting it, the Chief Justice stands well within a continuing reformist / Gandhian tradition of interpreting Hinduism so as to expunge it of 'untouchability'. For example, see Gandhi (1954): Sundarananda (1946).

involves a normative view of religion. Certain aspects of what is claimed to be religion are given recognition, support, and encouragement; others are the subject of indifference; finally, some are curtailed and proscribed. Religion, then, is not merely a datum for constitutional law, unaffected by it and independent of it. It is, in part, the product of that law. The American legal setting, for example, has made and presumably will continue to make a profound contribution toward shaping religion in the United States.

The Indian constitutional stance toward religion is more explicit and more complex. The Constitution attempts a delicate combination of religious freedom in the present with a mandate for active governmental promotion of a transformation of India's religions. For example, religions are to be divested of their character as sources of legal regulation of family life.[57] The Constitution also propounds equality among religions, but as we have seen, the State's reforming power with respect to Hinduism is more extensive. The law may be used to abolish caste distinctions by *inter alia* conferring rights of religious participation.

The broad constitutional mandate disposes of the notion that the law might confine itself to ascertaining and respecting a preordained religious sphere as implied by the 'separation-of-powers' model of secularism. There is a clear commitment to what Anthony Blackshield calls the 'overall arbitral role' in which the law exercises 'the function of regulative oversight and adjustment of the working and of the interlocking of particular social controls. . .'.[58] But how is this 'arbitral control' to be exercised? Here we may distinguish two alternative modes for the exercise of the law's 'regulative oversight' of religious controls. We might call them the mode of limitation and the mode of intervention. By limitation I refer to the shaping of religion by promulgating public standards and by defining the field in which these secular public standards shall prevail, overruling conflicting assertions of religious authority. By intervention I refer to something beyond this — to an attempt to grasp the levers of religious authority and to reformulate the religious tradition from within, as it were.[59]

At the risk of ignoring elements of overlap and mixture, we might visualize these alternative modes as related on two dimensions: first, the superior or overriding character attributed to the legal norms in cases of overlap, conflict, and characterization; second, the

[57] Article 44.

[58] Blackshield (1966: 60).

[59] For a temple-entry position delicately poised on the border between limitation and intervention, see Gandhi's (1965: 140), 1931 depiction of his dream of independent India: 'There will be no untouchability. The 'untouchables' will have the same rights as any other. But a *Brahmin* will not be *made* to touch anybody. He

asserted competence and mastery of the legal specialists in authoritative exposition of the religious norms. The relationship between our three modes of secular control may then be depicted in tabular form. The division of the Table into four distinct boxes for purposes of emphasis should not obscure the point that the dimensions in question are continuation rather than dichotomous distinctions. Our present table portrays only a part of the range along each dimension. Presumably, by extending it we could encompass various forms of 'religious state'.

Table

Alternative modes of secular control

External Superiority of Legal Norms	Internal Competence of Legal Specialists	
	low	high
high	Mode of Limitation	Mode of Intervention
low	'Separation of Powers' Mode	*

* The situation of high internal competence of legal specialists in religious norms combined with low external superiority of legal over religious norms is not another fom of secularism, but defines a condition that would be called a religious state.

Clearly the Constitution gives government power to promulgate certain reforms, irrespective of Hindu usage. There is a deliberate abandonment of part of Hindu tradition to State regulation. It is clearly the task of the courts to delimit that abandoned part and to interpret the secular public principles that are to apply in its stead. But does the Constitution give the court a mandate to go beyond this, to participate actively in the internal reinterpretation of Hinduism, to interpret Hinduism so as to accommodate these governmentally sponsored changes, and to legitimize them in terms of Hindu doctrine? Religious notables, publicists, and scholars are presently engaged in reinterpreting Hindu tradition. Does the Constitution empower the courts to participate actively in this

---

will be free to make himself untouchable and have his own well, his own temple, his own school and whatever else he can afford, so long as he uses these things without being a nuisance to his neighbours. But he will not be able, as some do now, to punish untouchables for daring to walk on public streets or using public wells. There will be under *Swaraj* no such scandal as that of the use of public temples being denied

reformulation? Is the Supreme Court a forum for promulgating official interpretations of Hinduism? Is it a Supreme Court of Hinduism?[60]

A number of factors in the Indian religious and political setting may impel judges to take this active interventionist role: the desire to reconcile, to avoid explicit disregard of religious authority, to make reforms palatable, to propagate strongly held views of religion, to teach the unenlightened, and to entertain pleasant images of one's past. There are other factors too that make such judicial activism acceptable and appealing to the educated reformist elite. Members of this elite are distressed by much of popular and traditional Hinduism which they feel lacks the dynamism, the concern for welfare and development, that they feel is necessary for India's progress. They are also distressed by its diffuse and fragmented character which in their eyes obstructs national unity and by its lack of coherence and organization which make the masses unavailable to mobilization for reform and development. These very shortcomings lend legitimacy to judicial activism. The sprawling, disjointed, unorganized character of Hinduism and the parochialism of its spokesmen disqualifies it from a right to self-definition. Since it is so organizationally fragmented and diffuse, there are no religious leaders who have a mandate to define it for the entire religious community. To permit each religious dignitary to define it for himself would subvert any attempt at integration. Thus, in the absence of credible spokesmen, only the judges can speak to and for Hinduism as a whole. It is assumed that judicial intervention will have a salutary unifying as well as a reforming influence. Indeed, it is only when the state intervenes to promote unity and infuse modernity that it can create in Hinduism a capacity for self-definition.

There are, I submit grounds for serious doubt whether such judicial intervention could actually be productive of the values to

---

to untouchables when it is allowed to all other Hindus. The authority of the *Vedas* and the other *Shastras* will not be denied, but their interpretation will not rest with individuals but will depend upon the courts of law, in so far as these religious books will be used to regulate public conduct. Conscientious scruples will be respected, but not at the expense of public morals or the rights of others. Those who will have extraordinary scruples will themselves have to suffer inconvenience and pay for the luxury. The law will not tolerate any arrogation of superiority by any person or class in the name of custom or religion. All this is my dream' (Cf. Derrett, 1968: 510ff).

[60] It might be argued that the area of untouchability/temple entry called for distinctive and interventionist treatment, because this area is carved out of Hinduism by the Constitution — that is, that in effect an ecclesiastical jurisdiction is conferred on the court to interpret untouchability out of Hinduism. Such assertiveness is not evident in other judicial encounters with untouchability (see Galanter (1969)). Nor is the interventionist stance taken in the Satsangi case explicitly limited to matters concerning untouchability.

which the elite subscribe. First, there is a tendency to underrate the strain of such a role on the capacities, energies, and persuasiveness of the judiciary. What, it may be asked, equips judges to prescribe the nature and content of Hinduism? It could be and is argued that the higher judiciary, many of whom are accomplished in Hindu learning or at least steeped in a Hindu atmosphere, are well equipped to filter and refine Hindu tradition and to arrive at assessments of Hinduism that combine fidelity to essentials with a progressive, modern outlook which recognizes the need for reform and growth.

But how is a common-law judge to do this? Is he to confine himself to making an assessment solely on the basis of the record before him? Or may be draw upon his own experience and prepossessions? How about the judge—and there are many, only meagerly acquainted with Hinduism or holding an idiosyncratic view of it? How about the non-Hindu judge? Is he to disqualify himself? Or ought he to sit, but to defer to the opinion of his Hindu colleagues? Or is he equally entitled, as an Indian, to voice the true meaning of Hinduism?

Once qualified to sit, how is the judge to proceed? An exponent of a tradition of textual exegesis, the common-law judge employs certain techniques of selecting authorities, interpreting and reconciling texts, and introducing innovations. Is it open to him to apply these techniques to the Hindu textual tradition? Should not the judge enter into that tradition to ascertain its own internal rules and techniques, its methods of assessing the relative importance of its various elements, and the admissibility of innovations?[61] In the Satsangi case, the court did not consult contemporary Hindu learning. It draws a picture of Hinduism based on western or western-inspired scholarly sources[62] and elicits principles from Hindu tradition by common-law techniques. But if the court cannot enter into Hindu tradition and work within it, how persuasive can it be to the living exponents of that tradition and to their followers?

This is not to suggest that what the courts say is of no effect. Whatever the courts do cannot help but have some impact on Hinduism. The kind of religion that the government and the elite favour may gain some acceptance as a result of judicial promul-

[61] Cf. the impact on Hindu law of its administration by common law courts which could not satisfactorily exercise the discretionary techniques that give it its flexibility see chap. 2; Derrett (1968: chaps. 8,9).

[62] The opinion is remarkable for the extent to which (though written by a judge himself learned in Sanskrit and a descendant of a family of eminent *dharmasastrins*) it approached Hinduism through Western and Western-inspired scholarly sources. Apart from the brief quotation from Tilak and another from the Bhagavad Gita the principal authors referred to are Monier-Williams, Max Muller, and S. Radhakrishnan.

gation. Like many an earlier reform movement it may succeed in being accepted as a sect alongside all the other movements in Hinduism. But is it likely to have the kind of influence that the elite desire and sway the whole Hindu community? One of the distinctive features of modern India is that there are, for the first time, levers for bringing changes to bear 'across the board' for the entire Hindu community, among them legislation and the courts. And it is the Westernized elite who grasp these levers. The question is: Will it get more thrust from them by using them under the banner of true Hinduism than under that of secular modernism?

It might be argued that the courts should not be shy of prescriptive assessments of Hinduism since their statements will inevitably carry normative overtones. If, in dealing with religious questions, the courts scrupulously avoid prescriptions based on reform views they in effect will give powerful support to existing forms and ossify present practice.

But if the courts deliver "reformist' decisions, does the addition of religious justifications enhance their effect? Here we need empirical information and we have none. From what little is known about the influence of the decisions of higher courts upon behaviour in other settings, we may surmise that when court decisions are influential it is not through their doctrinal pronouncements but through the rechanneling of major institutional opportunities and controls and by their liberating effect. The court may provide dissident and progressive elements with institutional support and with rationalization for their nontraditional behaviour or beliefs. But these effects may be relatively independent of whether the justifications provided are secular or religious. Those with serious traditional commitments are unlikely, in any event, to be persuaded by judicial exegesis. Indeed, it might be argued that the masses will do better at reconciling reforms with their religious understandings if the courts disclaim any competence in religious matters and reiterate instead their claim to overriding authority in whatever touches public life.

Quite apart from these problems of effectuating an interventionist approach, there is the question of the cost that might be entailed by its 'success'. The paradoxical character of the interventionist stance is revealed if we inspect the common argument in favour of Hindu unity. If judicial intervention did assist in making Hinduism unified and organized, would the modern reformist elite be as influential as they are at present? Their anxiety to see Hinduism organized is not to have a spokesman for it as it is, but so that it might be more readily mobilized to be what they think it ought to be. Indeed, government measures may succeed in precipitating a unified structure out of Hinduism. But the lack of unity and organization of traditional society is not only an obstacle to the greater influence of the elite, it is the condition of their present influence. Again, it is assumed that the unification and organization of Hinduism will somehow contribute to national integraton. But this is not a self-evident proposition. Perhaps the disunity of Hinduism contributes to national unity. The successful breaking down of Hinduism's capacity to generate and tolerate internal differences may well lessen India's capacity to sustain pluralistic democracy.

Of course, the 'success' of interventionism is unlikely to be more than partial and localized. In reading the Satsangi judgment one gets the impression that the public being addressed is not the unenlightened mass but the elite itself. The opinion is an occasion for intraelite debate, using the Satsangis as an object lesson. The concern of the court's judgment is to reduce the dissonance of the elite —a dissonance which derives from their own ambivalence about Hinduism and about secularism. But however successful this approach may be in assuaging the feelings of the educated and resolving their ambivalence, the use of religious justifications is not without danger. For it projects certain illusions about Indian society which increase the capacity of the elite to be seduced by their own tendency to concentrate on illusory verbal reforms. The lawyer's fallacy that behaviour corresponds to legal rules offers powerful reinforcement of the elite's fallacy that the masses are following them—a coincidence of illusions that can lead to dangerous miscalculation about popular sentiment and about the efficacy of legally enacted reforms.

Conclusion

Lawyers have tended to view the relationship of 'law' and 'morals' as a problem of the sources of legal norms and the scope of legal regulation, a problem, that is, whether law should express or enforce 'morals' (usually taken to mean some generally accepted normative rules). As our example of secularism indicates, there is another side to the relationship—the problem of the autonomy and authority of the various other traditions of normative learning with which the law coexists in society. The law must face the question of the mode in which it should recognize and/or supervise them. The alternatives emerge vividly in the law-religion relationship because religions are systems of control with complex learned traditions, expounded by their own specialists and even with their own doctrines concerning their relationship to law. But the same general problems of relative authoritativeness and competence are present in the relation of law to various simpler and unlettered traditions (for example, the custom of a trade, the usage of a caste) as well as to complex and learned traditions like Hinduism. The 'law and morals' problem reappears in the relation of law to every body of normative learning.

Our analysis of Indian secularism as presenting alternative possibilities for managing the relationship of law to Hindu tradition suggests the possibility of a more generally applicable typology of the relations between law and other bodies of normative learning. We may reformulate our alternative modes of secularism as possible modes in which the law can relate to various 'lesser' traditions, for example: (1) delegative recognition of an autonomously defined realm of authority, corresponding to the 'separation of powers' mode of secularism; (2) regulative management of boundaries, conflicts, and characterization of the 'lesser' tradition, corresponding to the 'limitation' mode of secularism; and (3) internal management (for example, interpretation, innovation) of the 'lesser' tradition by legal specialists, corresponding to the 'intervention' mode of secularism.

This crude typology not only invites refinement, but it points to a range of questions for empirical investigation, questions whose exploration might illuminate the 'law and morals' debate (which has been heavily analytic and prescriptive) by adding a descriptive and comparative dimension. For example, it suggests that we ought to ask:

(1) Do particular legal systems have predominant characteristic modes in which they relate to other normative traditions?

(2) What conditions—demographic, economic, cultural, structural political—are associated with such differences in style?

(3) Do particular areas (religion, economic activity, etc.) inspire similar treatment by diverse legal systems?

(4) What are the implications of different modes of doctrinal justification for different modes of practice? This gives us another way to look at the relation of law as authoritative doctrine and law as patterns of institutional practice. Does the relation of the 'law on the books' and the 'law in action' vary with the predominant mode of dealing with other normative traditions?

(5) Is there a tendency for modes to change over time? Is there a prevailing direction of change?

(6) What are the potentialities and problems of various modes in deliberately using law as an instrument of social change?

In the view put forth here, both 'church and state' and 'law and morals' are instances of a wider class, the relation of law to the whole array of normative traditions with which it coexists in society. Descriptive and comparative study of the forms of that coexistence would provide a focus of interest to both social, scientific and philosophical inquiry into those tensions which lie at the heart of legal reality—tensions between unity and plurality, stability and change, norm and practice.

APPENDIX

Relevant Constitutional and Statutory Provisions
Constitution of India
(1950)

Article 17 *Abolition of Untouchability.* — 'Untouchability' is abolished and its practice in any form is forbidden. The enforcement of any disability arising out of 'Untouchability' shall be an offence punishable in accordance with law.

Article 25 *Freedom of conscience and free profession, practice and propogation of religion.* (1) Subject to public order, morality and health and to the other provisions of this Part, all persons are equally entitled to freedom of conscience and the right freely to profess, practice and propagate religion.
(2) Nothing in this article shall affect the operation of any existing law or prevent the State from making any law —

(a) regulating of restricting any economic, financial, political or other secular activity which may be associated with religious practice;

(b) providing for social welfare and reform or the throwing open of Hindu religious institutions of a public character to all classes and sections of Hindus.

Explanation I. — The wearing and carrying of *kirpans* shall be deemed to be included in the profession of the Sikh religion.

Explanation II. — In sub-clause (b) of clause (2), the reference to Hindus shall be construed as including a reference to persons professing the Sikh, Jaina, or Buddhist religion, and the reference to Hindu religious institutions shall be construed accordingly.

Article 26 *Freedom to manage religious affairs.* — Subject to public order, morality and health, every religious denomination or any section thereof shall have the right—

(a) to establish and maintain institutions for religious and charitable purposes;

(b) to manage its own affairs in matters of religion;

(c) to own and acquire movable and immovable property; and

(d) to administer such property in accordance with law.

The [Central] *Untouchability (Offences) Act, 1955*
No. 22 of 1955

Section 3 *Punishment for enforcing religious disabilities.* — Whoever on the ground of 'untouchability' prevents any person —

257

(a) from entering any place of public worship which is open to other persons professing the same religion or belonging to the same religious denomination or any section thereof, as such person; or

(b) from worshipping or offering prayers or performing any religious service in any place of public worship, or bathing in, or using the waters of, any sacred tank, well, spring or water-course, in the same manner and to the same extent as is permissible to other persons professing the same religion, or belonging to the same religious denomination or any section thereof, of such person;

shall be punishable with imprisonment which may extend to six months, or with fine which may extend to five hundred rupees, or with both.

Explanation. — For the purposes of this section . . . persons professing the Buddhist, Sikh, or Jaina religion or persons professing the Hindu religion in any of its forms or developments including Virashaivas, Lingayats, Adivasis, followers of Brahmo, Pararthana, Arya Samaj and the Swaminarayan Sampraday shall be deemed to be Hindus.

The Bombay Hindu Places of Public Worship (Entry Authorization) Act, 1956
No. 31 of 1956

Section 2 *Definitions.* — In this Act, unless the context otherwise requires,

(a) 'place of public worship' means a place, whether a temple or by any any other name called, to whomsoever belonging, which is dedicated to, or for the benefit of, or is used generally by, Hindus, Jains, Sikhs or Buddhists or any section or class thereof, for the performance of any religious service or for offering prayers therein; and includes all lands and subsidiary shrines appurtenant or attached to any such place, and also any sacred tanks, wells, springs and water courses the waters of which are worshipped, or are used for bathing or for worship;

(b) 'section' or 'class' of Hindus includes any division, sub-division, caste, sub-caste, sect or denomination whatsoever of Hindus.

Section 3 *Throwing open of Hindu temples to all classes and sections of Hindus.*— Notwithstanding any custom, usage or law for the time being in force, or the decree or order of a court, or anything contained in any instrument, to the contrary, every place of public worship which is open to Hindus generally, or to any section or class thereof, shall be open to all sections and classes of Hindus; and no Hindu of whatsoever section or class shall, in any manner be prevented, obstructed or discouraged from entering such place of public worship, or from worshipping or offering prayers thereat, or performing any religious service therein, in the like manner and to like extent as any other Hindu of whatsoever section or class may so enter, worship, pray or perform.

11

Symbolic Activism: A Judicial Encounter with the Contours of India's Compensatory Discrimination Policy*

Pursuing Substantive Equality: the Classic Compromise

In the text of the Constitution, the general principle of compensatory discrimination is established as a Directive Principle, but the specific provisions authorizing it are framed as exceptions to general Fundamental Right. This arrangement expresses the basic tension between the broad purposes to be achieved and the commitment to confine the device and make it comport with other constitutional commitments, especially that to formal equality.

The exceptional character of compensatory discrimination has frequently been noted by the courts. That Article 15(4)[1] 'has to be

---

* This essay, drawn from Galanter, (1984), is reprinted from R. Dhavan, S. Kurshid and R. Sudarshan, eds., *Judges and the Judicial Power*, London: Sweet & Maxwell, Bombay, N.M.Tripathi (1985).

[1] The Equality provisions of the Indian Constitution are:

14. Equality before the law. The State shall not deny to any person equality before the law or the equal protection of the laws within the territory of India.

15. Prohibition of discrimination on grounds of religion, race, caste, sex or place of birth. (1) The State shall not discriminate against any citizen on grounds only of religion, race, caste, sex, place of birth or any of them. (2) No citizen shall, on grounds only of religion, race, caste, sex, place of birth or any of them, be subject to any disability, liability, restriction or condition with regard to (a) access to shops, public restaurants, hotels and places of public entertainment; or (b) the use of wells, tanks, bathing ghats, roads and places of public resort maintained wholly or partly out of State funds or dedicated to the use of the general public. (3) Nothing in this article shall prevent the State from making any special provision for women and children. (4) Nothing in this article or in clause (2) of article 29 shall prevent the State from making any special provision for the advancement of any socially and educationally Backward Classes of citizens or for the Scheduled Castes and Scheduled Tribes.

16. Equality of opportunity in matters of public employment. (1) There shall be equality of opportunity for all citizens in matters relating to public employment or appointment to any office under the State. (2) No citizen shall, on grounds only of religion, race, caste, sex, descent, place of birth, residence or any of them, be ineligible for, or discriminated against in respect of any employment or office under the State. (3) Nothing in this article shall prevent Parliament from making any law prescribing, in regard to a class or classes of employment or appointment to any office under the Government of, or any local or other authority within, a State or Union territory,

259

read as a proviso or an exception to Articles 15(1) and 29(2)' was evident from its history, according to the Supreme Court in *Balaji v. State of Mysore*.[2] This has been the received characterization of Article 15(4). Thus in *State of Andhra Pradesh v. Sagar* the Supreme Court emphasizes that as an exception, Article 15(4) cannot be extended so as in effect to destroy the guarantee of equality in Article 15(1).[3]

In *Devadasan v. Union of India* the Supreme Court emphasized that Article 16(4) 'is by way of a proviso or an exception' to Article 16(1) and 'cannot be so interpreted to nullify or destroy the main provision.' Thus its 'over-riding effect' is only to permit a 'reasonable number of reservations . . . in certain circumstances. That is all.'[4] The characterization as an exception was challenged by Subba Rao, J., dissenting in *Devadasan*. In his view, Article 16(4) 'has not really carved out an exception, but has preserved a power untrammeled by the other provisions of the Article.[5] An extension of this argument was put forward in *C.A. Rajendran v. Union of India*, where it was urged that Article 16(4) was not merely an exception engrafted on Article 16, but was itself a Fundamental Right granted to the Scheduled Castes and Tribes and untrammeled by any other provision of the Constitution.[6] The Supreme Court's response was that Article 16(4) imposed no duty on the government to make reservations for these classes but 'Article 16(4) is an enabling provision and confers a discretionary power on the State to make a reservation. . . .[7]

But if the particular means are discretionary, the object is not: the Constitution explicitly declares it 'the duty of the state' to promote the interests of the 'weaker sections' and to protect them.[8] And if

---

any requirement as to residence within that State or Union territory prior to such employment or appointment. (4) Nothing in this article shall prevent the State from making any provision for the reservation of appointments or posts in favour of any backward class of citizens which, in the opinion of the State, is not adequately represented in the services under the State. (5) Nothing in this article shall affect the operation of any law which provides that the incumbent of an office in connection with the affairs of any religious or denominational institution or any member of the governing body thereof shall be a person professing a particular religion or belonging to a particular denomination.

[2] A.I.R. 1963 S.C. 649, 657.

[3] A.I.R. 1968 S.C. 1378. The characterization of Article 15(4) as an exception reappears in *State of A.P. v. Balaram*, A.I.R. 1972 S.C. 1375, 1394 and *Janki Prasad v. State of J. & K.*, A.I.R. 1973 S.C. 930, 937.

[4] A.I.R. 1964 S.C. 179, 187.

[5] Id., at 190.

[6] A.I.R. 1968 S.C. 507. The argument was based upon Subba Rao, J.'s dissent in *Devadasan v. Union of India*, A.I.R. 1964 S.C. 179.

[7] A.I.R. 1968 S.C. at 513.

[8] Articles 37, 46.

these provisions are exceptions, they are exceptions of a peculiar sort. They do not merely carve out an area in which the general principle of equality is inapplicable. Rather, they are specifically designed to implement and fulfill the general principle.[9]

Article 15(4) and 16(4) are undoubtedly exceptions to the constitutional prohibition of State employment of the otherwise forbidden criteria of caste, religion and so forth. But it does not follow that they are exceptions to the policy of equal treatment mandated by Articles 14, 15, and 16. In respect to the general policy of equality they represent an empowerment of the State to pursue substantive equality in respect to the disparities between the backward classes and others. It might be argued that in a state generally committed to formal equality this commitment to reduce these disparities is an exception. Thus the State could pursue a policy of overcoming the inequalities of backward groups that would lie outside its mandate in dealing with the poor, the handicapped, veterans, or victims of disasters. But it seems clear that these disparities of circumstances and fortune may be addressed by policies utilizing reasonable classifications. So what these clauses do is to ensure State power to pursue substantive equality *vis à vis* certain historic formations in Indian society.

It was the realization that mere provision of formal equality would not suffice to bring about the desired 'EQUALITY of status and of opportunity' that led to the adoption of these provisions. As a Full Bench of the Kerala High Court observed:

> It has however been realized that in a country like India where large sections of the people are backward socially, economically, educationally and politically, these declarations and guarantees [of equality] would be meaningless unless provision is also made for the uplift of such backward classes who are in no position to compete with the more advanced classes. Thus to give meaning and content to the equality guaranteed by Articles 14, 15, 16, and 29, provision has been made in Articles 15(4) and 16(4) enabling preferential treatment in favour of the 'weaker sections.'[10]

Indeed, as the Supreme Court has observed, guarantees of equality might by themselves aggravate existing inequalities. If taken literally,

> instead of giving equality of opportunity to all citizens, it will lead to glaring inequalities. . . . In order to give a real opportunity to [the backward to] compete with the better placed people . . . [Article 16(4) is

---

[9] Cf. Gupta (1969: 83), who notes that these provisions'. . ; are only to be viewed as defining the principle of equality as contained in Article 14. They do not derogate from the principle of equality in any manner'.

[10] *Hariharan Pillai v. State of Kerala*, A.I.R. 1968 Ker. 42 at 47–8.

included in the Constitution.] The predominant concept underlying [Article 16] is equality of opportunity in the matter of employment; and, without detriment to said concept, the State is enabled to make reservations in favour of backward classes to give a practical content to the concept of equality.[11]

The tension between these commitments to non-discrimination and to substantive equalization was poignantly expressed by Prime Minister Nehru when he remarked in the course of the First Amendment debate that '. . . we arrive at a peculiar tangle. We cannot have equality because in trying to attain equality we come up against some principles of equality.'[12] The textual juxtaposition of guarantees of equality and authorization of compensatory discrimination reflects a deeper conflict between different views of equality and divergent notions of the goal and scope of protective discrimination. While in practice these views tend to merge and overlap, it may be helpful here to isolate in pure form what we may conveniently label the horizontal and the vertical perspectives on equality and compensatory discrimination.

In the horizontal view, the relevant time is the present. Equality is visualized as identical opportunities to compete for existing values among those differently endowed, regardless of structural determinants of the chances of success or of the consequences for the distribution of values. One of the dissenting judges in the *Thomas* case sums it up neatly when he cautions that Article 16(1) 'speaks of equality of opportunity, not opportunity to seek equality'.[13] In this view, preferential treatment is accepted as a marginal adjustment to be made where results of complete equality are unacceptable. Compensatory discrimination detracts from equality: it amounts to a kind of social handicapping to insure fair present distribution among relevant units. Thus in *Devasan v. Union of India* the Supreme Court emphasized that in combining these provisions it might strike a fair balance between the claims of the backward and the claims of other communities.[14]

The relation of equality and compensatory discrimination is viewed very differently in what we may call the vertical perspective. In this view the present is seen as a transition from a past of inequality to a desired future of substantive equality; the purpose of compensatory discrimination is to promote equalization by offsetting historically accumulated inequalities. Thus, compensatory discrimination

---

[11] *Triloki Nath Tiku v. State of Jammu and Kashmir*, A.I.R. 1967 S.C. 1283 at 1285.

[12] Parl. Deb. XII–XIII (Part II col. 9617.

[13] Gupta, J., dissenting in *State of Kerala v. N.M. Thomas*, A.I.R. 1976 S.C. 490 at 543.

[14] *Devadasan v. Union of India*, A.I.R. 1964 S.C. 179, Cf. *C.A. Rajendran v. Union of India*, A.I.R. 1968 S.C. 507.

does not detract from equality in the interest of present fairness; rather, it is seen as a requisite to the fulfilment of the nation's long-run goal of substantial redistribution and equalization. Not only present claims but historical deprivations and national aspirations are relevant. Such a view was given its most clear judicial expression in *Viswanath v. State of Mysore*, where Hegde J. repulsed the argument that reservations should be confined with the observation that counsel

> did not appear to be very much alive to the fact that there can be neither stability nor real progress if predominant sections of an awakened Nation live in primitive conditions, confined to unremunerative occupations and having no share in the good things of life, while power and wealth are confined in the hands of only a few and the same is used for the benefit of the sections of the community to which they belong. . . . Unaided many sections of the people, who constitute the majority in this State cannot compete with the advanced sections of the people, who today have a monopoly of education and consequently have predominant representation in the Government services as well as in other important walks of life. It is cynical to suggest that the interest of the Nation is best served if the barber's son continues to be a barber and a shepherd's son continues to be a shepherd. . . . We have pledged ourselves to establish a welfare State. Social justice is an important ingredient of that concept. That goal cannot be reached if we overemphasize the 'merit theory'.
>
> Advantages secured due to historical reasons cannot be considered a fundamental right guaranteed by the Constitution. The nation's interest will be best served — taking a long-range view — if the backward classes are helped to march forward and take their places in a line with the advanced sections of the people.[15]

From this 'long-range' perspective, Justice Hegde later elaborated, the 'immediate advantages of the Nation [in the effective utilization of talent] have to be harmonized with its long-range interests.'[16]

Rarely do these contrasting views of equality and compensatory discrimination appear with such purity and clarity; more common is the attempt to harmonize them in what we might call 'the classic compromise': advancement of the backward permits exceptional means and enjoys special priority, but these means and priorities must be balanced against other rights and other interests. Thus in *Balaji v. State of Mysore*, Gajendragadkar, J. observes that:

> It is obvious that unless the educational and economic interests of the weaker sections of the people are promoted quickly and liberally, the ideal

---

15 A.I.R. 1964 Mys. 132 at 136. A more summary and less vivid version of these observations was delivered by Justice Hegde after his ascension to the Supreme Court in *Periakaruppan v. State of Tamil Nadu*.

16 *Periakaruppan v. State of Tamil Nadu*, A.I.R. 1971 S.C. 2303 at 2309.

of establishing social and economic equality will not be attained. . . .[17]

Surely the State is authorized to take 'adequate steps' toward that objective, but these special, exceptional provisions do not override the Fundamental Rights of others. Furthermore, there are other crucial national interests which have to be taken into account:

> The interests of weaker sections of society which are a first charge on the States and the Centre have to be adjusted with the interests of the community as a whole.[18]

The inevitable weighing and balancing is rendered particularly difficult because the constitutional provisions set up another tension as to the relevant units whose interests are to be balanced or who are ultimately to be equalized, as the case may be. The Constitution confers Fundamental Rights on individual citizens in their personal capacity, not as members of communal groups. All citizens have a fundamental right that another, excepting a member of a backward class, shall not be preferred by the State on the basis of his membership in a particular group. However the government is obliged to advance the weaker sections of backward classes. Therefore the scope of compensatory discrimination involves tension between individuals or groups as objects of State policy.

The constitutional embrace of the antagonistic principles of equal treatment and compensatory discrimination, individual rights and group rights, confronts both government and courts with the problem of reconciling them in specific settings. The sweeping language of Articles 15(4) and 16(4) indicates that their framers relied primarily on the discretion of the politicians and administrators of the future, rather than on the courts, to effect such a reconciliation. But while these provisions give the executive and the legislatures broad discretion in their application, judicial intervention is not entirely excluded. Leverage for judicial oversight is supplied by the placement of these provisions as exceptions to the judicially enforceable Fundamental Rights. These rights can only be vindicated to the extent that the courts scrutinize the government's designation of beneficiaries to see that only the backward are included, that the extent or method of operation does not prejudice others unduly, that the schemes are designed and administered to work in favour of the intended beneficiaries and not to their detriment. Such review is necessary not only to vindicate individual rights but to effectuate the policy of these provisions — by preventing unwarranted dilution of benefits (for the more unrestrained the inclusion of beneficiaries.

[17] A.I.R. 1963 S.C. 649 at 661.
[18] Id., at 663.

the less the assistance the intended beneficiaries will receive) and abuses that undermine public support for these measures.

The courts have indeed played a major role in shaping policy in this area by defining the constitutional boundaries of preferential treatment. Recent developments in the constitutional doctrine of equality indirectly raise the question of whether courts might play an even more central role in the design and implementation of compensatory policies.

A New Constitutional Vista: the Thomas Case

For the first quarter-century of constitutional development, the approaches discussed in the previous section bounded the discourse about the way in which competing commitments to formal equality and compensatory discrimination might be combined. But much of the earlier understanding of constitutional policies of compensatory is cast into doubt by a remarkable 1975 decision of the Supreme Court in *State of Kerala v. N. M. Thomas*.[19]

Employees of the Registration Department of the State of Kerala were divided into Lower Division Clerks and Upper Division Clerks.[20] The former could be promoted to the higher position on a 'Seniority-cum-merit' basis. To qualify for promotion it was necessary to pass some tests ('Account Test . . . Lower Kerala Registration Test and . . . Test in the manual of office procedures') — this was the merit requirement. Among those who satisfied this prerequisite, the promotions went to the most senior Lower Division Clerks. The rules allowed for temporary appointments to the higher posts for a two-year period during which the clerk would have to pass the required tests; the two years was extended to four years in the case of Scheduled Caste and Scheduled Tribe clerks. Nevertheless a number of Scheduled Caste clerks had not satisfied the test qualifications within the extended period and were facing reversion to the lower posts.

In 1972 the Government promulgated a new rule:

19 A.I.R. 1976 S.C. 490.
20 The version of the facts here is a composite drawn from the accounts provided at various places in the seven opinions. I noted no major discrepancies in these accounts. See A.I.R. 1976 S.C. at 493–5, 496–7 to 601, 502–4, 520, 523. One factual question puzzled me: what would happen to these clerks at the expiration of the extension period if they had not passed the tests? My initial reading was that during the period that they were incumbents in the higher posts, they were eligible for promotion on senirotity and since they were exempt from passing the tests, they would be confirmed in the higher posts on grounds of senirotity. But the judges are unanimous that this is not the case, that the extension is only temporary, and that the tests must be passed before they can be confirmed in the higher posts. Ray, C.J. at 502, Mathew, J. at 520, Beg, J. at 523, Fazl Ali, J. at 544.

13AA. Notwithstanding anything contained in these rules, the Government may, by order, exempt for a specific period, any member or members, belonging to a Scheduled Caste or a Scheduled Tribe, and already in service, from passing the tests. . . .

On the same day the State promulgated an Order granting Scheduled Castes and Scheduled Tribes already in service 'temporary exemption . . . from passing all tests . . . for a period of two years'. In 1974 this was extended for a further period to ensure each employee two chances to appear for the required tests. This time the government ordered that 'these categories of employees will not be given any further extension of time to acquire the test qualifications.' Because of earlier difficulties with the test barrier, there was a heavy concentration of Scheduled Caste lower division clerks with high seniority. When the test barrier was removed temporarily, many of them were promoted to the higher posts. Thus in 1972, of 51 vacancies in the category of Upper Division Clerks, 34 were filled by Scheduled Castes who had not passed the tests and only 17 were filled by persons who had passed the tests. A writ petition was filed by N. M. Thomas, a lower division clerk who did pass the test and would have been promoted but for the extensions authorized by Rule 13AA.

A Division Bench of the Kerala High Court concluded that: 'What has been done is not to reserve . . . posts . . . [R]eservations had already been made. . . . What has been attempted by Rule 13AA is to exempt persons from possessing the necessary qualifications.' Such exemption lies beyond the scope of Article 16(4)'s authorization of reservations, and on the scale it is done here, directly violates Article 335's directive that claims of Scheduled Castes and Tribes may be taken into account in government employment 'consistently with the maintenance of efficiency of administration. . . .'[21] On appeal to the Supreme Court, counsel for the State took an innovative tack and argued that the extension need not be subsumed under Article 16(4)'s provision for reservations, but could be justified as a reasonable classification under Article 16(1).[22] A seven judge bench decided five to two to reverse, issuing seven separate opin-

---

[21] *Thomas v. State of Kerala*, I.L.R. 1974 (1) Ker. 549 at 556–7.

[22] In the High Court, counsel for the state had argued unsuccessfully that the state's action fell within the meaning of 'reservation' in Article 16(4), I.L.R. Ker. 1974 (1) at 561. Counsel had also advanced an ingenious textual argument that if the reservations were covered by Article 16(4), then preference in government employment apart from reservations was authorized by Article 15(4), but the Court held that Article 16(4) covered the employment area and its provisions exhausted the government's power to deviate from the guarantees of Articles 16(1) and 16(2). Id., at 557. There is no indication that counsel urged that Articles 16(1) or 16(2) themselves conferred power to make such arrangements.

ions.[23] One of the majority judges thought that Article 16(4) rightly interpreted would authorize the State's provision; the other four accepted some version of the broad classification argument advanced by the State.[24]

Chief Justice Ray's opinion for the majority sketches the outline of the classification argument: providing equal opportunity in government employment is a legitimate objective; Article 46 directs the State to promote the economic interests of Scheduled Castes and Tribes with special care; Article 335 directs the State to take into its consideration their claims regarding service under the State. Thus the classification of employees belonging to these groups to afford them an extended period to pass tests for promotion is 'a just and reasonable classification having rational nexus to the object of promoting equal opportunity . . . relating to [public] employment. . . .'[25] The difference in condition of these groups justifies differential treatment; just as rational classification is permissible under the general equal protection provision of Article 14, so it is permissible to treat unequals unequally under Article 16.[26]

The extent of the doctrinal innovation here can be appreciated by considering the opinion of Justice Beg, the only member of the majority who does not participate in the reconceptualization of Article 16. He helpfully restates a conventional understanding of the constitutional provisions. In this view 'the guarantee contained in Article 16(1) is not by itself aimed at removal of social backwardness due to socio-economic and educational disparities pro-

---

[23] *State of Kerala v. N.M. Thomas*, A.I.R. 1976 S.C. 490. The appeal was argued just after the onset of Mrs. Gandhi's emergency rule and the decision was announced on 19 September 1975, at the end of three months of emergency rule and at the height of optimism that it heralded an 'egalitarian break-through'.

[24] Although *Thomas* marks a sharp departure from earlier doctrinal analysis of the special provisions for Backward Classes, it was not without precursors and anticipations. Thus classification standards were used to permit reservations for various deserving groups in medical admissions. (Cf. *Chanchala v. State of Mysore*, A.I.R. 1971 S.C. 1762.) For example, *Sardool Singh v. Medical College*, A.I.R. 1970 J. & K. 45, applied general Article 14 classification standards to permit reservations in medical admissions for children of defence personnel, noting that Article 14 permits such classification quite apart from the specific provisions for reservations in Articles 15(4) and 16(4). Reservations of government posts for ex-servicemen were upheld on a similar ground in *Daya Ram v. State of Haryana*, A.I.R. 1974 P. & H. 279. The classification argument might be made where the beneficiaries would come under Article 15(4) or 16(4), as in *Raju v. Chief Electoral Officer*, A.I.R. 1976 Guj. 66, decided nine months before *Thomas*. The High Court had no difficulty in justifying a lower electoral deposit for Scheduled Castes and Tribes on Article 14 grounds. The extent to which classification principles were perceived as available to justify compensatory preference is suggested by the remarks of Chief Justice Hidayatullah in *Golak Nath v. State of Punjab*, A.I.R. 1967 S.C. 1643 at 1706.

[25] A.I.R. 1976 S.C. at 500.

[26] Id., at 502.

duced by past history of social oppression, exploitation, or degradation of a class of persons.[27] Instead 'it was in fact intended to protect the claims of merit and efficiency . . . against incursions of extraneous considerations.'[28] And efficiency tests, in turn, 'bring out and measure . . . existing inequalities in competency and capacity or potentialities so as to provide a fair and rational basis for justifiable discrimination between candidates.'[29] Thus provisions for equality of opportunity are meant to insure 'fair competition' in securing government jobs; they are not directed to 'removal of causes for unequal performances. . . .'[30] But such provisions do not stand alone: they are juxtaposed with Articles 46 and 335 which imply 'preferential treatment for the backward classes' to mitigate the rigour of equality in the sense of strict application of uniform tests of competence.

> Article 16(4) was designed to reconcile the conflicting pulls of Article 16(1), representing . . . justice conceived of as equality (in the conditions of competition) and of Articles 46 and 335, embodying the duties of the State to promote the interests of the economically, educationally and socially backward, so as to release them from the clutches of social injustice.[31]

Thus Article 16(4) may be thought to 'exhaust all exceptions made in favour of backward classes'.[32] Yet the effect of the Kerala promotion rules here is 'a kind of reservation', for it is a temporary promotion that would be confirmed only if the appointee satisfied specified tests within a given time. These rules may be viewed as 'implementation of a policy of qualified of partial of conditional reservations' which could be justified under Article 16(4). . . .'[33]

Dismissing the interpretation of Article 16(4) as an exception to Article 16(1), Justice Matthew articulates the view of equality that implies the doctrinal shift. The equality of opportunity guaranteed by the Constitution is not only formal equality with fair competition, but 'equality of result'.[34] In order to assure the disadvantaged 'their due share of representation in public services' the Constitutional equality of opportunity was fashioned 'wide enough to include . . . compensatory measures. . . .'[35] Thus the guarantee of equality

[27] Id. at 522.
[28] Id. at 522.
[29] Id. at 522.
[30] Id. at 522.
[31] Id. at 522.
[32] Id. at 522.
[33] Id. at 524.
[34] Id. at 518.
[35] Id. at 518.

'implies differential treatment of persons who are unequal'.[36] Article 16(1) is 'only a part of a comprehensive scheme to ensure equality in all spheres'.[37] It implies 'affirmative action' by government to achieve equality — that is, 'compensatory State action to make people who are really unequal in their wealth, education or social environment . . . equal'.[38]

> If equality of opportunity guaranteed under Article 16(1) means effective material equality, then Article 16(4) is not an exception to Article 16(1). It is only an emphatic way of putting the extent to which equality of opportunity could be carried *viz*, even up to the point of making reservation.[39]

Thus 'the state can adopt any measure which would ensure the adequate representation in public service of the members of the Scheduled Castes and Scheduled Tribes and justify it as a compensatory measure to ensure equality of opportunity provided the measure does not dispense with the acquisition of the minimum basic qualification necessary for the efficiency'.[40]

Justice Krishna Iyer propounds a complex vision of the constitutional commitment to equality. Interpreting the Constitution by 'a spacious, social science approach, not by pedantic, traditional legalism',[41] he proposes to erect a 'general doctrine of backward classification' to pursue 'real, not formal, equality'.[42] According to the doctrine of backward classification, the State may, for purposes of securing genuine equality of opportunity, treat unequals equally. Thus Article 16(4)

> serves not as an exception [to the strictures of Article 16(1) and (2)] but as an emphatic statement, one mode of reconciling the claims of backward people and the opportunity for free competition the forward sections are ordinarily entitled to. . . . Closely examined it is an illustration of a constitutionally sanctified classification.[43]

So, in addition to reservations provided by Article 16(4), the State may also confer 'lesser order(s) of advantage' on the principle of classification under Article 16(1).[44]

---

[36] Id. at 516.

[37] Id. at 519.

[38] Id. at 516.   The idea of affirmative action is attributed to the United States Supreme Court, citing cases involving the unconstitutionality of the poll tax and the right of criminal defendants to a free transcript and to counsel on appeal.

[39] Id. at 519.

[40] Id. at 519.

[41] Id. at 525.

[42] Id. at 529.

[43] Id. at 535.

[44] Id. at 536.

At this point in the argument there is a crucial divergence between the views of Justice Krishna Iyer and those of Justice Mathew. For the latter, the compensatory measures authorized by Article 16(1) might be extended to 'all members of the backward classes',[45] not only to the Scheduled Castes and Scheduled Tribes. But Justice Krishna Iyer's more complex vision of Article 16 contains a second layer: the power of classification outside the boundaries of Article 16(4) for purposes of overcoming inequality may be used only on behalf of Scheduled Castes and Tribes.

> Article 16(4) covers all backward classes, but to earn the benefit of grouping under Article 16(1) based on Articles 46 and 335 . . . the twin considerations of terrible backwardness of the type Harijans have to endure and maintainance of administrative efficiency must be satisfied.[46]

'Not all caste backwardness is recognized' as a basis for differential treatment under Article 16(1).[47]

> The differentia . . . is the dismal social milieu of harijans. . . .The social disparity must be so grim and substantial as to serve as a foundation for benign discrimination. If we search . . . we cannot find any large segment other than the Scheduled Castes and Scheduled Tribes . . . no class other than harijans can jump the gauntlet of 'equal opportunity' guarantee. Their only hope is in Article 16(4).[48]

This is perplexing, for it appears that the stronger measures of reservation may be taken on behalf of all the backward, yet those who suffer the most terrible backwardness are the only ones entitled to measures which confer 'a lesser order of advantage'.

Later he suggests that to allow the Other Backward Classes to participate in these benefits may be detrimental to those who are most deserving

> . . . no caste, however seemingly backward . . . can be allowed to breach the dykes of equality of opportunity guaranteed to all citizens. To them the answer is that . . . equality is equality. . . . The heady upper berth occupants from backward classes do double injury. They beguile the broad community into believing that backwardness is being banished. They rob the need based bulk of the backward of the . . . advantages the nation proffers.[49]

---

[45] Id. at 519.
[46] Id. at 536.   Although here and elsewhere in his opinion, Justice Krishna Iyer uses the term 'harijans', his explanations suggest that he is employing this as a shorthand term for both Scheduled Castes and Scheduled Tribes.
[47] Id. at 537.
[48] Id. at 537.
[49] Id. at 539.

This distinction is justified because 'the Constitution itself makes a super-classification between harijans and others, grounded on the fundamental disparity in our society and the imperative social urgency of raising the former's sunken status'.[50] From the provision of Articles 330, 332, 335, 338 and others, we may deduce that 'the Constitution itself demarcates Harijans from others. . . . This is based on the stark backwardness of this bottom layer of the community'.[51] This constitutional differentiation of Harijans is specifically extended to the area of government employment as part of the State's obligation to 'promote the economic interests of harijans and like backward classes'. Articles 14 and 16 are, according to Justice Krishna Iyer 'the tool kit' to carry out the 'testament' of Articles 46 and 335.[52]

An attenuated form of this 'super classification' argument is found in each of the four majority opinions that embrace the classification argument. That argument posits a general authorization flowing from Article 16(1) to adopt reasonable classifications for purposes of securing equality of opportunity. But it is conceded that this does not include a power to employ those classifications specifically forbidden in Article 16(2). Therefore each of these four 'classification' opinions argues that the Scheduled Castes and Tribes do not comprise a classification on the basis of 'caste'.

Thus Justice Krishna Iyer argues that the ban of 16(2) doesn't arise in connection with measures for the Scheduled Caste and Scheduled Tribes for they are 'no[t] castes in the Hindu fold but an amalgam of castes, races, groups, tribes, communities or parts thereof found . . . to be the lowliest and in need of massive State aid and notified as such by the President'.[52] Article 16(4) allows use of these forbidden grounds to identify the Backward Classes that may be recipients of reservation. The four judgements imply that the other kinds of compensatory treatment justified directly by Article 16(1) are available only to classes which avoid the classification forbidden by Article 16(2), including Scheduled Castes and Tribes since they are not castes. But if the idea is to confine compensatory classification under Article 16(1) to Scheduled Castes and Tribes, this argument proves too much. For there are innumerable categories — e.g., on the basis of income, occupation, physical handicap, etc. — that are not based on the classifications forbidden in Article 16(2). The argument for Scheduled Castes and Tribes as a 'super-classification' cannot be sustained on the basis of the structure of

---

[50] Id. at 532–3.
[51] Id. at 533.
[52] Id. at 533.
[53] Cf. Chief Justice Ray, id. at 501; Justice Mathew, id. at 519; Justice Fazl Ali, id. at 549, 552. But cf. Justice Gupta, id. at 542.

Article 16, but only on the basis of their special recognition in the Constitution. But if that is the argument for their distinctiveness, the fact that they are not 'castes' within Article 16(2) is a distracting irrelevance.

In this and other matters, the *Thomas* opinions leave open many perplexing questions. Are equalizing measures permissible even if they employ the categories forbidden by Article 16(2)? Or, conversely, is the power to adopt equalizing classifications under Article 16(1) to be used exclusively to address disparities along the dimensions listed in Article 16(2)? How about differences in class? Income? Is the doctrine of classification merely permissive, so that government may make such compensatory classifications, but need not? Suppose it fails to do so? Do classifications that do not take account of inequalities violate the Article 16(1) guarantee of equality of opportunity?

Yet *Thomas* is welcome because it makes unavoidable reflection on compensatory treatment. By revealing the constitutional indeterminacy and doctrinal disarray in this field, it poses new challenges for jurisprudence and policy. Where courts could rely without much thought on quotations from *Balaji* or *Sagar*,[54] they must now articulate their choices. Actors, governmental and private, can no longer assume that the categories of compensatory policy are immutable.

A Symbolic Breakthrough, but . . . Who Gets What, When, How?

In terms of doctrinal housekeeping, there was little need for the *Thomas* reconceptualization of Article 16. 'Reservation' in Article 16(4) could readily have been construed to accomodate the Kerala scheme (as it was by Justice Beg). Such a construction would usefully have clarified the status of the age-waivers, fee concessions, travel allowances, coaching schemes, lowering of minimum marks and other provisions that typically accompany reservations and often exist apart from reservations *per se*. If the state enjoys the plenary power to reserve places for a beneficiary group, does this not imply a power to take lesser measures to help the members of that group obtain places? In order to join issue with the large questions, the *Thomas* majority (the four) had to accede to a very narrow definition of 'reservation' so that all of these practices are now exposed to challenge and possibly to the necessity of securing justification on broad classification principles, a course which may be strewn with conceptual landmines. Thus the radical reconceptualization of

---

[54] *Balaji v. State of Mysore*, A.I.R. 1963 S.C. 649 and *State of Andhra Pradesh v. Sagar*, A.I.R. 1968 S.C. 1379, are dutifully and copiously cited in cases involving the Backward Classes, but rarely subjected to critical analysis.

Article 16 arrives on the agenda at the cost of an unimaginatively narrow reading of what common sense would regard as the most relevant constitutional provision, Article 16(4).

A generous construction of Article 16(4) would not only have addressed the problem in *Thomas*, but would have been readily comprehensible to all the different groups concerned with reservations in government posts — officials in charge of appointments and promotions, government servants and aspirants to government posts. *Thomas* reaches the same immediate outcome by elaboration of ambiguous and recondite doctrine that is not readily accessible to officials who must design and manage programmes of preference, or to the beneficiaries of such programmes, or to disappointed non-beneficiaries — or in large measure even to lawyers and judges.

Perhaps, though, the accomplishment of *Thomas* lies in its providing government with more ample means to pursue compensatory policies. The doctrinal obstacles that might have impeded governmental policies in this area stood for the most part on weak ground, unlikely to withstand sustained conceptual attack.[55] Openings for favourable reinterpretation were plentiful. To bestow on government more ample authority to do something does not automatically mean that more of it will get done. Was the critical shortage one of doctrine favourable to such policies? Or, the will, energy and competence to implement them?

Consider the situation of the Kerala Registration clerks in *Thomas*. The State's scheme here seems less a carefully calculated modification of job requirements than a desparate improvization amounting to a confession of failure of its earlier policies. Extensions hadn't worked before, so another one is tried. There is no indication that any thought was given to (1) some way of helping the Scheduled Caste lower division clerks prepare to pass these tests; nor to (2) modifying the tests to eliminate cultural biases or extraneous matters and measure qualities genuinely needed for the job; or (3) modifying the job to make it suitable to candidates with the qualifications of these applicants. The immediate thrust of the decision is to enlarge the State's authority to confer preferential treatment. But the failure of the earlier measures was not due mainly — perhaps even at all — — to lack of State authority, but to lack of will or capability to make its schemes work. Enlarging State authority will not necessarily supply the lack. Indeed, it may allow the State to substitute easy short-cuts (like the one here) for thinking about how to do the job effectively. Since it may now in effect decree the result of keeping these

---

55 For example, the quantitative restrictions imposed by *Devadasan v. Union of India*, A.I.R. 1964 S.C. 179, or the restrictions on the use of layers of preferential treatment found in *Balaji v. State of Mysore*, A.I.R. 1963 S.C. 649, and elsewhere.

clerks in the higher posts, it may have less incentive to devise ways of motivating or enabling them to grasp these opportunities and improve their performance, thus prolonging their lack of qualification and reinforcing their dependence. And, as the State's authority is more broadly defined, there will be fewer occasions for the courts to observe State programmes and to monitor and energize government performance.

Of course, it is not only a question of the capability of the State, but also of the recipients of preferential treatment. The argument in favour of the extension of the Scheduled Caste clerks in the upper division posts is that it will somehow motivate or equip them to pass the test eventually. And unless they are motivated and enabled to take the initiative and improve their own capacity to perform the extension (and any compensatory measure for that matter) is only a temporary palliative. There is no indication that the State of Kerala in any way enhanced its ability to elicit a satisfactory performance from these clerks.[56]

Thus Thomas does not offer any hope of breaking out of the pattern of patronage and dependence. Both majority and minority judgements visualize the Scheduled Castes as the passive recipients of governmental largesse, rather than as active participants in their own improvement. And for all its apparent radicalism, *Thomas* enlarges State power in a way that may jeopardize the future of compensatory preference for the Backward Classes. In two senses this enlargement is a false victory for the Scheduled Castes.

First, the court passed up the chance to make the modest contribution of usefully expanding the meaning of reservations to align doctrine with existing practice about age waivers, coaching schemes, etc. In taking the more radical course, the court attempts to make legal doctrine yield up benefits that it cannot yield. Doctrine can give only authority, not power. And the crucial shortage was not of State authority, but of the will and capacity of the State to deliver benefits (and of the receipients to utilize those opportunities to enlarge their capabilities). The result is a symbolic breakthrough in which Scheduled Castes, their well-wishers and wider publics are beguiled into thinking that much (or too much) is being done for Scheduled Castes.

Second, the new reading of equality may detract from the attention and priority accorded to the Backward Classes. The new equality doctrine is so ample that it sweeps in its path all that confines the commitment to compensatory treatment to specific historic

---

[56] I am indirectly informed that by June 1978, most of the Scheduled Caste clerks whose promotion sparked the *Thomas* case have been provisionally certified as Upper Division Clerks upon passing the departmental test.

groups. If Articles 14–16 proclaim a regime of substantive equality
and if the State may employ classification to remedy any falling
short of the equality thus mandated, the government's responsibility
to confer compensatory preference is vastly larger than it has hitherto been understood. It is a responsibility that runs not only to
Scheduled Castes and Tribes and to Backward Classes more widely
conceived, either in terms of social groups or the poor. It also includes those who suffer difficulties as a result of personal misfortune
(disaster victims), accidents of personal history (the physically handicapped) or as a result of meritorious service to the nation (ex-
servicemen or dislocated children of diplomats).

In a setting of chronic shortage, an enlarged commitment to
remedy all undeserved difficulties betokens a commendable generosity of spirit. But it also raises the question of priorities and of allocation of scarce resources, including attention. The Government's
authorization to pursue substantive equality is vastly greater than
the resources that will conceivably be available to it. Among the
claimants on its compensatory powers will be many who are better
placed to press their claims on the attention and sympathies of
government. Will not the commitment to the lowest social groups
—especially where these are perceived to receive massive benefits—
be overwhelmed by governmental response to better-placed claimants on its compensatory attention?

Of course, the State could always take account of these difficulties
under Article 14 in terms of reasonable classification. But the notion
that the State has a general obligation to produce substantive
equality means that the kinds of disadvantage that afflict the better
off — diplomats, central government employees, retired army
officers, physically handicapped children in well-to-do, educated
families — are now elevated, as far as compensatory responsibilities
go, to parity with the government's commitment to overcome disparities associated with the traditional social hierarchy. The earlier
sense of a regime of formal equality qualified by a singular exception,
to alleviate disparities derived from position in the traditional social
hierarchy, is now liquidated or dissolved into a general and unful-
fillable commitment to substantive equality.

The distributive potential of this new dispensation is dramatically
realized in *Jagdish Rai v. State of Haryana* where the State reserved
a substantial portion of government posts for ex-servicemen on the
ground that they were handicapped because 'over the years [they]
have lost opportunities for entering government service and have
also lost contact with ordinary civilian life'.[57] The Full Bench sweeps

---

[57] A.I.R. 1977 P. & H. 56, 61. Since India has no conscription, these were volun-
teers. Two separate reservations figured in this case: the Division Bench addressed

aside the notion that reservations have to be justified by Article 16(4)
as 'a relic of the old way of thinking. . . . The old idea has now given
way to the idea that [Article 15(4) and Article 16(4)] . . . are themselves aimed at achieving the very equality proclaimed and guaranteed by Article 14 and other clauses of Articles 15 and 16'.[58]
After extensive citation from *Thomas*, which it commends for having
'got rid of the old sterility' and 'introduced a new dynamic and a
new dimension into the concept of . . . equality of opportunity'
the Court observes that '[i]t is no longer necessary to "apologetically"
explain laws aimed at achieving equality as permissible exceptions.
It can now be boldly claimed that such laws are necessary incidents
of equality'.[59] Reservation of posts for ex-servicemen is justified for
they suffer difficulties in competing with civilians for civilian job
and the State has an obligation to provide them employment. Thus
the State is justified in classifying them separately as a source of recruitment.[60] To secure a just proportion of posts to those who suffer
a peculiar handicap in competition is 'an extension of the principle
of Article 16(4) to those that do not fall under Article 16(4)'.[61]

In effect, other deserving groups are now entitled to reservations
along with the Backward Classes specified by Article 16(4) and in
effect, all the kinds of preferential treatment which seemed to be
allowed exclusively to the Backward Classes may now be bestowed
on other groups regarded as deserving by the State. It is a fitting
symbol that the losing petitioner in *Jagdish Rai* was herself a Scheduled Caste.

Thus *Thomas* opens Pandora's box: compensatory classification
is available — perhaps incumbent — to succour all the disadvantaged. The earlier notion that 'the interests of the weaker sections
are a first charge'[62] on government is dissolved into a diffuse, and
in the nature of things, largely symbolic, generalized egalitarianism.
Justice Krishna Iyer's doctrine of super-classification may be understood as an attempt to close the lid again, confining compensatory
classification to Scheduled Castes and Tribes. Were this doctrine to
gain acceptance, state authorization to confer benefits on these groups

---

a 50 per cent reservation for 'Ex-Emergency Commissioned Officers' as demonstra-
tors in a Dental College. Id. at 57. On reference to a Full Bench, attention shifted
to a reservation of 28 per cent of posts of sub-inspectors in the Food and Supplies
Department for 'released Army Personnel'. Id. at 58.

[58] Id. at 58.

[59] Id. at 60.

[60] Id. at 61. The 'source' language is borrowed from *Chanchala v. State of Mysore*,
A.I.R. 1971 S.C. 1762. This borrowing confirms the suspicion that the distinction
between making reservations and specifying sources is a thin one indeed.

[61] Id. at 61.

[62] *Balaji v. State of Mysore*, A.I.R. 1963 S.C. 649 at 651.

might end up not very far from where liberal reading of Article 15(4) and 16(4) would have left it. Superclassification would re-establish the priority of Scheduled Caste claims by emphasizing a picture of Indian society as riven by an unbridgeable dichotomy between Scheduled Castes and Tribes and the rest of the population. This dichotomous picture, by drawing a rigid line between Scheduled Castes and Other Backward Classes, could impede programme administration and obstruct the eventual dismantling of preferential treatment by preventing merger of these categories. More immediately, it would further stigmatize Scheduled Castes and Tribes by portraying them as uniquely hapless and helpless specimens, a potential demonstrated in the overtones of condescension found in some of the *Thomas* opinions.

But it would be shortsighted to judge *Thomas* only in terms of what it does to governmental power in this area. There is another side to the coin and one that is potentially of even greater significance. If Articles 14, 15, and 16 are read as mandating the pursuit of substantive equality, then to what extent is substantive equality — or at least governmental efforts to promote it — an enforceable Fundamental Right? Since those denied Fundamental Rights may resort to the courts to enforce them, would the potential beneficiaries of such equalizing measures have a right to resort to the courts to secure governmental compliance with this right of equalization?

It would not be surprising if the courts would shrink from affirmative enforcement of these reconceptualized rights to equality. But imagine for a moment that they were willing to do so. Scheduled Castes and others would not have to wait for government preference on their behalf, but could take the initiative in the courts to secure the 'enforcement' or implementation of their rights to substantively equalizing measures. Scheduled Castes movements could mobilize around these issues, generating the kind of political movement that would make government responsive to judicial proddings.

We shall in a moment mention the doubts that surround this scenario. But to emphasize it once more, in this reading the ultimate significance of *Thomas* is not the enlargement of State authority to confer preferential treatment, but the acknowledgement of a Fundamental Right to substantive equality and the possibilities for affirmative litigation by disadvantaged groups to force the State to fulfil its responsibilities. This would amount to an ironic reversal of the meaning of *Thomas*. We started with the obvious view that *Thomas* loosened judicially-imposed restraints on government, allowing it to patronize the least advantaged; we end by viewing *Thomas* as suggesting the imposition on government of a new and onerous accountability to these disadvantaged, an accountability mediated through the courts.

This scenario is subject to a number of contingencies. First, do the litigants and their lawyers have the legal imagination to devise these claims? Second, do they have the capacity to organize to sustain such litigation and press for the implementation of favourable rulings? (Of course, the capacity for organization might be enhanced by creative deployment of legal services.) Third, are the judges likely to be responsive to such claims, claims that ask them to depart from deeply held notions about the judicial role? They would have not only to innovate original standards of what is appropriate movement towards equality, they would have to undertake a sustained activist monitoring of government that they may find both ideologically uncongenial and institutionally discomfiting.[63] (Were they to shrink from this, while maintaining an interpretation of equality rights as substantive, *Thomas* would, in the guise of exalting the Directive Principles, have helped to demote these Fundamental Rights to the non-enforceable status that led to the disdain of the Directive Principles.) Fourth, even if claimants come forward and courts are responsive, do the realities of litigation in India, the delay and manoeuver and cost, make it possible for groups to improve their position through litigation against powerful adversaries? This is related to a fifth contingency: are the real barriers to improvement of the conditions of these groups reachable by even the most well-disposed and capable judiciary? To what extent is the notion of remedying these problems through litigation yet a further tempting illusion?

Conclusion

It is not clear yet in what ways and how deeply *Thomas* will transform earlier doctrine and practice. It was pronounced by a divided court; it expresses the heated symbolic egalitarianism that was both institutionalized and discredited during the Emergency Rule. It liberates the courts conceptually from what we tagged the 'classic compromise' on compensatory discrimination, but it does so at the cost of obscuring and diffusing the commitment to remedy the historical plight of the lowest groups in Indian society. But if it provides no satisfying answers, its presence invites and requires some fresh thought about the principles underlying compensatory discrimination policies and about the role of courts, lawyers, and citizens in effectuating them.

[63] The accomplishments, problems, and discomforts of the American counterpart of this kind of activist 'public law' judging have spawned an immense literature. A useful introduction is provided by Chayes (1976); D. Horowitz (1976); Galanter, Palen and Thomas (1979).

12
New Patterns of Legal Services in India

India is a society in which law (in the conventional sense of author-itative general rules propounded by official agencies) is called upon to play a major role in maintaining order, effectuating social control. implementing deliberate change, and adjusting the accompanying strains and tensions. There seems to be every prospect that law (as opposed, for example, to party cadres or charismatic mobilization) will continue to play a central role, both symbolically and substan-tively. In its adherence to legal forms and loyalty to legal procedures, India is quite unusual among Third World countries. Upon closer observation, this reliance upon and loyalty to law turns out to be more ambiguous than it first appears. In place of the image of India as a *Rechtsstaat*, we might substitute a double vision in which law in India is at once elaborated but attenuated, pervasive but precarious.

Notwithstanding recurring complaints that India's legal system is unsuitable to her conditions, courts and lawyers are a highly visible part of Indian life, frequently resorted to in matters ranging from great public issues to village disputes. Although almost everyone professes a preference for harmony and conciliation. Indians tend to perceive many situations as involving violations of legal entitlements for which appropriate remedy is vindication in a public forum. Although it is hard to document the common assertion that the rate of litigation is high, it is significant that it is widely perceived to be high.

India has a numerous legal profession — both absolutely and re-latively. A recent estimate places the number of lawyers at 228,000 — approximately 336 for each million people, a ratio far larger than almost all Third World countries.[1] In spite of differences of region, language, caste and religion, these lawyers share a common legal culture which they put into the service of a wide variety of interests. The lawyer and the legal system provide familiar techniques and standards for forwarding private interests and conducting public business. The law forms part of the familiar idiom of public life.

The pervasive attachment to law in India is evidenced by the extraordinary respect accorded the judiciary, whose exalted consti-

[1] The Indian figure is taken from a personal communication from Dr N.R. Madhava Menon of the Bar Council of India Trust.

tutional position is matched by enormous popular regard. Judges are on the whole seen as honest, independent and immune to the partiality and venality that are attributed to most actors in public life. In spite of some outspoken complaints about the formalism of law and the conservatism of judges, leading to obstruction of progressive reforms, attitudes towards the higher judiciary remain remarkably respectful. Among those who deplore particular decisions, there has been little talk of outright defiance: even open advocacy of evasion is not respectable in Indian public life. During Emergency rule (1975–7) the government took care to curtail the authority of the courts by formal legal means, rather than by breaching the fabric of legal authority. (Although the Supreme Court disappointed many critics of Emergency rule, the High Courts emerged with an enhanced reputation as guardians of citizens' rights.) The extraordinary regard in which the higher judiciary is held is evidenced by reliance on them for extracurricular 'integrity jobs' and the frequent calls for judicial inquiries.

Loyalty to Anglo-Indian legal institutions and legal forms was strikingly demonstrated during the 1975–7 Emergency. Although there were a number of drastic incursions into the legal order (curbing the power of courts, denying civil liberties, etc.), there was a very consistent adherence to superficial legal proprieties and a striking unwillingness to depart from the major institutional forms of Anglo-Indian law. There was no serious consideration of replacing the familiar brand of Anglo-Indian courts with either 'people's courts' or with indigenous *panchayats*. Indeed, it was during the Emergency that government undertook new commitments to support a massive legal aid programme, presumably to increase access to and use of the legal system.

At the same time, the role of law and lawyers in India seems superficial and precarious. The prestige of the legal profession has declined since the pre-independence days when lawyers were key spokesmen for national aspirations. The emphasis on economic development and the burgeoning of elective politics enhanced the power and prestige of other callings, accelerating the long term drift away from the bar as a focus of ambition for the educated. It is clear that lawyers play a much less prominent role in public life than they did before independence. Lawyers are often seen as parasites and mere word-mongers who have little to contribute to national development, when they are not obstructing it by over-zealous protection of vested interests. Recently there has been some erosion of the position of the judiciary *vis à vis* the government. Institutional disarray within is matched by a greater readiness to direct sharp criticism at the courts and an increasing potential for government interference with judicial independence.

Trust in law has been underminded by the dramatic increase in lawless violence, both private and public. Lawlessness is perceived as widespread and officials as both unable to control the violators and as susceptible to pressures, both legitimate and otherwise. There is an increasing tendency for many sorts of groups to take direct action to vindicate their claims and the success of direct action confirms cynicism about legal methods. The visibility of dramatic instances of legal ineffectiveness produces widespread cynicism about the possibilities of social change through law. Government has engaged in mass production of rights and entitlements that it cannot easily fulfil. Implementation is lax and malleable. The profusion and complexity of laws and the difficulties of implementing them create broad possibilities for manipulation and manoeuvre. For the poor and disadvantaged the law and its personnel are present as agents of oppression; its promises of improving their lot and protecting them from their oppressors seem empty, if they are known at all.

Legal ineffectiveness gives new life to persisting doubts about the suitability of current legal arrangements. There are enduring loyalties to the notion that the adversary conflict institutionalized in the law is destructive and counter to India's indigenous traditions. Yet no promising alternatives have been formulated.[2]

Thus the portrait is a very mixed one. India's legal institutions have endured and flourished, but they are beset by a growing sense of despair. They arouse very high expectations, but are widely seen as failing to fulfil these. Promises of liberation and protection are seen as illusory. Current concern with legal aid in its broad sense of access to justice (*vide* the Bhagwati Report)[3] is a manifestation of discomfort with the gap between promise and performance. Present patterns of delivering legal services are, I argue here, a key element maintaining the persistent tension between high expectations and disappointing performance.

Organization of Legal Services in India

Legal aid is often approached simply as a matter of providing the prevalent type of legal services to those who are unable to afford to purchase them on the private market. Instead I shall adopt the more ample perspective of the Bhagwati Report which sees legal aid as implicating larger questions of access to justice and of the effectiveness of law. In this wider view, the problems of delivering legal services to the poor and to other unrepresented groups cannot be

2 See chaps. 3 and 5.
3 Ministry of Law, Justice & Company Affairs (1978).

addressed without considering the general question of the way in which the delivery of legal services is organized in India.

Cost barriers are not the only impediment to the delivery of legal services. The structure and organization of professional life itself affect what services are provided and to whom. Extending legal services to the poor and to unrepresented interests provides an occasion for examining the wider problems of what services get produced, whether they are really the ones that clients or society need most and whether legal work could be so organized as to provide a more appropriate range of services. For example, the range of legal services presently available to those who can afford it is (with only limited exceptions) confined largely to advocacy in court. A programme of legal aid that aimed at extending the present mix of legal services to the entire population would be limited to token coverage or would immensely overburden the judicial system. The current concern to extend benefits to presently unserved groups points to questions of enlarging the creative role of the profession in meeting India's needs for new forms of protection and participation, and for systematic but flexible regulation.

Existing studies portray a profession characterized by a cluster of mutually reinforcing characteristics which to a large extent inhibit the growth and adaptation invited by these new demands. Among the prominent features of Indian lawyers are their orientation to courts to the exclusion of other legal settings; their orientation to litigation rather than advising, negotiating or planning; their conceptualism and orientation to rules; their individualism; and their lack of specialization.[4]

Public and profession in India concur in visualizing the lawyer in the role of courtroom advocate rather than advisor or negotiator, much less business or social planner. Lawyers see themselves this way and clients typically come to them at a relatively late stage of a dispute, already committed to go to court. Ties with clients (and with legal agencies other than courts) tend to be episodic, not enduring. Lawyers have little incentive to develop new expertise in matters relevant to the client's affairs; nor do they enjoy opportunities to pursue the clients' interests in arenas other than courts.

The lawyer's business is usually at a court; he typically spends his working life at a particular level of the system. But in spite of the stratification that this entails, the profession is relatively undifferentiated in character. Within each level, lawyers are stratified by skill, influence, prestige and wealth. Lawyers at higher levels enjoy more of these on the whole, but lawyers at each level do much the same

---

[4] See the various studies on the Indian legal profession in *Law & Society Review* (1968–9).

sort of thing. There is little division of labour by specialization (beyond civil-criminal) and little co-ordination in the form of partnerships or firms. Basically, all lawyers offer the same narrow range of services under conditions of chronic oversupply. Competitiveness limits solidarity and the capacity for corporate action.[5]

The emphasis on litigation and the barrister role reinforces lawyer's rule-mindedness. Where the lawyer's task is to win disconnected battles rather than to pattern relationships, there is little to induce the practising lawyer to go beyond the kind of arid conceptualism that has been characteristic of much Indian legal scholarship and which pervades Indian legal education. Writing and teaching about the law are, with significant exceptions, confined to textual analysis on a verbal level with little consideration either of underlying policy on the one hand or problems of implementation on the other.

The Indian lawyer, in spite of his estimable record of public service, has thus remained restricted in his professional rule. The intermediary, negotiator, trustee, planner, advisor, and spokesman functions that might have gravitated to the lawyer are performed, if at all, by others — clerks and touts, village notables and businessmen, politicians and administrators — and frequently in a manner that excludes the influence of authoritative legal commitments to the disadvantaged.

Compared, for example, to his protean, expansionist American counterpart, the Indian lawyer remains restricted in his role. It is only a slight exaggeration to think of him as a somewhat enlarged, popularized version of the barrister half of his British counterpart. (But in Britain barristers make up less than 10 per cent of the divided legal profession!)

Thus we get a picture of legal services supplied by relatively unspecialized lawyers, involved in little co-ordination of effort, offering a relatively narrow range of services. Relations with clients are episodic and intermittent. The lawyer addresses discrete problems in isolation from the whole situation. of the client and uninformed by considerations of long-range strategy. This kind of *atomized* legal service accentuates the disadvantages of the poor and disadvantaged in using the legal system. Vindication of the entitlements of the poor is not part of the standard repertoire of lawyers, which is oriented to the needs of the recurrent users of the system. The ways the poor could use the legal system are more inchoate and

---

5 Prestige and power do not serve to co-ordinate activity within the profession. Koppell observed that the leaders of the metropolitan bar "did not have a following of young lawyers who would implement plans and projects developed by their seniors" —an observation that parallels Morrison's account of the isolation of "leading lawyers" in his study of the Ambala bar. Koppell (1966); Morrison (1968–9).

require investment in innovative research and investigation. But lawyers are disinclined to make these investments because of lack of imagination, resources and models and because the poor and disadvantaged are not organized to provide a sustained market for such expertise if it were developed.

Indian lawyers are reminiscent of the Indonesian bazaar merchants, described by Clifford Geertz in *Peddlers and Princes*,[6] who displayed the calculating rationality deemed requisite for economic development, but remained locked into a hyper-individualized, undifferentiated and inefficient bazaar economy. This example of producers with the 'right' culture, but unable to navigate the transition to more productive forms of organizing their work, points to the larger questions about the delivery of legal services in India that must frame inquiry into possibilities for legal aid.

Can the Indian legal profession expand its role to meet the new demands? Will lawyers detach themselves from the courts and learn to operate in a wider range of legal settings? Will they overcome their individualism to find forms of sustained collaboration that will permit the development of expertise in the substantive problems of those they serve? Will they temper their rule-mindedness into a more flexible and pragmatic problem-solving approach? The problem of adapting to broader functions is not primarily a problem of lack of skills or drive, but of lack of appropriate organizational forms for mobilizing skills and channelling them to meet emerging needs. If such a transformation depends upon the adaptive capacity of the profession, it also requires that the demand for more differentiated, complex and widely distributed legal services be made effective. Thus, the development of large scale legal aid programmes provides an opportunity for promoting new organizational forms for legal services. Indeed, I shall argue that only through the development of new forms of collaboration, specialization, and service delivery, can the promise of legal aid be fulfilled.

Styles of Delivering Legal Services

The term 'public interest law' is often invoked in discussions of extending effective representation to interests and constituencies which have not enjoyed effective participation in the legal process. If it is not a defensible analytic construct, it does have heuristic value — even in a comparative context. In order to extract that value it is necessary to detour slightly, referring to the organization of legal

6 Geertz (1963).

services in the United States — the original locus of public interest law.

I should preface this detour by emphasizing that the reference to the American legal profession is not offered in a spirit of invidious comparison. If the American profession has many magnificent achievements, it cannot be held up as a model for the provision of legal services to the poor or to diffuse but unorganized interests. The American profession has afforded comprehensive, highly skilled and wide ranging services to large enterprises and organizations but it has failed to provide the same kind of comprehensive service to the poor and unorganized.

Law practice in the United States is a bifurcated structure, organized around different kinds of clients. The upper strata consists mostly of large firms whose members are recruited largely from elite schools and who serve corporate clients: the lower strata practice as individuals or small firms, are drawn from less pestigious schools, and serve individual clients. Much of the variation within the profession, as one recent study concludes, is accounted for by 'one fundamental difference . . . a difference between lawyers who represent large organizations (corporations, labor organizations or government) and those who represent individuals or individuals' small businesses.'[7]

The corporate or organizational segment of modern American law practice is characterized by a cluster of features, which (though each can be found apart from the cluster) taken together constitute a very distinctive kind of lawyering.[8] These lawyers typically practice in large units with considerable specialization both among firms and by individual lawyers within firms. These firms are characterized by co-ordination among lawyers of different specialities to collaborate on particular problems. There is considerable training within the firm: the work of juniors is supervised and reviewed by seniors.

These firms tend to have more enduring and continuous relations with clients, to encounter problems at an earlier stage when preventive and planning work is still possible, and to offer very wide range of services. Such firms typically can engage in more painstaking investigation and more elaborate research. They can explore the options more exhaustively and are on the whole more innovative in their tactics. Of course this kind of legal service is much more costly than the more routine and less elaborate episodic legal services delivered to individuals. The client is not only represented in court

---

[7] Heinz and Laumann (1978: 1117).

[8] This style of lawyering is described in Galanter (1982).

but in a variety of other forums, including administrative and legislative ones. The lawyer negotiates with both actual and potential collaborators, antagonists and regulators.

Although much of this kind of lawyering, like virtually all ordinary lawyering, is routine and stereotyped, there is more scope for individualized response to the clients problem. What is more striking in this kind of law practice is that the lawyer is not confined to addressing specific problems of the client that have already developed into troubles, but can take a larger view and try to use the law strategically to pursue the long range interest of the client. The lawyer looks at the total situation of the client and the opportunities and problems that surround him, not just at one isolated trouble. The lawyer participates in planning the operations of the client to take optimum advantage of the legal environment.

By comparison the legal services supplied to individuals tend to be more episodic and intermittent, more confined to particular troubles, less comprehensive and less informed by considerations of long-term strategy. These services are provided (typically) by less specialized lawyers, involved in less co-ordination of effort; research, investigation, and the use of experts are less elaborate. Traditional legal aid programmes extend this kind of service to persons who otherwise could not afford it. Typically (and often fallaciously) the legal problems of the poor are thought to be simple and routine, readily manageable with only minimal attention and effort. Thus, lawyers who do 'service' work are thought to be 'quite justified in handling large numbers of cases and are not expected to raise 'test case issues' or spend too much time on any particular case.'9 Too often such programmes respond to high caseloads and restricted aspirations by handling cases routinely and unaggressively, dispensing service that is 'shallow, cautious and incomplete'.

Many 'law reform' and 'public interest law' efforts in the United States can be seen as attempts to put the more expansive 'corporate' style of lawyering at the service of non-established interests: the poor, minorities, consumers, the handicapped, inmates of mental institutions and so forth. What is pursued here is not just the services of skilled and dedicated lawyers, but something more — operation of a scale, scope, and continuity to reap the advantages conferred by this sort of large-scale lawyering. These advantages include the ability to pursue a long run strategy by acquiring specialized expertise, co-ordinating efforts on several fronts, selecting targets and forums, managing the sequence, scope and pace of litigation, monitoring developments, and deploying resources to maximize long

---

9 Bellow (1977: 114).

term advantage (including educational and organizational effects as well as favourable awards).

Models of Legal Services Delivery

This American example is not offered to identify an exclusive or optimal method for effectively serving the legally deprived. Its value is that it suggests the intimate connection between the range and complexity of services provided and the organization of legal work. These connection are not fixed and invariant. It is not necessary nor even possible to copy American arrangements in the Indian setting. But considering them should alert us to the range of possibilities encompassed in such terms as 'legal aid' or 'public interest law'.

To map these possibilities, I shall postulate two contrasting types of legal services programmes which I call the Service type and the Strategic type.[10] (My typifications depict them as polar opposites, but of course there are in the real world many intermediate cases in which these characteristics are mixed. However, I think there is a tendency for these various characteristics to cluster.)[11] It will be immediately apparent that these types are legal aid analogues to the contrasting styles of law practice ('corporate' and 'individual') described above.

The Service legal aid programme confines its attention to the discrete claims brought to it by individuals. Typically these are claims which are already clearly defined as vindicatable under existing law. The object at hand is some tangible gain (or avoidance of loss) for a specific individual. The service programme is reactive in its mobilization of cases — that is, it chooses its cases from those presented to it by prospective clients. Typically, such programmes make little effort to reach out to potential clients by publicizing the availability of services, by educating clients about potential claims,

10 In formulating these types, I draw upon the helpful presentation of Trubek (1978).

11 The Strategic versus Service distinction summarizes a number of features. A more elaborate treatment might begin by depicting as separate dimension (1) the product delivered (service vs. strategic) and (2) the organizatin of professional work (complex vs. undifferentiated). In the present discussion, which relies on the strong correlation between complex organization and the capacity to deliver strategic lawyering. I have not pursued this distinction. But I recognize that the correspondence is not exact: there are undoubtedly instances of service lawyering provided by complex organizations and probably instances of strategic lawyering by undifferentiated providers. For purposes of the present discussion, I have used the term innovative legal services to include programmes that aspire to use new means and/or deliver a wider range of services.

or providing them with supportive services to facilitate claim-making. Hence case intake is limited by the profile of existing information, initiative and resources in the community. Since individuals must overcome the various obstacles to bringing a case (lack of information, psychic cost, the transaction cost of taking a day off and travelling to the law office, fear of reprisal, etc.) it is not likely that cases will be brought to the service programme where individual stakes are concealed or small, even though large numbers of people may share a similar problem.

Typically, the Service programme focuses on representation in court. Lawyers are seen as courtroom advocates and legal services are viewed as primarily courtroom representation by lawyers in litigation. Work in other arenas (administrative, legislative, or mere advising) are seen as ancillary or subordinate parts of the legal services effort. The tasks that the lawyer does are confined to the traditional modes of advocacy in the litigation setting — i.e. oral and written argumentation. In this setting of individual claims being argued in court, clients are typically encouraged to assume passive and dependent roles in their relations with attorneys. They are typically treated as incapable of participating in decisions about the pursuit of their goals.

To this Service type of legal services delivery — reactive, paternalistic, limited to the pursuit of individual claims in the courtroom setting by formal legal advocacy — is contrasted what I shall call the Strategic type of legal services delivery. Such a programme is oriented to the longer run as well as to immediate advantage or relief and to matters that effect groups rather than isolated individuals. Such programmes focus on those complaints with a public dimension where a whole class of persons are potential beneficiaries. They are willing to invest in establishing favourable precedents by test-case litigation and interested in aggregating claims too small to pursue thoroughly into larger aggregates which can be economically pursued on behalf of a whole group. In order to assemble such claims, they will undertake pro-active strategies of mobilization, reaching out into relevant constituency groups, informing potential clients of their services, searching for test cases and aggregating diffuse claims into formidable units. Rather than operate exclusively through litigation, the Strategic programme pursues the interests of its clients in legislative, administrative, media, educational and political arenas. In addition to the modes of advocacy characteristic of courtroom settings the Strategic programme engages in research, negotiation in a variety of settings, citizen education, media relations and so forth. The strategic programme's aspiration to comprehensive attention to the larger (and changing) interests of the client groups leads to client participation in decision-making both for its educative value and because the client participation is needed to choose between competing versions of the goals to be pursued.

The distinction of Strategic and Service types should not be taken to describe mutually exclusive and opposite types. There are in practice many combinations of the two themes. In particular, the Strategic mode should not be misinterpreted as implying an exclusive concentration on test cases and law reform efforts, subordinating the interests of individual clients to the reform goals of the professionals. (This can be dispelled by recalling that the paradigm case of strategic lawyering was drawn from a setting — large firm corporate practice in the United States — in which lawyers were highly responsive to the interests of their clients.) At its best strategic lawyering is a way of enhancing service to client groups.

But the juxtaposition of these types points to contrasting dangers that threaten the legal services enterprise. Lawyering involves translation of the problems and goals of clients into a legal idiom of claims. The classical problem of service lawyering is that the client's problem is reduced to a sterotyped legal claim which is then treated in isolation from his total position. Conversely, the classical problem of strategic lawyering is that the goals that inform the strategy are those of the lawyers rather than the clients. (If service lawyering offers a band-aid, strategic lawyering may offer unnecessary surgery!) The problem is to find ways to combine the larger strategic vision with attentiveness to client needs and responsiveness to client goals.[12]

Styles of Lawyering and Legal Aid in India

Most lawyering in India is 'atomistic'. The lawyer addresses discrete problems in isolation from the whole situation of the client: there is very little planning or preventive work. Relations with clients tend to be episodic. The range of services offered is narrow. There is little specialization and little professional collaboration. Lawyers contend with a constant struggle to mobilize a clientele. (In these respects,

---

12 A very interesting linkage of strategic elements with service to poor clients is proposed by Bellow (1977) as the 'focussed case strategy.' Lawyers in a poverty law office co-ordinate and compare the way they handle ordinary day-to-day cases. This comparison identifies 'Target' institutions whose illegal practices affect significant numbers of the client group. Through referrals, in-depth interviews, and solicitation the programme discovers and represents large numbers of clients victimized by these institutions. By co-ordination and communication within the office, the lawyers will crystallize strategies and goals, directed against these target institutions, including legislative efforts, direct negotiation with the target institutions, support for community organization and class action litigation, as well as the pressing of individual claims.

lawyering in India resembles the segment of American practice which services individuals.)

Legal Aid has long been a declared aspiration of the Indian bar, honoured largely in the breach until interest quickened duing the Emergency. (The Bhagwati Committee on Judicare was established in May 1976 to propose implementation of the 1973 Krishna Iyer report.[13] The 42nd. Amendment, while curtailing the powers of the higher judiciary, deposited in the Constitution's chapter on Directive Principles a provision directing the State to provide free legal aid.[14] Several states now have such programmes in operation others are in the planning stage. Various groups of lawyers have schemes makes generalization hazardous and although new variants sponsored legal aid programmes of their own. Although the variety of are still appearing, I think it is possible to venture some surmises about the central tendency of such efforts.

Not surprisingly, most discussion of legal aid envisions programmes close to the Service end of the spectrum. A typical programme is oriented toward pursuit (or defence) of individual claims in court; it is passive in case selection and paternalistic in relation to clients. It eschews any deliberate impact on the design or administration of government policy. It envisions attorneys entering cases after the dispute has already taken shape and focusing their efforts on courtroom advocacy. There is little provision for follow-up, for aggregating similar claims, for operating in other arenas, etc.

One significant departure from this pattern — infrequent, but symbolically important — is what is called in India 'public interest litigation'. This refers to cases in which the Supreme Court or a High Court permits volunteer lawyers to bring a case on behalf of some victimized group — prisoners, bonded labourers, inmates of a home for neglected girls, pavement dwellers threatened with relocation. Typically there is no active participation by the would-be beneficiaries. Often the court actively invites (or induces) the bringing of the case. This public interest litigation is undertaken by a small number of judges. (The persistence of such enclaves of judicial activism is made possible by the attentuation of hierarchic control in Indian courts, rendering them colleges within which can flourish many inconsistent strands). Traditional notions of remedy are often stretched to devise appropriate relief.

---

[13] Ministry of Law, Justice & Company Affairs (1973).

[14] Article 39A reads:

Equal justice and free legal aid — The State shall secure that the operation of the legal system promotes justice, on a basis of equal opportunity, and shall, in particular, provide free legal aid, by suitable legislation or schemes or in any other way, to ensure that opportunities for securing justice are not denied to any citizen by reason of economic or other disabilities.

Public interest litigation departs from typical Indian lawyering in its proactivity and its orientation to large questions of policy, particularly vindication by government of its commitments to welfare and the relief of the oppressed. Although broader in scope than typical legal aid schemes, in crucial ways it replicates the prevailing atomistic style. Public interest litigation too is initiated and controlled by elites and is responsive to their sense of priorities. It carries no accountability to a specific client constituency nor does it imply a sustained commitment to such a constituency. Typically, it is an episodic response to a particular outrage. It does not mobilize the victims nor help them to develop capabilities for sustained effective use of law.

The Indian legal services scene does contain a scatter of programmes that break from the prevailing atomistic style of legal services delivery. These innovative programmes are small and not family institutions. They are located in the voluntary sector. They are not sponsored by the government. For the most part they depend on short term funding, very often from foreign sources. I would like to briefly sketch a few of these programmes, not because they are in any way representative, but because they suggest the range of possible rearrangements of the ingredients present in the Indian legal setting.

Rajpipla. Rajpipla Free Legal Aid is part of a wider developmental and service programme in a tribal area in Gujarat, run by Jesuits and supported by overseas charities. It has trained 'barefoot lawyers' or 'paralegals' who disseminate legal information, mobilize cases and provide assistance and support to clients. Clients are brought to an office staffed by three lawyers who screen, select and prepare cases; select, brief and pay local counsel to do courtroom advocacy. This arrangement enables them to discern patterns in the caseload, aggregate cases, devise campaigns focused on suitable targets, and monitor the performance of outside lawyers.

Rangpur. The Legal Support Scheme for the Poor operated by the Anand Niketan Ashram is also only one component of a wider developmental and service programme in a tribal area of Gujarat, supported largely by foreign funds, run by a charismatic Gandhian. Like Rajpipla there is use of paralegals for intake and support. The key difference is the establishment of offices staffed by salaried lawyers at the taluk towns and the district headquarters (and the retention of a clerk and part time advocate at the High Court). With some 12 lawyers, 43 paralegals and 18 clerks and administrators, this Scheme is no doubt the largest law firm in the state. It is distinctive in its vertical integration of operations from the village through the High Court. During its initial year of operation (1978–9), the first centre rendered advice in about 5000 cases, and 259 cases were filed in court. A long-established Lok Adalat (people's court) hears disputes among the tribals themselves and serves as an additional intake mechanism for disputes with outsiders and as a training ground for paralegals. Finally, it should be mentioned that this Ashram can generate considerable political clout and legal initiatives are consciously co-ordinated with approaches through other channels.

Aware is a development programme for tribals which operates in nine districts of Andhra Pradesh. Its legal work emphasizes legally informed direct action rather than litigation. There are about thirty trained 'barefoot lawyers' who, on the basis of investigation, make representations to force government to implement legislation favourable to the tribals. They emphasize legal education and knowledge rather than use of legal proceedings, encouraging collective efforts to bring pressure to enforce the law. The barefoot lawyers are backed up by a legal department consisting of three retired magistrates. When they decide to litigate, they retain outside counsel but the investigation is done by AWARE's legal department.

CERC. A very different kind of innovation is the Consumer Education and Research Centre in Ahmedabad. This is a public interest law firm (in the American sense) which has vigorously pursued a variety of issues from the performance of the Life Insurance Corporation to food adulteration. It has a full-time paid staff of 10, including a professional manager, researchers, a librarian and a journalist. It has developed working relations with technical experts in a number of fields. The lawyers are not full-time. In addition to innovative litigation, the Centre publishes research reports, handles complaints, runs training programmes and sponsors public meetings. It is oriented to the media and has displayed great skill in focusing attention on its issues. It has no organized constituency to whom it is responsible. Funding has come from foundations and to a slight extent from the government.

Notwithstanding the unrepresentative character of my sample, some patterns suggest themselves. For example, all the rural groups are in tribal areas — there seems to be no instance of innovative legal service in a caste village setting. Each of the tribal programmes involves the leadership of charismatic outsiders who supply not only technical know-how but overall direction and day to day leadership. All seem weakly institutionalized in their dependence on outside funding. But what I want to emphasize here is the profusion of innovations and the various ways that the deficiencies of atomistic lawyering are overcome: the development of proactive mobilization and investigatory capacity by use of paralegals; the development of capacity for discovering patterns in the caseload and aggregating cases afforded by full-time attorneys or by the 'solicitors' office' at Rajpipla; the vertical linkage from village mobilization through the high court, overcoming the separation of levels typical of Indian law practice; the use of cumulative educational efforts (through a variety of media) to support legally informed direct action; the continuing linkage with technical experts; and so forth.

What do we learn from this profusion of innovations? Can these programmes devise ways to provide strategic lawyering in the Indian setting? Can they be successfully enlarged and securely institutionalized? Can they be replicated? Can their example induce other legal service deliverers, including the massive state legal aid schemes, to incorporate some features of the strategic type and deliver more effective lawyering? The uncertainty of the answers increases as we proceed to ask about wider and more enduring effects.

Although these questions remain unanswered, I think the developments sketched here allow us to draw some soft but important conclusions and to pose a further set of questions.

First, we may conclude that although the typical features of legal life in India are firmly institutionalized, they are not inalterably fixed. There are possibilities for re-arrangement of the familiar ingredients that produce strikingly different patterns. Innovators have hit on a variety of ways to reassemble them and it is likely that still others will be forthcoming.

Second, the example of these strategic programmes may provide an impetus to the strategic use of law by groups far removed from e.g., tribals and consumers. The existence of these programmes may contribute to a wider appreciation of the possibilities of using law strategically and to organizational innovations to realize these possibilities. And there is no reason to suppose that such appreciation will be confined to the poor or 'have not' groups.[15] The prevailing 'atomistic' pattern is a low level equilibrium in which resources are used inefficiently and opportunities foregone. The organizational innovations of lawyers for the poor may contribute to the development of the Indian version of the law firm.

Third, one may anticipate more strategic use of law both by 'have not' and 'have' groups. Law has always been used instrumentally in India[16] but now it will be used in a more comprehensive fashion in sustained campaigns utilizing various *fora*, etc.

Fourth, what happens to the courts (which are already in some

[15] Cf. the reborrowing by large corporate economic interests of the American public interest law format for large-scale lawyering — the tax exempt foundation free of responsibility to the short-term needs of specific clients and free to pursue its notion of the public interest. See Galanter (1982).

[16] E.g. Kidder (1973); Mendelsohn (1981).

institutional disarray) if many groups start using them strategically as part of protracted campaigns? Can courts produce the goods in terms of devising new remedies, enforcing compliance with their judgements? Will this restore their institutional vigour or will they be consumed by the political heat that will be generated?[17]

Finally, even if there is a growth of innovative providers of strategic legal services at the disposal of the poor and the courts are responsive, can law really be liberative for the disadvantaged? In many ways India is a propitious setting for innovative and aggressive legal services. Government has distributed 'rights' and 'legal entitlements' broadside to the poor and unrepresented. There is no shortage of authoritative policy favourable to these groups but unenforced or unimplemented — e.g. land reforms, anti-discrimination measures, food adulteration laws, regulation of the working conditions of contract labourers, etc. Even when unfulfilled these commitments may provide useful resources to social action groups.[18] Legal action can push for fulfilment of existing legal commitments, without carrying the burden of securing them. The pursuit of such entitlements through legal action is familiar and acceptable — if often inaccessable — to wide sections of the population.

But of course there are other barriers to the fulfilment of these commitments: retaliatory pressures, lack of resources and competing priorities. Those who will be discomfited by implementation have other resources they can use to frustrate legal gains. Good legal outcomes secured by good lawyering will not automatically transform Indian society. Whether legal institutions have enough leverage

[17] That is if courts enjoyed relative autonomy so long as they addressed issues of power in a fragmented and intermittent way, will their independence be tolerated if they are used by parties to thwart or pressure power holders in a sustained and systematic way? Cf. Toharia (1975).

[18] Cf. the observations of Upendra Baxi (1982) that . . . information about the state law may often help ORP [Organizations of Rural Poor] to fashion more effective strategies for attainment of social change objectives over period of time. If the state law imposes prohibition, or reckless felling of trees, or pollution of environment or provides for free and compulsory education of children or family planning, the ORP following the same objectives can attain several concrete results by using these information packages. First they can resist delegitimation from adversaries by stressing the commonality of the development objectives. Second, they can negotiate bureaucratic contingencies as more effectively (e.g., from obtaining major resource allocation decisions in their favour to even getting support for seemingly small steps as use of a public place in a village for night-school for landless labourers, and getting priority quota for kerosene lamps for this purpose from public distribution systems). Third, ORP can by reference to the commonality with the national objective embodied in state law add to their capabilities to withstand repression. Fourth, ORP can also thereby obtain political space and time — both valuable resources in themselves — needed for their growth and viability. Last but not the least, insofar as ORP are funded, especially from overseas, they need to have legal habitation (a bank account at the very least') and a name: and this process is facilitated by invocation of commonality.

on administrative and private action so that legal determinations would actually affect distributive outcomes is always an open question: the answer is contingent on many factors. But India remains a place with strong commitment to legal forms. There is reason to think that law can exert some pressure and influence on the uses of power and the distribution of benefits. Enhanced legal services is one way to increase the chances of mobilizing the resources needed to follow through and realize the benefits of legal gains.[19]

Whether courts and the poor can use law to force the government to carry out its commitments points to a distinction between the higher state and the local state.[20] Innovative lawyers and courts offer themselves as an instrument for carrying out the redistributive and welfare policies of the higher state against the resistance of locally dominant groups allied with the local state. Whether enhanced legal services will make a difference depends in part on how the incumbents of the higher state pursue their commitments to equality and welfare in the light of their competing goals and their relation to those who control the local state. The law cannot abolish the ambitions and alliances that temper and dilute those commitments. But the law provides a channel through which persistent and imaginative players can reinforce those commitments while using them to secure redress.

---

[19] Ironically, the low state of organization of most legal service delivery creates a setting in which sustained 'strategic' legal services programmes may provide previously unrepresented groups with an organizational edge, at least temporarily, in their ability to use law to produce significant change.

[20] This distinction is usefully adumbrated in R. Dhavan (1981).

Epilogue
Will Justice Be Done?

When I first travelled around India, observing lawyers and courts, I was told a joke about the fresh junior lawyer who was despatched by his senior to argue his first case in some remote mofussil town. At the hearing the young lawyer prevailed. He was elated and immediately proceeded to the telegraph office and wired his senior 'Justice is done'. Exhausted, he retired to his hotel and fell into a deep sleep. He was rudely awakened by a pounding on the door. It was a messenger with a return telegram, which he tore open hastily, to find a message 'Appeal at once'.

This story seemed to capture many facets of law in contemporary India. It records the widely felt disjunction of law and justice. It reflects a resigned acknowledgement of the inability of the system to render justice. At the same time, it observes ruefully the perpetual renewal of faith — here that of the young aspirant — doomed to be betrayed by experience. It portrays veteran lawyers as cynical scoundrels and identifies them as the active perverters of the system. Having cherished this as a characteristically Indian story, I was bemused when some years later I heard it told (in America) as a story about American lawyers. This turned out to foreshadow the larger realization that much of what I once thought peculiar to India turns out to be pertinent as I look closer to home.

In 1957, a year after graduating from law school, I went off to India to study the abolition of untouchability. I encountered a world vastly different than my training had led me to expect or equipped me to deal with. (This would no doubt have been true had I remained at home and gone into law practice.) Immersion in another culture is a celebrated method for exposing our presuppositions and showing us that familiar arrangements are problematic rather than the way things have to be. And India, with its profound dissociation of legal norms and social reality, and its spectacular display of familiar elements arrayed in dramatically different shapes was especially well-suited to expose the contingency of my picture of what law was like.[1] This picture, I should add, was not the product of

---

[1] India is especially suited to nourish that robust sense of possibility whose possessors Robert Musil called 'possibilitarians.'

Anyone possessing it does not say, for instance: Here this or that has happened, will happen, must happen. He uses his imagination and says: Here such and such

296

extensive experience with the American or any other legal system;
it was the picture I had acquired in the course of going to law school.
It portrayed the law as an integrated purposive system, residing in a
hierarchy of agencies, moved by and applying a hierarchy of norms.
Law, in this view, expresses the central values and concerns of the
society; it draws on the power of the State but disciplines that power
by its own autonomous and internally derived norms. The central
legal institution is the court and the central and the typical activity
of courts is adjudication. With some slippage and friction, social
behaviour is aligned with and guided by legal rules. Moreover, this
behaviour can be deliberately modified by appropriate alterations
of these rules. Hence law can be used to re-design social arrangements; it can guide and monitor the transformation of social life.
If, a generation after the assaults of Legal Realism, few took this as
a literal description, it still provided a picture of the normal shape
of legal reality that was sufficiently resilient to withstand dislodge-
ment by ordinary counter-instances.[2]

But in the Indian crucible 'all that is solid melts into air. . . .'[3]
The study of India provided a series of lessons that undermined that
picture by violating my expectations of continuity and correspondence between law and society. To the naive and curious student,
India suggested a set of bewildering possibilities, namely that:

> a legal system may flourish and become deeply rooted although
> dissonant with central cultural values, embodied in other
> institutions;
> a legal system may be internally disparate, embodying dis-
> cordant norms and institutionalizing conflicting practices in
> different levels and agencies;
> the divergence of legal norm and social practice is not transient
> and exceptional, but normal and institutionalized;
> deliberate legal changes do not ordinarily produce the effects
> their proponents avow; and the effects they do produce are
> largely unanticipated;

might, should or ought to happen. And if he is told that something *is* the way it is,
then he thinks: Well, it could probably just as easily be some other way. So the
sense of possibility might be defined outright as the capacity to think how everything could 'just as easily' be, and to attach no more importance to what is than
to what is not. Musil (1953: 12).

2 For a fuller sketch of this paradigm, see Galanter (1974) Trubek and Galanter
(1974).

3 Marx and Engels (1947: 17).

— the effects of legal regulations depend not only on the conduct of legal authorities, but on the consumers or users of the law and their differential capability to use it;
— The law as a system of symbols diverges from the law as a system of operative controls

The papers in this collection may be viewed as a series of attempts to appreciate and respond to these perplexities. They attempt to develop a view of Indian legal reality that steers between the 'classic' fallacy of treating India as a flawed pathological specimen of law and the 'romantic' fallacy of thinking that beneath Indian law's false 'foreign' surface lies a more authentic, ineffably Indian legal genius waiting to be released. The first ('Harvard of the East') approach leads to prescriptions for transplanting missing institutions — law schools using the case method, research institutes, public interest law firms and so forth. From the latter comes the enduring allegiance to the idea of *panchayats*. I should emphasize that Indians are not exempt from the first nor do they have a monopoly on the latter.

'The Uses of Law in Indian Studies' was written in the early 1960s; the most recent of these essays in the early 1980s. The first proposed to use law as a path to understanding India; the latter attempt to assess Indian experience in the light of more general understanding of the dynamics of legal life. In part the swerve is one of biography — but it is also part of a larger trajectory. For between my early contact with India and now there has grown up a vigorous scholarly community devoted to social research on law. This multi-disciplinary enterprise goes by several names — 'law and society', the sociology of law, socio-legal studies, and law and behavioural science. It cultivates a second kind of learning about law, that seeks explanation rather than justification, that emphasizes process rather than rules, and that tries to appreciate the distinctiveness of law against the background of larger patterns of social behaviour rather than as something autonomous and self-contained. The second kind of learning about law, only dimly anticipated in the earliest of these essays, has in the intervening years become a substantial intellectual presence.[4]

Looking back at the extended absorption in India that constituted my 'Journey to the East', I am reminded of two contrasting readings of such a pilgrimage. In one reading, the seeker abandons

---

[4] On the major themes of social research on law, see Galanter (1985).

familiar scenes and comforts to undertake an arduous journey. He encounters wondrous creatures and curious customs, penetrates to the heart of inaccessible mysteries and is in the process indelibly transformed. But there is a second version: it is of the seeker who pursues his quest to a distant city where he uncovers a puzzling map which finally reveals to him that the treasure he seeks is buried in his own back yard — which he then returns to unearth. In the first reading, the protagonist penetrates mysteries and is transformed in a way that separates him from the quotidian world from which he embarked and joins him to something finer and more profound. In the second reading, the hero also undergoes a process of discovery: the vicissitudes of the journey equip him to discover the wondrous lurking beneath the familiar from which he started out. His village remains unchanged during his absence, but he returns to it with eyes enabled to see layers and patterns within it that were not discernible before his journey.

During the past dozen years, most of my research has been on the United States. Although India is mentioned only infrequently in this work, it is infused by perspectives developed in my encounter with Indian law. Each of my two major theoretical papers, 'Why the "Haves" Come out Ahead'[5] and 'Justice in Many Rooms',[6] could be thought of as a transposition of Indian materials. The former paper starts from the differences between the capabilities of disputing parties and traces the pervasive effects of these differences in patterns of litigation, legal services, legal institutions and legal rules. The immediate impulse for the paper was the juxtaposition of observations from the handful of then available empirical studies of litigation in the United States. But the attempt to examine law in the light of the disparate capabilities and opportunities of its various users was an approach I had developed in studying litigation under the Untouchability (Offences) Act.[7] Not legal rules, but the structures, strategies, and capabilities of the contending parties were seen to shape the patterns of law in action. The 'Haves' paper, elaborated this perspective and applied it to a wide range of instances in American society.

'Justice in Many Rooms' visualizes modern society as an arena of multiple and overlapping normative orderings, in which governmental

---

[5] Galanter (1974a).

[6] Galanter (1981).

[7] Galanter (1972). Although the final version was not published until 1972, this paper was prepared for a 1968 conference and a preliminary version published in 1969 [Galanter (1969)].

law is surrounded by rival and companion regulative ideas and institutions. It documents the persistence of 'indigenous law' through-out modern societies and argues that the effects of legal change depend on its interaction with these 'lesser' normative orderings. This normative and regulatory multiplicity is vividly displayed in India, where the coexistence of a multitude of distinctive styles of life is everyday reality. I was powerfully attracted to this pluralism, and intellectually engaged by the problem of how the autonomy and integrity of groups could be combined with the voluntarism and equality that animated India's constitutional regime.[8] My paper on 'Hinduism, Secularism and the Indian Judiciary'[9] examines the various modes in which the official legal system monitors and re-gulates the indigenous normative orderings. Its direct descendant, 'Justice in Many Rooms', elaborates this theme to comprehend the role of courts in the United States and their relations to the various indigenous regulatory orderings with which they coexist.

An interest in the micro-politics of litigation and a sense of the inherent multiplicity of legal ordering were both things I acquired in India. I like to think that my long pursuit of the elusive 'reality' of Indian law, which entailed unlearning much of the received view of how legal systems work, enabled me to develop a fresh perspective from which to view the legal process in America. Indian law has

[8] Of course exploration of India was not just a disinterested quest for scientific truth. If the initial appeal was a heady mixture of philosophical quest for the most profound unity, the allure of bejewelled dancers, and Nehru's aloof neutralism and patrician socialism, the enduring attraction was to the rich proliferation of human variety and the gracious accomodation of difference, not as a precarious political accomplishment but as an effortless instinctive style of life and thought.

Only when I read Rajeev Dhavan's introduction did it strike me how much my focus on pluralism and accomodation reflects a (then) barely articulated concern about Jewish identity and continuity. My first major paper on Indian pluralism, 'The Problem of Group Membership' was published in 1963 — the same year as 'A Dissent on Brother Daniel', criticizing the Supreme Court of Israel for failing to adopt a suitably pluralistic solution to the question of Jewish identity. Similarly, I criticized Indian judges for paying undue respect to hierarchy and rigid boundaries, deviating from the 'pragmatic' view of Indian society. This pragmatic view discerned beneath the rigidity and domination of the caste system a possibility of melioration in which the system could be turned on its side, so to speak, de-hierarchized, drained of coerciveness, infused with voluntarism. India seemed to contain the possibility of a benign compartmentalization which would permit to flourish a variety of less than universal fellowships. This vision of a non-hierarchic pluralistic India also represented a fantasy solution to the Jewish problem of how to maintain continuity and integrity without isolation, how to attain participation and acceptance without assimilation.

[9] Chap. 10. Although it was not published until 1971, this paper was largely written in India in 1966.

been a source of ideas and perspectives that have been assimilated into the repertoire of the 'law and society' research community that has emerged over the past 25 years.[10] But the flow of ideas from India to the American law and society community has been only a minor trickle compared to the surging tide of influence of American law on India.

Since independence, legal intercourse between India and the United States has been heavily one-directional. Indian courts have been influenced by American constitutional doctrine. Legal education and legal scholarship have been influenced by American models.[11] But this influence has been largely inadvertent. India has been invisible to the American legal academy and the world of law practice.

That a country with almost one-sixth of all the people on earth, with a legal system akin to ours (and that operates in English, no less!) has pulled off the improbable feat of establishing a regime of constitutional liberty, with vigorous judicial protection of human rights, in an extremely poor and extraordinarily diverse society ought I would think, to have attracted the attentions of American legal intellectuals. But the American legal world has been totally oblivious. In the two hundred law schools in the United States, there is not a single academic programme that focuses on Indian law; among their 6000 law teachers, there is not to my knowledge a single one paid to specialize in the law of India.[12] In the midst of this indifference, American law schools hosted as graduate students a whole generation of young Indian legal scholars who today populate the upper echelons of India's legal academy. Since this training was not rooted in any significant American scholarly involvement in Indian matters, it contained little that addressed the realities of the setting to which these students would return.

Only twice have events briefly penetrated American parochialism to spark an interest in Indian law — during the Emergency and in the aftermath of the Bhopal disaster. In the year and a half since the gas leak at Bhopal there has been more attention in America to the reality and potential of the Indian legal system, than in the

[10] This is epitomized by Vol. 9, No. 1 of the *Law & Society Review*, containing the first part of a symposium on 'Litigation and Disputes Processing' that had a profound effect on the direction of social inquiry about law. This number contained four articles — two field studies of Indian litigation, and two theoretical papers by authors whose work was informed by extensive experience in India.

[11] Baxi (1985); Dhavan (1985).

[12] The state of studies of South Asian law in American universities is reviewed by Robert Kidder in Beer, *et al.*, (1978).

thrity-seven years from independence to the tragedy in Bhopal.[13] Whether this surge of interest will lead to anything enduring remains to be seen. The effect of Bhopal on the flow of information was of course far greater in India. An unprecedented barrage of information about torts, damages, and procedure reached an expanded and attentive audience. New models and new perspectives appeared just when Bhopal had raised both public expectations of accountability and remedy and public awareness of the lamentable state of the Indian legal system. The system's troubles and failures were palpable and publicly acknowledged. Thus Chief Justice Bhagwati warned in late 1985 that 'the judicial system in the country is almost on the verge of collapse' and condemned the failure of the legal system to serve as an instrument for succoring the poor and disadvantaged and for transforming Indian society.[14] Issues of procedures and remedies, the structure of courts, and the organization of legal services gained a new salience. By dramatizing the gap between aspiration and performance, Bhopal has presented a singular opportunity for wide-ranging public debate on the structure of justice in Indian society.

In closing, I would like to return to the story at the beginning of this Epilogue, because I think it invites a re-reading. If we confine ourselves to what it says, we are left with a counsel of resignation and despair. But if we add what it leaves unsaid, we see a more complex message.

The story is permeated by a feeling that law ought to be the institutionalized pursuit of justice. Its divergence from this proper state, suspected but painful to acknowledge, is the 'dirty little secret' that the joke reveals. But the joke is even more grim, for it shows us that youthful ignorance in eager pursuit of justice can itself be manipulated to serve the schemes of the unjust.

Yet in depicting this nasty state of affairs, we are given a glimmer of hope. For a moment the veil is lifted; the injustice that parades as justice reveals itself. Presumably the young lawyer gains insight into the nature of the system in which he is enmeshed. Perhaps he succumbs to despair and joins the army of the disillusioned. But we are invited to hope that he is transformed by the experience. Armed with this hard-won cynical knowledge, he can carry on the struggle in a more measured and self-aware way.

[13] This is exemplified by the devotion of the plenary session at the 1986 Annual Meeting of the Association of American Law Schools to a discussion of 'Bhopal, the Good Lawyer, and American Law Schools'. The papers delivered on that occasion have been published in the *Journal of Legal Education*, Vol.36, No. 3 September 1986).

[14] Speech of Chief Justice P.N. Bhagwati on Nov. 26, 1985, on the 36th. anniversary of the adoption of the Indian Constitution.

This little drama, then, tells us that learning how the system works equips us to resist it and perhaps even to transform it. It teaches us '[t]o recognize the absence of justice where justice is postponed or deformed, without succumbing, therefore, to a belief in its non-existence. . . .'[15] In the midst of entrenched injustice, it echoes Gramsci's admonition to 'be a pessimist of the intellect, an optimist of the will.'

Just this combination of bitter knowledge and vibrant action is visible in the post-Emergency Indian legal world. A saving remnant of judges, lawyers and academics, together with a few social activists, journalists and social workers have taken bold and inventive initiatives to use the law to provide justice for the most deprived and oppressed. Although still 'token' in scale, these initiatives point to the possibility of effective use of law by the oppressed and the unrepresented. While exposing and discrediting the present operations of the legal system, they affirm its potential for transformation. India has an immense pool of talent, a great fund of skills, and vast institutional resources. But legal institutions and legal actors remain locked in stultifying patterns. This leaven of activism and innovation demonstrates how these ingredients can be reassembled in more productive and beneficent combinations. Ahead lies the challenge of the arduous transition from such experiments to general and institutionalized reforms.

---

15 Sugarman (1979: 221).

REFERENCES

Acharya, Bijay Kisor (1914), *Codification in British India*. Calcutta: S.K. Banerji & Sen.

Adler, Chaim, Reuven Kahane and Amy Avgar (1975), *The education of the disadvantaged in Israel: comparisons, analysis and proposed research*. Jerusalem: The Hebrew University of Jerusalem, School of Education.

Agarwal, Shriman Narayan (1946), *Gandhian constitution for free India*. Allahabad: Kitabistan.

Ahmad, M.B. (1941), *The administration of justice in medieval India*. Aligarh: The Aligarh Historical Research Institute.

Aiyar, P.S. Sivaswami (1965), 'A great liberal', in K.A. Nilakanta Sastri, ed., *Speeches & writings of P.S. Sivaswami Aiyar*. Bombay: Allied.

Ali, Hamid (1938), *Custom and law in Anglo-Muslim jurisprudence*. Calcutta: Thacker, Spink and Company.

All-India Congress Committee (1954), *Report of the Congress village panchayat committee*. New Delhi.

Ambedkar, B.R. (1946), *What Congress and Gandhi have done to the Untouchables*, second edition, Bombay: Thacker & Co, Ltd.

Angell, James W. (1934), 'Reparations', in *Encyclopaedia of the Social Sciences*, 13:300–8 New York: MacMillan.

Altekar, A.S. (1958), *State and government in ancient India*, third edition, Delhi: Motilal Banarsidass, 1958.

Austin, Granville (1966), *The Indian Constitution: a cornerstone of a nation*. Oxford: Clarendon Press.

AVARD [Association of Voluntary Agencies for Rural Development]. (1962), *Panchayat raj as the basis of Indian polity: an exploration into the proceedings of the constituent assembly*. New Delhi: AVARD.

Ayyar, A.S.P. (1958), *Rambles in literature, art, law and philosophy*. Madras: Madras Law Journal Office.

Baden-Powell, B.H. (1886), 'English legislation in India', *The Atlantic Quarterly Review*, October, 2 (4), 365–80.

Bailey, F.G. (1965), 'Decisions by consensus in councils and committees, with special reference to village and local government in India', in Banton, L.M. ed., *Political systems & distribution of power*. London: Tavistock for Association of Social Anthropologists.

——. (1957), *Caste and the economic frontier*. Manchester: Manchester University Press.

Bastedo, T. (1969), 'The judiciary in Bihar'. Unpublished Doctoral Dissertation, Duke University.

Basu, K. (1967), *Reform of panchayati raj in West Bengal*. Garia, 24 Parganas: Mahendra Nath De.

Baxi, U. (1967), 'The little done, the vast undone: some reflections on reading Granville Austin's 'The Indian Constitution', *Journal of the Indian Law Institute*, 9: 323–40.

———.(1968), 'Directive Principles and sociology of Indian law', *Journal of Indian Law Institute*, 10: 245–72.

———.(1975), 'Legal assistance to the poor: a critique to the Expert Committee Report', *Economic and Political Weekly*, 10(27): 1005–13.

———.(1976), 'From Takrar to Karar: the lok adalat at Rangpur––A preliminary study', *Journal of Constitutional and Parliamentary Studies*, 10: 52.

———.(1982), 'Legal mobilisation and the needs of the rural poor' (Mimeo).

———.(1985), 'Understanding the traffic of "ideas" in law between America and India', in Robert M. Crunden, ed., *The traffic of ideas between India and America*. Delhi: Chanakya Publications, pp. 319–42.

Bayley, D. (1969), *The police & political development in India*. Princeton: University Press.

Beals, A. (1955), 'Interplay among factors of change in a Mysore village', in M. Marriott, ed., *Village India*. Chicago: University of Chicago Press, pp. 80–104.

———.(1970), 'Namahalli 1953–1966: urban influence and change in southern Mysore', in Ishwaran, ed., *Change and continuity in India's villages*. New York: Columbia University Press.

Beer, L.W. Clarence Dias, Randle Edwards, Robert Kidder, Daniel Lev and Barry Metzger (1978), 'Asian legal studies in the United States: a survey report', *Journal of Legal Education*, 29: 501–67.

Bellow, Gary (1977), 'Turning solutions into problems: the legal aid experience', *NLADA Briefcase* 34: 106.

Berreman, Gerald D. (1963), *Hindus of the Himalayas*. Bombay: Oxford University Press.

Bhat, K.S. (1970), 'The role of political parties', in R.N. Haldipur and V.R.K. Paramhansa, eds., *Local government institutions in rural India: some aspects*. Hyderabad: National Institute of Community Development, pp. 217–25.

Bittker, Boris (1973), *The case for Black reparations*. New York: Random House.

Blackshield, A.R. (1966), 'Secularism and social control in the west the material and the ethereal', in G.S. Sharma, ed., *Secularism: its implications for law and life in India*. Bombay: N.M. Tripathi Pvt. Ltd., pp. 9–85.

Borale P.T. (1968), *Segregation and desegregation in India*. Bombay: Manaktalas.

Bose, Nirmal Kumar, ed. (1960), *Data on caste: Orissa*. Calcutta. Anthropological Survey of India.

Bose, Pramatha Nath (1917), 'The legal exploitation of the Indian people', *Modern Review*, 21:30–7.

Brass, Paul R. (1969), 'The politics of ayurvedic education', in S. Rudolph and L. Rudolph, eds., *Education and politics in India: studies in organization, society and policy*. Delhi: Oxford University Press.

Bryce, James (1901), *Studies in history and jurisprudence*. New York: Oxford University Press.

Chakraverti, Suranjan (1965), *Domestic tribunals and administrative jurisdictions*. Lucknow: Eastern Book Company.

Chatterjee, B., S. Singh & D. Yadav (1971), *Impact of social legislation on social change*. Calcutta: The Minerva Associates.

Chattopadhyay, G. (1964), *Ranjana: a village in West Bengal*. Calcutta: Bookland Private Ltd.

Chayes, Abram (1976), 'The role of the judge in public law litigation', *Harvard Law Review*, 89: 1281.

Cohn, B.S. (1955), 'The changing status of a depressed caste', in M. Marriott, ed., *Village India*. Chicago: University of Chicago Press, pp. 54–79.

——. (1959), 'Some notes on law & change in north India', *Economic Development & Cultural Change*, 8:79–93.

——. (1960), 'The initial British impact on India: a case study of the Benares region', *Journal of Asian Studies*, 24: 418–31.

——. (1961), 'From Indian status to British contract', *Journal of Economic History*, December, 21(4) 613–28.

——. (1962), 'The British in Banaras: a nineteenth century colonial society', *Comparative Studies in Society and History*, 4: 169–99.

——. (1965), 'Anthropological notes on disputes and law in India', in L. Nader, ed., *The ethnography of law* (*American Anthropologist* Special Publication 67 (No.6), part 2 82–122.

Cohn, H. (1958), 'The spirit of Israel law', *International Lawyers Convention in Israel*, Jerusalem.

Conlon, Frank F. (1963), 'The Khojas and the courts'. An unpublished manuscript, Department of History, University of Minnesota.

Connell, A.K. (1880), *Discontent and danger in India*. London: Kegan, Paul & Co.

Coulson, N.J. (1963), 'Islamic family law: progress in Pakistan', in J.N.D. Anderson, ed., *Changing law in developing countries*'. London: George Allen and Unwin, 114–53.

Coupland, R. (1944), *The Constitutional problem in India*. Oxford: Oxford University Press.

Dalal, M.N. (1940), *Whither minorities?* Bombay: D.B. Taraporevala Sen & Co.

Davies, H.O. (1962), 'The legal and constitutional problems of independence', in Peter Judd, ed., *African independence*. New York: Bell.

Das, Govind (1967), *Justice in India*. Cuttack: Shangon & Shangon.

De Bary, W.T. (1958), *Sources of Indian tradition*. New York: Columbia University Press.

Derrett, J.D.M. (1956), 'Hindu law: the Dharmasastra and Anglo-Hindu law—scope for further comparative study', *Zeitschrift für vergleichende Rechtswissenschaft*, 58: 199–245.

———. (1957), *Hindu law past and present*. Calcutta: A Mukherji and Company.

———. (1958), 'Statutory amendments of the personal law of the Hindus since Indian independence', *American Journal of Comparative Law*, 7:380–93.

———. (1959a), 'Sir Henry Maine and law in India', *The Judicial Review*, 40–55.

———.(1959b), 'The role of Roman law and continental laws in India', *Zeitschrift für ausländisches und internationales Privatrecht*, XXIV(4), 657–85.

———. (1961a), 'The administration of Hindu law by the British', *Comparative Studies in Society and History*, 4: 10–52.

———. (1961b), 'Sanskrit legal treatises compiled at the instance of the British', *Zeitschrift für vergleichende Rechtswissenchaft*, 63: 72–117.

———. and J.H. Nelson (1961), 'A forgotten administrator historian of India', in C.H. Phillips, ed., *Historians of India, Pakistan, Ceylon*. London: Oxford University Press, pp. 354–72.

———. (1963a), 'Justice, equity and good conscience', in J.N.D. Anderson, ed., *Changing law in developing countries*. London: George Allen and Unwin, 113–53.

———. (1963b), *Introduction to modern Hindu law*. Oxford: Oxford University Press.

———. (1964a), 'Law and the social order in India before the Muhammadan conquests', *Journal of Economic and Social History of the Orient*, 8 1: 73–120.

———. (1964b), 'Aspects of matrimonial causes in modern Hindu law', *Revue du sud-est Asiatique*, 3: 203–41.

———. (1968), *Religion, law and the state in India*. London: Faber & Faber.

———. (1969), 'Law'. Paper presented to Study Conference on tradition in Indian politics and society', 1–3 July 1969, University of London (School of Oriental and African Studies).

———. (1978), *The death of a marriage law: An epitaph for the rishis*. New Delhi: Vikas.

Desika Char, S.V. (1963), *Centralised legislation: a history of the legislative system of British India from 1834 to 1861*. London: Asia Publishing House.

Dewey, John (1946), *The public and its problems*. Chicago: Gateway Books.

Dhavan, R. (1981), 'Public interest litigation in India. An investigative report'. New Delhi: Committee for Implementing Legal Aid Schemes.

———. (1985), 'Borrowed ideas: on the impact of American scholarship on Indian law', *American Journal of Comparative Law*, 33: 505–26.

Dhavan, S.S. (1964), 'The Indian judicial system I', in M.G. Gupta, ed., *Aspects of Indian Constitution*, second edition, Allahabad: Central Book Depot.

———. (1966), 'Secularism in Indian jurisprudence', in G.S. Sharma, ed., *Secularism: its implications for law and life in India*. Bombay: N.M. Tripathi Pvt. Ltd, pp. 102–38.

Dickinson, John (1853), *Government of India under a bureaucracy*. London. [Reprinted and published by Major B.D. Basu, Allahabad, 1925].

Durkal, J.B. (1941), *Conservative India*. Ahmedabad: 2:99.

Dushkin, Lelah (1957), 'The policy of the Indian National Congress towards the depressed classes'. Unpublished Master's thesis, University of Pennsylvania.

—— (1961), 'The Backward Classes: special treatment policy', *Economic and Political Weekly*, 13:1665–8, 1695–1705, 1729–38.

Eldersveld, Samuel J., and Bashiruddin Ahmed (1978), *Citizens and politics: mass political behavior in India*. Chicago: University of Chicago Press.

Fox, Richard G. (1967), 'Resiliency and change in the Indian caste system: the Umar of U.P', *Journal of Asian Studies*, 26: 575–87.

Fyzee, Asaf A.A. (1955), *Outlines of Muhammadan law*. London: Oxford University Press.

Gadbois George H.Jr. (1970), 'The Supreme Court of India: a preliminary report of an empirical study', in *Journal of Constitutional and Parliamentary Studies*, 41:33.

——. (1977), 'The Emergency: Mrs Gandhi, the judiciary and the legal culture'. Paper presented to the 16th. Annual Meeting of the Southeast Conference of the Association for Asian Studies.

Gajendragadkar, P.B. (1954), *The Hindu Code Bill*. Karnataka University Extension Lecture Series, No.2. Dharwar: Karnataka University.

——. (1965), *Law, liberty and social justice*. Bombay: Asia Publishing House.

Galanter, M. (1961a), 'Caste disabilities and Indian federalism', *Journal of the Indian Law Institute*, 3: 205–34.

——. (1961b), 'Equality and "protective discrimination" in India', *Rutgers Law Review*, 14: 42–74.

——. (1962), 'The problem of group membership: some reflections on the judicial view of Indian society', *Journal of the Indian Law Institute*, 4: 331–58.

——. (1963a), 'A dissent on brother Daniel', *Commentary*, 36 (1): 10–17.

——. 8 (1963b), 'Law and caste in modern India', *Asian Survey 1963*, 3: 544–59.

——. (1964a), 'Hindu law and the development of the modern Indian legal systems'. Paper delivered at the 1964 annual meeting of the American Political Science Association, Chicago.

——. (1964b), 'Temple entry and the Untouchability (Offences) Act, 1955', *Journal of the Indian Law Institute*, 6: 185–95.

——. (1965), 'Secularism east and west', *Comparative Studies in Society and History*, 7: 133–59.

——. (1966a), 'The religious aspects of caste: a legal view', in Donald E. Smith, ed., *South Asian politics and religion*. Princeton: Princeton University Press, pp. 277–310.

——. (1966b), 'The modernization of law', in Myron Weiner, ed., *Modernization*, New York: Basic Books, pp. 153–65.

——. (1968–9), 'The study of the Indian legal profession', *Law Society Review*, 3: 201–17.

———. (1969), 'Untouchability and the law', *Economic and Political Weekly*, 4: 131–70.

———. (1972), 'The abolition of disabilities—Untouchability and the law', in M. Mahar, ed., *The Untouchables in India*. Tucson: University of Arizona Press.

———. (1973), 'A note on contrasting styles of professional dualism: law and medicine in India', in M. Singer, ed., *Entreprenership and modernisation of occupational cultures in South Asia*. (Durham, Duke University Comparative Studies on Southern Asia., 1973), 310:3.

———. (1974a), 'Why the "haves" come out ahead: speculations on the limits of legal change', *Law & Society Review*, 9:95–160.

———. (1974b), 'The future of law and social sciences research', *North Carolina Law Review*, 52:1060–8.

———. (1978), 'Remarks on family law and social change in India', in David C. Buxbaum, ed., *Chinese family law in historical and comparative perspective*. Seattle: University of Washington Press, pp. 492–7.

———. (1981), 'Justice in many rooms: courts, private ordering and indigenous law', *Journal of Legal Pluralism*, 19:1–47.

———. (1982), 'Mega-law & megalawyering in the contemporary United States', in R. Dingwall and P. Lewis, eds., *The sociology of the professions: lawyers, doctors and others*. London: Macmillan, pp. 152–76.

———. (1984), *Competing equalities: law and the Backward Classes in India*. Berkeley, University of California Press and New Delhi: Oxford University Press.

———. (1985), 'The legal malaise; or, justice observed', *Law & Society Review*, 19: 537–56.

———, F. Palen & J. Thomas (1979, 'The crusading judge: judicial activism in urban trial courts', *Southern California Law Review*, 52: 699–741.

Gandhi, M.K. (1954), *The removal of Untouchability*. Bharatan Kumarappa, ed., Ahmedabad: Navajivan Publishing House.

———. (1962), *The law and the lawyers*. S.B. Kher Ahmedabad: Navajivan Publishing House.

———. (1965), *My picture of free India*, A.T. Hingorani, ed., Bombay: Bharatiya Vidya Bhavan.

Gangrade, K.D. and C.G. Sanon (1969), 'Panchayat elections in a Punjab village: changing political status of a depressed caste', *Eastern Anthropologist* 22:38–54.

Geertz, Clifford (1963), *Peddlers & princes: social development & economic change in two Indonesian towns*. Chicago: University of Chicago Press.

Ghouse Mohammed (1965), 'Religious freedom and the Supreme Court of India', *Aligarh Law Journal*, 2: 60–85.

Gledhill, A. (1954), 'The influence of common law and equity on Hindu law since 1800', *International and Comparative Law Quarterly*, 3: 576.

———. (1960), 'The compilation of customary law in the Punjab in the nineteenth century', in *La Redaction des Coutumes Dans le Passe et Dans le*

Present. Colloque Des 16–17 Mai, 1960. Centre d'Histoire et l'ethnologie Juridiques. Bruxelles: Editions de l'Institute de Sociologie.

Gordon, Philip R.T. (1914), *The Khasis*, London.

Gough, E. Kathleen (1960), 'Caste in a Tanjore village', in E.R. Leach, ed., *Aspects of caste in south India, Ceylon and North-West Pakistan*. Cambridge: Cambridge University Press, pp.11–60.

Gould, Harold, A. (1963), 'The adaptive functions of caste in contemporary Indian society, *Asian Survey*, 3: 427–38.

Govinda Menon, P. (1951), 'The bench and the bar in India—A survey of their problems', *All-India Reporter Journal*, 1951:90.

Grimshaw, Allen, D. (1959), 'The Anglo-Indian community: the integration of a marginal group', *Journal of Asian Studies*, 18: 227–40.

Grossman, Kurt Richard (1954), *Germany's moral debt: the German-Israel agreement*. Washington: Public Affairs Press.

Gune, V.T. (1953), *The judicial system of the Marathas*. Poona: Deccan College Post-Graduate and Research Institute.

Haldipur, R.N., and V.R.K. Paramahansa, eds. (1970), *Local government institutions in rural India: some aspects*. Hyderabad: National Institute of Community Development.

Harper, Edward B. (1964), 'Ritual pollution as an integrator of caste and religion', *Journal of Asian Studies*, 23:151–97.

Hattori, T.D. (1963), 'The legal profession in Japan: its historical and present state', in A.T. von Mehren, ed., *Law in Japan*. Cambridge: Harvard University Press, pp. 111–52.

Hazari (1951), *An Indian outcaste*. London: The Bennisdale Press.

Heinz, John P., and Edward O. Laumann (1978), 'The legal profession: client interest, professional roles and social hierarchies, 76 *Michigan Law Review*, 1111.

Hitchcock, John D. (1960), 'Surat Singh, head judge', in Joseph B. Casagrade, ed., *The company of man*. New York: Harper & Bros., pp. 234–72.

Hoebel, E. Adamson (1954), *The law of primitive man*. Cambridge: Harvard University Press.

——. (1965), 'Fundamental cultural postulates and judicial law-making in Pakistan', in L. Nader, ed., *The ethnography of law* (*American Anthropologist* Special Publication V 67: 6, part 2), pp. 43–56.

Horowitz, D. (1976), *The courts & social policy*. Washington: The Brooking's Institute.

Hunter, W.W. (1897), *Annals of rural Bengal*, seventh edition, London: Smith Elder.

Hutton, J.H. (1961), *Caste in India*, Bombay: Oxford University Press.

Ibert, Sir Courtenay (1907), *The government of India,* Oxford: The Clarendon Press.

Indian Institute of Public Opinion Evaluation of Panchayat Elections in Punjab (1960), 6 *Monthly Public Opinion Survey*, 8–9, 1961.

Ishwaran, K. (1964), 'Customary law in village India', *Internatonal Journal of Comparative Sociology*, 5:228–43.

Issacs, Harold R. (1965), *India's Ex-Untouchables*. New York: John Day.

Iyengar, G. Aravamuda (1935), *The temple entry by Harijans*, second edition, Nellore: Sanatana Dharma Printing Agency.

Jagannadh Das, B. (1955), 'The judiciary in India', *All-India Reporter Journal*, 1955:42.

Jain, M.P. (1963), 'Custom as a source of law in India', *Jaipur Law Journal*, 3: 96–130.

Jain, S. (1968), *Community development and panchayati raj*. Bombay: Allied Publishing House.

Jennings, Sir Ivor (1955), *Some characteristics of the Indian Constitution*. Oxford: Oxford University Press.

Jones & Lewalla (1969), 'Power factions in Kabuapur panchayat', in Mathur & Narain, eds., *Panchayat raj, planning and democracy*. New Delhi, Asia Publishing House.

Joshi, Rao Bahadur P.C. (1913), 'History of the Pathare Prabhus and their gurus or spiritual guides', *Journal of the Anthropological Society of Bombay*, 10: 100–38.

Journal of the Indian Law Institute, 8 (No.4) October–December [Special Issue on Work of Justice P.B. Gajendragadkar.]

Kahana, K. (1960), *The case for Jewish civil law in the Jewish state*. London: The Soncino Press.

Kane P.V. (1930–62), *History of Dharmasastra* (5 Vols. in 7). Poona: Bhandarkar Oriental Research Institute.

——. (1950), *Hindu customs and modern law*. Bombay: University of Bombay.

Kantowsky, D. (1968), Indische Laiengerichte, 'Die nyaya panchayats in Uttar Pradesh', 1, *Verassung und Recht in Ubersee*, 140.

Karve, Irawati (1961), *Hindu society—an interpretation*. Poona: Deccan College.

Katju, K.N. (1948), Speech reported in the *All-India Reporter Journal*, 1 1948: 22.

Keer, D. (1962), *Dr Ambedkar: life & mission*, second edition, Bombay.

Khare, Ravindra, S. (1972) 'Indigenous culture and lawyers' law in India', *Comparative Studies in Society and History*, 14: 71–96.

Khosla, G.D. (1949), *Our judicial system*. Allahabad: University Book Agency.

Kidder, Robert (1973). 'Courts and conflict in an Indian city: a study in legal impact', *Journal of Commonwealth Political Studies*.

—— (1978), 'Western law in India: external law and local response', in Harry M. Johnson, ed., *Social system and legal process*. San Francisco: Jossey-Bass, pp. 155–80.

Kikani, L.T. (1912), *Caste in courts or rights and powers of caste in social and religious matters as recognised by Indian courts*. Rajkot: Ganatra Printing Works.

Koppell, G. Oliver (1966), 'Legal aid in India', 8, *Journal of the Indian Law Institute*, 224.

Krishnamacharya, U.P. (*c.* 1936), *Temple worship and temple entry*, second edition, Nellore.

Kumar, Ravinder (1965), 'The Deccan riots of 1875', *Journal of Asian Studies*, 4: 613–35.

Kushawaha, R. (1977), *Working of nyaya panchayats in India: a case study of Varanasi District*. New Delhi: Popular Prakashan.

Law Society Review (1968–9), A special issue devoted to lawyers in developing societies with particular reference to India, 3: (2–3), November 1968–February 1969.

Lawson, F.H. (1953), *A common lawyer looks at the civil law*. Ann Arbor: University of Michigan Law School.

Lecky, Robert S., and H. Elliott Wright, eds. (1969), *Black manifesto—religion, racism and reparations*. New York: Sheed and Ward.

Leslie, Charles (1969), 'Modern India's ancient medicine', *Transaction*, June: 46–55.

—— (1970), 'The professionalisation of indigenous medicine'. Paper presented to the Conference on Occupational Cultures in Changing South Asia, May 15–16, 1970, University of Chicago.

Lev, Daniel S. (1965), 'The politics of judicial development in Indonesia', *Comparative Studies in Society and History*, 3: 173–99.

Levy, H.L. (1961), 'The Hindu Code Bill in British India'. Unpublished M.A. thesis, Department of Political Science, University of Chicago.

——. (1968–9), 'Lawyer-scholars, lawyer-politicians and the Hindu Code Bill, 1921–1956', *Law and Society Review*, 3: 303–16.

Lipstein, K. (1957), 'The reception of western law in India', *International Social Science Bulletin*, 85–95.

Llewellyn, Karl N. (1960), *The common law tradition: deciding appeals*. Boston: Little, Brown & Co.

Luchinsky, Mildred S. (1963a), 'The impact of some recent Indian government legislation on the women of an Indian village', *Asian Survey*, 3: 573–83.

——. (1936b), 'Problems of culture change in the Indian village', *Human Organisation*, 22: 66–74.

Luthera, Ved Prakesh (1964), *The concept of the secular state and India*. London: Oxford University Press.

Lynch, Owen M. (1967), 'Rural cities in India: continuities and discontinuities', in Philip Mason, ed., *India and Ceylon*, New York: Oxford University Press, pp. 142–58.

Mack, Elmar (1955), *Mainly scripts and touts*. Madras: Higginbothams.

Madan, T.N. (1969), 'Who chooses modern medicine and why', *Economic and Political Weekly*, 4: 1475–84.

Maddick, H. (1970), *Panchayati raj: a study in rural local government in India*. Harlow: Longmans.

Maine, Sir Henry (1890), *Minutes by Sir H.S Maine 1862*. Calcutta: Government of India, Legislative Department.

——. (1895), *Village communities in the east and west*, seventh edition, London: John Murray.

Maitland, Frederick W. (1957), 'Why the history of English law is not written', in V.T. Delany, ed., *The Maitland reader*. New York: Oceana Books.

Malaviya, H.D. (1956), *Village panchayats in India*. New Delhi: Economic and Political Research Department, All-India Congress Committee.

Malik, Suneila (1979), *Social integration of Scheduled Castes*. New Delhi: Abhinav Publications.

Marriott, M. (1955), 'Little communities in an indigenous civilisation', in M. Marriott, ed., *Village India*, Chicago: University of Chicago Press, pp. 175–227.

Marx, Karl and Friedrich Engels (1947), *The communist manifesto*, Chicago: Charles H. Kerr & Co.

——. (1959), 'Interactional and attributional theories of caste ranking', *Man in India*, 39: 92.

Mathur, R., and I. Narain, eds. (1969), *Panchayati raj, planning and democracy*. New Delhi: Asia Publishing House.

—— and A. Sinha (1966), *Panchayati raj in Rajasthan*. New Delhi: Impex India.

Maududi, Abula LA. (1960), *Islamic law and constitution*, second edition, Lahore: Islamic Publications Ltd.

Mayhew, Leon (1971), 'Stability and change in legal systems', in Bernard Barber and Alex Inkeles, eds., *Stability and social change*. Boston: Little, Brown, 187–210.

McCloy, John J. (1958), 'The extra-curricular lawyer'. Lecture given at Washington and Lee University, April 18, 1958. (Mimeo.)

McCormack, William (1963), 'Lingayats as a sect', *Journal of the Royal Anthropological Institute*, 93, Part 1: 57–71.

—— (1966), 'Caste and the British administration of Hindu law', *Journal of Asian and African Studies*, 1: 25–32.

Mendelsohn, Oliver (1981), 'The pathology of the Indian legal system', *Modern Asian Studies*, 15: 823.

Miller, Robert (1966), 'Button, button . . . great tradition, little tradition, whose tradition?' *Anthropological Quarterly*, 39: 26–42.

Misra, B.B. (1959), *The central administration of the East India Company, 1773–1834*. Manchester: Manchester University Press.

——. (1961), *The judicial administration of the East India Company in Bengal, 1765–82*. Delhi: Motilal Banarsidass.

Misra, K.L. (1960), 'Presidential address', 15th. Uttar Pradesh Lawyers' Conference. Allahabad: *All-India Reporter Journal*, 1960:38.

Misra, L.S. (1954), 'Inaugural address'. *All-India Reporter Journal*, 1954:48.

Monier-Williams, M. (1891), *Brahmanism & Hindism or religious thought and life-in India*, fourth edition, London: John Murray.

Mookerji, R.K. (1958), *Local government in ancient India*, second edition, Delhi: Motilal Banarsidass.

Moon, Penderel (1945), *Strangers in India*. New York: Reynal and Hitchcock.

Moran, Francis (1960), 'The republic of Ireland', in A.L. Goodhart ed., *The migration of the common law*. London: Stevens and Company, 31–5.

Morley, William H. (1850), '*An analytical digest of all the reported cases . . .*' Vol. 1, London: W.H. Allen and Company.

Morris-Jones, W.H. (1963), 'India's political idioms', in C.H. Philips, ed., *Politics and society in India*. London: Allen and Unwin. 133–54.

Morrison, Charles (1968–9), 'Social organisation at the district court: colleague relationships among Indian lawyers', *Law & Society Review*, 3: 251.

—— (1972), 'Kinship in professional relations: A study of north Indian lawyers', *Comparative Studies in Society and History*, 14: 100–25.

Mukherjee, S. (1974), *Local government in West Bengal*. Calcutta: Dasgupta & Co. Ltd.

Mulla, Dinshah Fardunjl (1901), *Jurisdiction of courts in matters relating to the rights and powers of castes*. Bombay: Caxton Printing Works.

Musil, Robert (1953), *The man without qualities*, Vol. I. trans. E. Wilkens and E. Kaiser, London: Secker & Warburg.

Myrdal, G. (1968), *The Asian drama: an enquiry into the poverty of nations*. New York: Pantheon.

Narain, J. (1969), 'The emerging concept', in R. Mathur and Narain, eds., *panchayat raj, planning and democracy*. New Delhi: Asia Publishing House.

Narayan, Shriman (1955), 'Socialist pattern and social revolution', in Myron Weiner, ed., *Developing India* (2 Vols.). Chicago: University of Chicago Press.

Nelson, J.H. (1877), *Review of the Hindu law as administered by the high court of judicature at Madras*. Madras: Negapatam.

——. (1880), *A prospectus of the scientific study of Hindu law*. London: Kegan Paul and Company.

——. (1886), *Indian usage and judge-made law in Madras*. London: Kegan Paul and Company.

Nicholas, Ralph (1965), 'Factions, a comparative analysis', in M. Banton, ed., *Political systems and distribution of power*. London: Tavistock for Association of Social Anthropologists.

—— and T. Mukhopadhyay (1962), 'Politics and law in two West Bengal villages', *Bulletin of the Anthropological Survey of India*, 11: 15–40.

Orenstein, Henry (1965), 'The structure of Hindu caste value: a preliminary study of hierarchy and ritual defilement', *Ethnology*, 4: 1–15.

Pannikar, K.M. (1961), *Hindu society at cross roads*, third edition, Bombay: Asia Publishing House.

Parekh, Manilal C. (1936), *Sri Swami Narayana: a gospel of Bhagwat-Dharma or God in redemptive action*. Rajkot: Sri Bhagwat Dharma Mission House.

Patnaik, N. (1960a), 'A gram panchayat election in Orissa', 12: *Economic Weekly*, 45–8.

——. (1960b), 'Outcasting among oilmen for drinking wine', *Man in India*, 40: 1–7.

Patra, Atul Chandra (1961), *The administration of justice under the East India Company in Bengal.* Calcutta: Firma K.L. Mukhopadhyay.

Pickett, J. Waskom (1933), *Christian mass movements in India.* New York: Abingdon Press.

Pillai, K. (1974), 'Criminal jurisdiction of nyaya panchayats'. Master's Dissertation, Mimeo.

Pillai, P. Chidambaram (1933), *Right of temple-entry.* Nagercoil.

Popkin, W.D. (1962), 'Prematurity and oliter dictum in Indian judicial thought', *Journal of the Indian Law Institute,* 4: 231–60.

Priestly, J.B. (1962), 'Wigs and robes', *New Statesman.* August 17, 1964, 196–7.

Purwar, Vijaya Lakshmi (1960), *Panchayats in Uttar Pradesh.* Lucknow.

Pye, Lucian (1966), *Aspects of political development.* Boston: Little, Brown.

Rai, Lala Lajpat (1915), *The Arya Samaj.* London: Longmans, Green & Co.

Radhakrishnan, S. (1923), *Indian philosophy* 2 Vols. New York: MacMillan & Co.

Rajamannar, P.V. (1949), Speech reported in the *All-India Reporter Journal,* 1949: 26.

Ramachandran, V.G. (1950), 'Re-organisation of the legal profession in India', *All-India Reporter Journal,* 1950:52.

Ramakrishna Aiyar, C.S. (1918), 'Caste customs, caste questions and jurisdiction of courts', *Hindu Law Journal,* 1 (May 1918–June 1919): Sec. IV: 33–68. Madras, 1921:

Rankin, G.C. (1940), 'Custom and the Muslim law in British India'. Transactions of the Grotius Society, 25: 89–118.

———. (1946), *Background to Indian law.* Cambridge: Cambridge University Press.

Rattigan, Sir William Henry (1953), *A digest of civil law for the Punjab, chiefly based on customary law,* thirteenth edition, Allahabad: University Book Agency.

'The reception of foreign law in Turkey' (1957), *International Social Science Bulletin,* 9: 7–81.

Retzlaff, Ralph H. (1962), *Village government in India: a case study.* London: Asia Publishing House.

Robins, Robert S. (1962), 'India: judicial panchayats in Uttar Pradesh', *American Journal of Comparative Law,* 11:239–46.

Rocher, Luco (1972), 'Schools of Hindu law', in J. Ensink and P. Caeffke, ed., *India major: congratulatory volume presented to J. Gonda.* Leiden: Brill, pp. 167–76.

Rowe, Peter (1968–9), 'Indian lawyers and political modernisation: observations in four district towns', *Law & Society Review,* 3:219–50.

Roy, Sarat Chandra (1928) *Oraon religion and customs.* Calcutta.

Roy, Sripati (1911), *Customs and customary law in British India.* Calcutta: Hare Press.

Rudolph, Lloyd I. and Rudoph, Susanne H. (1965), 'Barristers and Brahmins in India: legal cultures and social change', *Comparative Studies in Society and History* 8: 24–49.

——. (1960), 'The politics of India's caste associations', *Pacific Affairs*, 33: 5–22.

Samant, S.V. (1957), *Village panchayats* [with special reference to Bombay State]. Bombay: Local Self-Government Institutes.

Sampurnanand (1963), 'Presidential address'. *Jaipur Law Journal*, 3: 1–6.

Sankaran Nair, C. (1911), 'Indian law and English legislation'. *Contemporary Review*, 100: 213–26.

Sarkar, U.C. (1958), *Epochs in Hindu legal history*. Hoshiarpur: Vishveshvara and Vedic Research Institute.

Sastri, Subramanya, *Hindu law*, eighth edition.

Schuchter, Arnold (1970), *Reparations: the black manifesto and its challenge to white America*. Philadelphia: Lippincott.

Seminar (1965), 'Secularism, a symposium on the implication of national policy' (No. 67) March 1965.

Sen, A.K., M.C. Setalvad and G.S. Pathak (1964), *Justice for the common man*. Lucknow: Eastern Book Co.

Sen-Gupta, N. (1953), *Evolution of ancient Indian law*. Calcutta: Eastern Law House.

Setalvad, M.C. (1960), *The common law in India*. London: Stevens and Sons.

Shah, Vimal P. and Tara Patel (1977) '*Who goes to college: Scheduled Caste/Tribe post-matric scholars in Gujarat*. Ahmedabad: Rachana Prakashan.

Sharma, G.S., ed. (1966), *Secularism: its implications for law and life in India*. Bombay: N.M. Tripathi Pvt. Ltd.

——. (1967), 'Changing perceptions of law in India', *Jaipur Law Journal*, 7: 1–19.

Sharma, V.S. (1951), *The panchayat raj*. Simla: Udbodhan Mandal.

Shore, Frederick John (1837), *Notes on Indian affairs*, 2 Vols. London.

Siegal, B.J. and A. Beals (1960), 'Pervasive factionalism', *American Anthropologist*, 62: 394–417.

Singer, Milton (1961), 'Text and context in the study of contemporary Hinduism', *The Adyar Library Bulletin*, 25: 274–303.

Singh (1969), 'Social structure and panchayats', in Mathur & Narain, eds., *Panchayat raj, planning and democracy*. New Delhi: Asia Publishing House.

Singh, K. (1972), *Rural democratization x-rayed*. Ghaziabad, Uttar Pradesh: Vimal Prakasha.

Sinha, D.P. (1960), 'Caste dynamics: case from Uttar Pradesh', *Man in India*, 40: 19–29.

Sinha, V.K., ed. (1968), *Secularism in India*. Bombay: Lalvani Publishing House.

Smith, Donald, E. (1963), *India as a secular state*. Princeton, N.J.: Princeton University Press.

—— ed. (1966), *South Asian politics and religion*. Princeton: Princeton University Press.

Smith, Munroe (1927), *A general view of European legal history*. New York: Columbia University Press.

Smooha, Sammy (1978), *Isreal: pluralism and conflict*. Berkeley: University of California Press.

Sorabji, Cornelia (1933), 'Temple entry and Untouchability', *Nineteenth Century*, 113: 698–702.

Srinivas, M.N. (1962), *Caste in modern India and other essays*. Bombay: Asia Publishing House.

——. (1962a), 'The study of disputes in an Indian village', in M.N. Srinivas, *Caste in modern India and other essays*. 112–19. Bombay: Asia Publishing House.

——. (c.1964), 'A study of disputes'. Delhi: University of Delhi (Delhi School of Economics, Dept. of Sociology).

Stokes, Eric (1959), *The English utilitarians and India*. Oxford: The Clarendon Press.

Stokes, Whitley (1887–8), *The Anglo-Indian codes*. Oxford: The Clarendon Press.

Subrahmanian, N. (1966), 'Hinduism and secularism', *Bulletin of the Institute of Traditional Cultures*, Madras, Part I: 1–21.

——. (1961), 'Freedom of religion', *Journal of the Indian Law Institute*, 3, 323–50.

Sugarman, R. (1979), 'To love the Torah more than god', *Judaism*, 28: 216–23.

Sundaranda, Swami (1945), *Hinduism and untouchability*, second edition, Delhi: Harijan Sevak Sangh.

Takayanagi, K. (1963), 'A century of innovation: the development of Japanese law 1868–1961', in A.T. von Mehren, ed., *Law in Japan: the legal order in a changing society*. Cambridge: Harvard University Press.

Tinker, H. (1954), *Foundation of local self-government in India, Pakistan and Burma*. London: University of London Historical Studies.

Titus, Murray T. (1959), *Islam in India and Pakistan*. Calcutta: Calcutta YMCA Publishing House.

Toharia, Jose (1975), 'Judicial independence in an authoritarian regime: the case of contemporary Spain', *Law & Society Review*, 9: 475–96,

Trubek, David (1978), 'Unequal protection: thoughts on legal services, social welfare and income distribution in Latin America', *Texas International Law Journal*, 243.

—— and Marc Galanter (1974), 'Scholars in self-estrangement: some reflections on the crisis of law and development studies in the United States, *Wisconsin Law Review*, 1062–102.

Tucker, Henry St. George (1853), 'Memorials of Indian government', in J.W. Kaye, ed., London: Richard Bentley.

Twining, William (1964), 'The place of customary law in the national legal systems of East Africa'. Lectures delivered at the University of Chicago Law School in April-May 1963. Chicago: University of Chicago Law School.

Venkataraman, S.R. (1946), *Temple entry legislation reviewed with acts and bills*. Madras: Bharat Devi Publications.

Venkatrangiah, M. and M. Patabiraman, eds. (1969), *Local government of India*. Bombay: Allied Publishers.

Vepa, Ram K. (1970), 'Has panchayat raj any future?' in R.N. Haldipur and V.R.K. Paramhansa, eds. *Local government institutions in rural India: some aspects*. Hyderabad: National Institute of Community Development, pp. 247–55.

Von Mehren, Arthur T. (1963a), 'The judicial process with particular reference to the United States and to India', *Journal of the Indian Law Institute*, 5:271–80.

——. (1963b), 'Some observations on the role of law on Indian society and on Indian legal education', *Jaipur Law Journal*, 3: 13–18.

Weber, Max (1954), *Law in economy and society*, Max Rheinstein, ed., Cambridge: Harvard University Press.

—— (1958), 'Politics as a vocation', in H.H. Gerth and C.W. Mills, eds., *From Max Weber: essays in sociology*. New York: Oxford University Press.

Weiner, M. (1967), *Party building in a new nation: the Indian National Congress*. Chicago: University of Chicago Press.

Woodruff, Philip (1953), *The men who ruled India: the founders*. London: Jonathan Cape.

Yadin, U. (1962), 'Reception and rejection of English law in Israel', *International and Comparative Law Quarterly*. 2, 59–72.

Zelliot, Eleanor (1966), 'Buddhism and politics in Maharasthra', in Donald E. Smith ed. *South Asian politics and religion*. Princeton N.J.: Princeton University Press, pp. 191–212.

GOVERNMENT PUBLICATIONS

Central statistical organization (1968), Statistical abstract of the Indian Union. Delhi: Manager of Publications.

Government of India (1925), Report of the Civil Justice Committee, 1924–5 Calcutta: Government of India.

—— (1953), Report of the All India Bar Committee. Delhi: Government of India, Ministry of Law.

—— (1955), Backward Classes Commission, Report. 3 Vols., Delhi: Manager of Publications.

—— Commissioner for Scheduled Castes and Scheduled Tribes. Report 1951–6, 1957–8, to 1978–9. Delhi: Manager of Publications [Abbreviated at RCSCST].

—— (1957). (Balwantray Mehta Report). Report of the Study Team for Community Development and National Extension Service.

—— (1958), Law Commission of India. Fourteenth report (reform of judicial administration), 2 Vols. New Delhi: Government of India, Ministry of Law.

—— (1961–6, 1969–70), Report on the administration of justice, Uttar Pradesh Allahabad.

—— (1962), Report of the Study Team on Nyaya Panchayats. Delhi: Ministry of Law.

—— (1963), The nyaya panchayat road to justice.

—— (1966), Panchayat raj at a glance (as on 31 March, 1966). New Delhi: Ministry of Food, Agriculture, Community Development and Co-operation. (Department of Community Development).

—— (1974), Towards equality: Report of the Committee on the Status of Women. Ministry of Education and Social Welfare.

—— (1977), Panchayati raj at a glance. statistics 1975–6. Ministry of Agriculture and Irrigation, Department of Rural Development.

—— (1978) Ministry of Law, Justice and Company Affairs. (Bhagwati Report). Report on national juridicare: equal justice—social justice.

Government of Maharashtra (1971), Maharashtra Report. Report of the evaluation committee on panchayati raj.

Government of Manipur. (1972), Manipur Administrative Report.

Government of Rajasthan. (1963), Rajasthan Administrative Reports.

—— (1964), Panchayat and Development Department: Report of the Study Team on Panchayati Raj. Jaipur.

—— (1973), Rajasthan Report. Report of the High Powered Committee on Panchayati Raj.

General Index

Table of Cases